Sustainable Fishery S

Fish and Aquatic Resources Series

Series Editor: Professor Tony J. Pitcher
Director, Fisheries Centre, University of British Columbia, Canada

The *Blackwell Science Fish and Aquatic Resources Series* is an initiative aimed at providing key books in this fast-moving field, published to a high international standard.

The Series includes books that review major themes and issues in the science of fishes and the interdisciplinary study of their exploitation in human fisheries. Volumes in the Series combine a broad geographical scope with in-depth focus on concepts, research frontiers and analytical frameworks. These books will be of interest to research workers in the biology, zoology, ichthyology, ecology, physiology of fish and the economics, anthropology, sociology and all aspects of fisheries. They will also appeasl to non-specialists such as those with a commercial or industrial stake in fisheries.

It is the aim of the editorial team that books in the *Blackwell Science Fish and Aquatic Resources Series* should adhere to the highest academic standards through being fully peer reviewed and edited by specialists in the field. The Series books are produced by Blackwell Science in a prestigious and distinctive format. The Series Editor, Professor Tony J. Pitcher is an experienced international author, and founding editor of the leading journal in the field of fish and fisheries.

The Series Editor and Publisher at Blackwell Science, Nigel Balmforth, will be pleased to discuss suggestions, advise on scope, and provide evaluations of proposals for books intended for the Series. Please see contact details listed below.

Titles currently included in the Series
1. *Effects of Fishing on Marine Ecosystems and Communities* (S. Hall) 1999
2. *Salmonid Fishes* (Edited by Y. Altukhov *et al.*) 2000
3. *Percid Fishes* (J. Craig) 2000
4. *Fisheries Oceanography* (Edited by P. Harrison & T. Parsons) 2000
5. *Sustainable Fishery Systems* (A. Charles) 2001
6. *Krill* (Edited by I. Everson) 2000
7. *Tropical Estuarine Fishes* (S.J.M. Blaber) 2000

For further information concerning books in the series, please contact:
Nigel Balmforth, Professional Division, Blackwell Science, Osney Mead, Oxford OX2 0EL, UK
Tel: +44 (0) 1865 206206; Fax: +44 (0) 1865 721205
e-mail: nigel.balmforth@blacksci.co.uk

Sustainable Fishery Systems

Anthony T. Charles

Saint Mary's University
Halifax, Nova Scotia, Canada

**Blackwell
Science**

Editorial offices:
Blackwell Science Ltd, 9600 Garsington Road, Oxford OX4 2DQ, UK
    Tel: +44 (0) 1865 776868
Blackwell Publishing Professional, 2121 State Avenue, Ames, Iowa 50014-8300, USA
    Tel: +1 515 292 0140
Blackwell Science Asia Pty, 550 Swanston Street, Carlton, Victoria 3053, Australia
    Tel: +61 (0)3 8359 1011

First published 2001
Reprinted 2006

ISBN-10: 0-632-05775-0
ISBN-13: 978-0-632-05775-7

Library of Congress Cataloging-in-Publication Data
Charles, Anthony Trevor, 1956-
    Sustainable fishery systems / Anthony T. Charles.
        p. cm. -- (Fish and aquatic resources series ; no. 5)
    Includes bibliographical references (p.).
    ISBN 0-632-05775-0
    1. Sustainable fisheries   I. Title. II. Fish and aquatic resources series ; 10.

    SH329.S87 C53 2000
    333.95'616--dc21

                                                                        00-052959

A catalogue record for this title is available from the British Library

Set in 10/13 Times
by Sparks Computer Solutions Ltd, Oxford
http://www.sparks.co.uk
Printed and bound in Great Britain
by Marston Book Services, Oxford

For further information on Blackwell Publishing, visit our website:
www.blackwellpublishing.com

This book is dedicated to Beth, Gavin and Ivy

Contents

Series Foreword

Fish researchers (a.k.a. fish freaks) like to explain, to the bemused bystander, how fish have evolved an astonishing array of adaptations, so much so that it can be difficult for them to comprehend why anyone would study anything else. Yet, at the same time, fish are among the last wild creatures on our planet that are hunted by humans for food. As a consequence, few today would fail to recognise that the reconciliation of exploitation with the conservation of biodiversity provides a major challenge to our current knowledge and expertise. Even evaluating the trade-offs that are needed is a difficult task. Moreover, solving this pivotal issue calls for a multidisciplinary conflation of fish physiology, biology and ecology with social sciences such as economics and anthropology in order to probe new frontiers of applied science. The Blackwell Science series on *Fish and Aquatic Resources* is an initiative aimed at providing key, peer-reviewed texts in this fast-moving field.

While bony fish stem from a great radiation that followed the invention of the swimbladder in the Cretaceous period 100 million years ago, some fish groups, such as the sharks, lungfish and sturgeons, are more ancient beasts. Survivors from earlier eras may be more widespread than we think: the deep-sea coelacanths, formerly known only from the Indian Ocean, have recently turned up in Indonesia. Also, these fishes may be more effectively adapted to specialised niches than their ancient body plans would suggest. For example, rays and angel sharks have perfected the art of the ambush predator, while most cartilaginous fishes can detect electric discharges in the nerves of their prey.

Bony fish themselves have evolved into an amazing array of habitats and niches. As well as the open sea, there are fish in lakes, ponds, rivers and rock pools; in deserts, forests, mountains, the great deeps of the sea, and the extreme cold of the Antarctic; in warm waters of high alkalinity or of low oxygen; and in habitats like estuaries or mudflats, where their physiology is challenged by continuous change. Air-breathing climbing perch (regularly found up trees), walking catfish and mangrove mudskippers are currently repeating the land invasion of their Carboniferous ancestors. We can marvel at high-speed swimming adaptations in the fins, tails, gills and muscles of marlins, sailfish and warm-blooded tunas; gliding and flapping flight in several groups of fish; swinging, protrusible jaws providing suction-assisted feeding that have evolved in parallel in groupers, carps and cods; parental care in mouth-brooding cichlids; the birth of live young in many sharks, tooth carps, rockfish and blennies; immense migrations in salmon, shads and tunas; and even the so-called four-eyed fish, with eyes divided into upper air and lower water-adapted sections.

In addition to food and recreation (and inspiration for us fish freaks), it has, moreover, recently been realised that fish are essential components of aquatic ecosystems that provide vital services to human communities. But, sadly, virtually all sectors of the stunning biodiversity of fishes are at risk from human activities. In fresh water, for example, the largest mass-extinction event since the end of the dinosaurs has occurred as the introduced Nile perch in Lake Victoria eliminated over 100 species of endemic haplochromine fish. But, at the same time, precious food and income from the Nile perch fishery was created in a miserably poor region. In the oceans, we have barely begun to understand the profound changes that have accompanied a vast expansion of human fishing over the past 100 years.

It is a sad fact that most attempts at fishery management have failed. Thirty years ago, in the optimistic era of Peter Larkin's 'stained glass cathedral' of simple, elegant ecological theory, it would scarcely have been thought credible that in a future planned with such powerful tools, one would be making such a statement, or that it would be the majority view among practicing fishery scientists. Why has it failed? The science was not wrong (although it needed extending from single species to exploited ecosystems), and it was certainly necessary, but it was not sufficient to avert a long series of fishery collapses, disasters and serious depletion throughout the world today. Is there a way out of this sorry state of affairs?

An important component of the answer is contained in *Sustainable Fishery Systems*, a 15-chapter book by Anthony Charles from St Mary's University, Halifax, Nova Scotia, Canada that comprises the 5th volume in the Blackwell Science *Fish and Aquatic Resources* series. The book commences with definition and description of what the author means by fishery systems – the whole enterprise of catching and selling fish for food and profit that includes people and their economic, social and cultural activities as well as fish ecology. This integrated view of a fishery may not seem so revolutionary today, but, surprisingly recently, fisheries scientists were often trained with little or no appreciation of the human sciences. Needless to say, such gaps in training made for rather ineffective fishery managers. This book will redress the balance, and provides a thorough and stimulating overview of the human factors that must be taken into account if fisheries are to be sustained. The book includes case studies from Canadian Atlantic groundfish and from a mangrove fishery system in Costa Rica.

Using a full arsenal of clearly described social science techniques, Charles introduces the concept of resilience to natural or human-made perturbations and leads the reader towards an integrated 'biosocioeconomic' analysis of fisheries by way of a deep analysis of co-management of resources, and how conflicts can be described and resolved using analytical tools such as paradigm triangles. The role of traditional ecological knowledge in sustainable fisheries is emphasised. By the end of the book, the reader feels able to draw on a web-like analytical framework and look forward to a future of profitable, co-managed community fisheries in which people manage themselves and the fish.

This book should become essential reading for all concerned with the management of fisheries, and the rigour of its methodology should open the eyes of scientists who have not yet come into contact with such an interdisciplinary and integrated perspective on fishery resources. I am delighted to welcome it to the Blackwell Science *Fish and Aquatic Resources* series.

Professor Tony J. Pitcher
Editor, Blackwell Science Fish and Aquatic Resources Series
Director, Fisheries Centre, University of British Columbia, Vancouver, Canada
October 2000

Preface and Acknowledgements

I suspect that the writing of almost any book is a labour of love, and this is no exception. What you have before you is in some ways a culmination of my interest, over several decades, in the interdisciplinary and systematic analysis of fisheries, and in seeking out approaches to improving the sustainability of fisheries. I have sought, in writing this, to produce something accessible to everyone interested in looking at fisheries from an integrated perspective and in exploring the various routes to more sustainable fisheries. I hope that this would include undergraduate and graduate students from various disciplines, as well as professionals in the fishery field, whether on the academic side, in science and management, or within the fishery sector itself.

A wide range of topics is covered in this book, from the structure and dynamics of the natural and human components of fisheries, through the various pieces of the management system, to an examination of a set of key themes that dominate fishery discussions: sustainability, resilience, uncertainty, complexity, diversity, conflict and institutions. Key approaches that are emerging as critical to sustainability and resilience are discussed in some detail, notably the precautionary approach, the ecosystem approach, the co-management approach, rights-based approaches and robust management. Given the breadth of coverage in the book, those familiar with certain aspects of fishery systems will undoubtedly be able to skip over parts of the volume – and indeed may find some material presented at a simpler level than might be preferred – since ultimately the aim here has been to present a fairly comprehensive and accessible sense of what fisheries are all about, and where they are heading (or should be heading).

To that end, the style of presentation is generally informal and meant for relatively 'easy reading'. Technical details, mathematics and the like are omitted, or placed in an appendix, although abundant references are provided throughout for those wishing to explore topics in further depth. Considerable use is made of 'boxes' – as case studies or more in-depth illustrations of particular points, or as optional side-trips relative to the primary direction of the text. In most cases, the boxes are not referred to specifically in the text itself, but each box is titled, enabling the reader to make the decision whether to take the side-trip or not, depending on the topic at hand.

Of course, the author of a book such as this owes a debt of gratitude to many people. First, I would like to acknowledge two special individuals. Colin Clark has inspired me from the very beginning of my career, leading me to focus on fisheries and to do so in an interdisciplinary manner. Elisabeth Mann Borgese has become, to myself and undoubtedly to a great many others concerned about the ocean and ocean users, very much a role model, showing how to balance the local and the global, and the need to protect both the natural world and human communities.

This book would be far inferior if not for the insightful and constructive reviews of one or more of the book's chapters by the following individuals, whom I would like to acknowledge with gratitude, without implicating them in the results: Fikret Berkes, Kevern Cochrane, Parzival Copes, Brian Davy, Brad de Young, Michael Fogarty, Bruce Hatcher, Angel Herrera, Jeff Hutchings, John Kearney, Daniel Lane, Jon Lien, Gary Newkirk, Michael Sinclair, Ralph Townsend, David VanderZwaag and Melanie Wiber. I also thank an anonymous reviewer for helpful comments. The suggestions of all these individuals have been incorporated as far as possible into the book – again, any remaining errors are my own responsibility. I would also like to thank Dominique Levieil for helping me to learn more about European fisheries, and Ransom Myers for providing a map of North Atlantic fishing zones.

I am also grateful to my co-authors on journal papers, books and reports over the years, for their stimulation and insights. Many of those insights have undoubtedly found their way into this book, although any errors are my own and not those of: Renato Agbayani, Emelita Agbayani, Max Agüero, Carol Amaratunga, Tissa Amaratunga, John Beddington, Evelyn Belleza, Alicia Bermudez, Theo Brainerd, Yvan Breton, Mark Butler, Joseph Catanzano, Chen Hailiang, Colin Clark, Scott Coffen-Smout, Parzival Copes, Mel Cross, Brad de Young, Rod Dobell, Michael Fogarty, Exequiel González, John Helliwell, Michael Henderson, Angel Herrera, Hu Baotong, Jennifer Leith, Laura Loucks, Marc Mangel, Michael Margolick, Jack Mathias, Leigh Mazany, Hermie Montalvo, Gordon Munro, Jose Padilla, Randall Peterman, Evelyn Pinkerton, Robert Pomeroy, Bill Reed, Paul Starr, Bozena Stomal, Chris Taggart, Ralph Townsend, Jean-Yves Weigel, George White and Chiwen Yang.

I appreciate the support of Fishing News Books and Blackwell Science, notably Richard Miles and Nigel Balmforth, as well as that of series editor Tony Pitcher. I am very grateful to my research assistants, who have provided invaluable support in the preparation of this book. Cheryl Benjamin has been involved from the start, and has helped to fill in many of the gaps along the way. Nicole McLearn drafted the figures and bibliography, as well as assisting on a variety of research activities. Erin Rankin provided background research in several crucial subject areas, while Sherry Mills and Philip Myers helped to complete the index.

Finally, on a personal note, I take this opportunity to thank my parents for those many years of continuous support. Above all, I thank my wife, Beth Abbott, for her long-standing and patient support of my work, and my children – Ivy and Gavin – for being generally wonderful, and in particular, for tolerating their father during the course of writing this book.

Anthony T. Charles
Saint Mary's University
Halifax, Canada
(t.charles@stmarys.ca)

Introduction

The title of this book, *Sustainable Fishery Systems,* reflects a combination of two interrelated terms: 'sustainable fisheries' and 'fishery systems'. An underlying premise of the book is that success in the pursuit of sustainability (and the related goal of resilience) is closely linked to adoption of a sufficiently broad conception of the fishery as a 'system' of interacting ecological, biophysical, economic, social and cultural components.

This statement of purpose raises several obvious questions. What are sustainability and resilience, and why are they important? What might a sustainable, resilient fishery look like? What exactly is a 'fishery system' and how is a systems perspective connected with sustainability and resilience? These questions are explored in detail within the book, but it is worth introducing the discussion here.

First, consider the idea of sustainability. In recent years, it has become standard practice, in all sectors of economic activity, to emphasise the pursuit of *sustainable development,* through which the economy operates in such a way as to meet human needs now while safeguarding the future (World Commission on Environment and Development 1987). This concept is by no means new to fisheries, or to forestry and other renewable resource sectors, where the idea of achieving a *sustainable yield* from the resource, i.e. a level of output that can be maintained indefinitely into the future, has been central to discussion (if not action) for many decades. The sustainable development approach has, however, brought about an important evolution from a focus merely on 'sustaining the output' to a more integrated view in which sustainability is multifaceted, and emphasises the process as much as the output.

All this discussion of sustainability is timely, given the unfortunate reality that – despite the above-noted history within fisheries of discussing sustainability, and despite the current worldwide focus on sustainable development – many fisheries are in a state of crisis, and indeed the majority are considered to require urgent attention. Many international agencies and congresses (e.g. Hancock *et al.* 1997; OECD 1997; Caddy & Griffiths 1995) are focusing on this crisis, and the need for strategies to promote *sustainable fisheries.*

There is certainly a fundamental problem with the state of the natural resource base, as the Food and Agriculture Organisation (FAO 1997a) has noted:

'... about 35% of the 200 major fishery resources are senescent (i.e. showing declining yields), about 25% are mature (i.e. plateauing at a high exploitation level), 40% are still 'developing', and 0% remain at low exploitation (undeveloped) level. This indicates that around 60% of the major world fish resources are either mature or senescent ...

A strikingly similar conclusion was reached by [Garcia & Newton 1994] which concluded that 44% of the stocks for which formal assessments were available were intensively to fully exploited, 16% were overfished, 6% depleted, and 3% slowly recovering, concluding therefore that 69% of the known stocks were in need of urgent management.'

It must be emphasised, however, that critical concerns about sustainability arise not only in terms of catch levels or even biomasses, but in all aspects of the fishery from the ecosystem, to the social and economic structure, to the fishing communities and management institutions, as well as the fish stocks themselves. For this reason, the pursuit of *sustainable fisheries* is best seen as requiring more than keeping the catch of fish to a level that is 'not too large'. Instead, sustainability can be viewed comprehensively as the maintaining or enhancing of four key components (as elaborated in Chapter 10):

- *Ecological sustainability* – incorporating concerns about sustaining harvests, maintaining the resource base and related species, and maintaining or enhancing the resilience and overall health of the ecosystem.
- *Socioeconomic sustainability* – dealing at an aggregated 'macro' level by maintaining or enhancing long-term socioeconomic welfare, including generation of sustainable net benefits, reasonable distribution of benefits, and maintaining the system's overall economic viability.
- *Community sustainability* – emphasising the 'micro' goal of sustaining communities as valuable human systems, by maintaining/enhancing economic and sociocultural well-being, overall cohesiveness, and the long-term health of the relevant human systems.
- *Institutional sustainability* – the maintenance of suitable financing, administrative and organisational capability over the long term, as a prerequisite for the above components; in other words, ensuring that the relevant fishery institutions function well over the long term.

(A brief aside is required here. The term 'institution' as used above reflects common usage of the concept, referring to an organisation of some sort, within which people interact and manage themselves, such as a Department of Fisheries, or a fisher association/cooperative. A meaning more often adopted by social scientists is of an *institution* as a set of rules or 'norms' governing the behaviour of individuals in the system (North 1990). Examples of such institutions include the market-place, the legal system, municipal councils, and so on. In this sense, an organisation is a manifestation of the underlying institution, so that, for example, a fishery management agency reflects the rules and norms adopted by society to manage natural resources. In this book, the term 'institution' is used often and in either of the above senses, depending on the context.)

While sustainability is important, it should be pursued in conjunction with (or incorporate) the fundamental goal of *resiliency* – the ability of a fishery to absorb and 'bounce back' from perturbations caused by natural or human actions. Certainly, the widespread range of experiences with fishery collapses worldwide clearly suggest a lack of resilience in these cases. As will be discussed, the idea of resilience, while first formulated with ecosystems in mind (Holling 1973), is of great relevance to all parts of the fishery: resilience is needed in the human aspects of the fishery (fishing communities, for example) and in the management infrastructure, as well as in the ecosystem. For example, management must be designed with resilience in mind, so that if something unexpected happens (as is bound to be the case from time to time), the management processes still perform adequately.

Management and policy measures to promote sustainability and resilience in fisheries are certainly central to this book, but the above discussion also highlights the fact that any attempt to analyse aspects of sustainability and resilience requires us to take a broad, integrated view of the fishery. Specifically, sustainable development is not just a matter of protecting fish stocks, but rather involves all aspects of the fishery. We cannot assess the state of ecological sustainability if we fail to look at the ecosystem beyond individual fish stocks, and we cannot enhance community sustainability if we restrict our attention solely to those catching the fish.

This leads us to a 'systems focus' that looks comprehensively at the full *fishery system*. The idea of this approach is to envision fisheries as webs of interrelated, interacting ecological, biophysical, economic, social and cultural components, not as the fish separate from the fishers, separate from the processors, and so on. A *systems* perspective is an *integrated* one, facilitating the assessment of management and policy measures in terms of implications throughout the system. The need for such a perspective has been a key lesson of recent decades, as it has been realised that the reductionism at the heart of most scientific disciplines (the idea of dividing up the study of a system into small pieces for ease of analysis) has proved useful but not sufficient. To put it simply, we cannot lose sight of the forest while we study the trees.

With this rationale, a major focus of the book lies in developing an integrated view of the fishery system, i.e. exploring the nature, structure and dynamics of the various components of the fishery. The idea is to provide an idea of the 'pieces of the puzzle' in fisheries, and how they fit together to create a *fishery system*. Specifically, the discussion is organised according to the following set of key fishery components:

- *The natural system*
  - The fish
  - The ecosystem
  - The biophysical environment
- *The human system*
  - The fishers
  - The post-harvest sector and consumers
  - Fishing households and communities
  - The social/economic/cultural environment
- *The fishery management system*
  - Fishery policy and planning
  - Fishery management
  - Fishery development
  - Fishery research

A focus on systems avoids both an overly simplistic view of the fishery – 'fish in the sea, people in boats' – and the contrasting view of the fishery as an unintelligible mess of 'so many types of fish, so many types of fishers, so many conflicts and debates'. While fisheries certainly are complex, there is a pattern, a structure and a set of fundamental themes that arise repeatedly in fishery discussions. A systems perspective aims to look at this 'big picture', both for a better understanding of the unique nature of the fishery as a human activity, and through this, to help make the fishery 'work better' by developing appropriate management measures and policy approaches.

The emphasis of this book, on an integrated approach to research and management in fishery systems, and on major recurring themes such as uncertainty and complexity, reflects the directions advocated in Agenda 21 (see box below), the action document resulting from the United Nations Conference on Environment and Development, also known as the 'Rio Conference' (Borgese 1995).

Structure of the book

The book is grouped into two segments, Part I and Part II, with an overall format as shown graphically in Fig. I.1. The systems approach is developed in Part I, while its application to recurring themes arising across fishery systems, such as sustainability, conflict, uncertainty and complexity, is to be found in Part II. The content and organisation of this book seeks to reflect an integrated, interdisciplinary (or perhaps 'transdisciplinary') structure. The book is organised in a non-disciplinary manner, with no particular focus on any one discipline at any point. Instead, use is made throughout of material from a range of disciplines within the realm of fisheries analysis, ranging from biology, ecology, oceanography and stock assessment to economics, law, policy, sociology and anthropology.

Part I (Chapters 1–9) focuses on fishery systems, their structure and dynamics. This begins in Chapter 1 with an overview of fishery systems, emphasising how these systems are depicted, and how they are characterised. Chapter 2 provides an overview of the natural system:

Agenda 21, Chapter 17:
Protection of oceans and coastal areas, and their living resources

A Integrated management and sustainable development of coastal areas, including exclusive economic zones
B Marine environmental protection
C Sustainable use and conservation of marine living resources of the high seas
D Sustainable use and conservation of marine living resources under national jurisdiction
E Addressing critical uncertainties for management of the marine environment and climate change
F Strengthening international cooperation and coordination
G Sustainable development of small islands

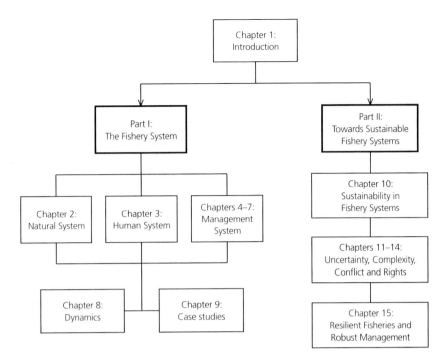

Fig. I.1 Structure of the book

the fish, the ecosystems and the biophysical environment. Analogously, Chapter 3 provides an overview of the human system, including the fishers, the post-harvest sector, households and communities, and the broader socioeconomic environment. It should be noted that most of Chapter 2 will not be new to those having a basic familiarity with biological and oceanographic aspects of fisheries, and similarly Chapter 3 might be safely omitted by those familiar with social and economic aspects of fisheries.

Chapters 4–7 focus on the fishery management system. Of the three major sub-systems in the fishery system, this one receives disproportionate attention in Part I, reflecting the emphasis later in the book on management and policy approaches to sustainability. Thus, individual chapters are provided here on each of the four parts of the management system: Chapter 4 concerns aspects of policy and planning, Chapter 5 covers fishery management *per se*, Chapter 6 examines fishery development, and Chapter 7 discusses fishery research.

The discussion in Chapters 2–7 is largely focused on structure, and is largely generic in nature, i.e. with general applicability, rather than applied to specific cases. To round out the discussion in Part I, therefore, the final two chapters complement this earlier material. Chapter 8 focuses on the *dynamics* of the various components and of the overall fishery system. Since the dynamics of each component are discussed sequentially in the chapter, readers have the option of either examining the chapter as a whole, or doing so one component at a time, in conjunction with corresponding earlier chapters. Chapter 9 presents two case studies that use the framework of system components to describe the structure of specific fishery systems: the groundfish fishery in Atlantic Canada, and the Gulf of Nicoya fishery in Costa Rica.

The second half of the book (Chapters 10–15) focuses on a set of key themes in fishery systems, all of which relate closely to the pursuit of sustainability and resilience. The discussion begins in Chapter 10 with an examination of the nature of sustainability and resilience, and the methodologies of sustainability assessment and sustainability 'indicators'. In Chapter 11, the ubiquitous presence of uncertainty in fisheries is discussed, with special attention to the various forms this uncertainty takes, the connection between uncertainty and risk, the relevance of these to the pursuit of sustainability and resilience, and the value of applying a precautionary approach to fishery decision making. Chapter 12 presents an analogous treatment of complexity and diversity in fisheries, i.e. the challenge of dealing with complex systems, the value of diversity and the benefits of an ecosystem approach.

Conflict plays a major role in fishery systems. A typology of fishery conflicts is presented in Chapter 13, together with an examination of the co-management approach that has emerged as a major strategic tool of management, helping to ameliorate some forms of conflict. The need for appropriate rights guiding the access to and use of fishery resources underlies the discussion in Chapter 14, which addresses concepts of property rights and use rights, analysing each of the various rights-based options. Finally, Chapter 15 synthesises preceding discussions of uncertainty, complexity, conflict and rights to derive policy and management measures that help in the process of developing sustainable, resilient fisheries, in part through measures to make fishery management more *robust*.

It should be noted that while Chapter 10 is a natural opening to Part II, and Chapter 15 wraps up the discussion, the text is designed so that it is perfectly feasible to read the chapters (10–15) independently of one another, in any order; linkages among them are relatively weak, particularly in the case of Chapters 11–14. On the other hand, the reader intending to read the whole of Part II would be advised to do so in sequence, in order to benefit from what linkages there are between the chapters.

Part I
The Fishery System:
Structure and Dynamics

Chapter 1
Fishery Systems

'... most fisheries problems are complex and contain human as well as biological dimensions. Too frequently we see the consequences of trying to deal with complexity in a fragmentary or narrow way. Management plans based on the soundest of biological information fail when it is discovered that fishing pressure cannot be controlled because of unforeseen political or economic constraints. Economic policies fail when unforeseen biological limits are exceeded. In short, fisheries represent dynamic (time varying) systems with interacting components ...'

Walters (1980: p. 167)

This chapter begins the examination of fishery systems with an overview of their nature, structure and characterisation, starting with a review of the various approaches available for depicting fishery systems in words or in graphical form. Following this, attention is focused on defining and characterising fishery systems, particularly in terms of spatial scale, and the dichotomy between small-scale and large-scale fisheries.

1.1 Depicting fishery systems

What does a fishery system look like? How can we describe such a system verbally or graphically? These questions are the focus of this section. First, recall from the Introduction that the presentation of fishery systems in this book is organised around a certain set of components:

- *The natural system*
  - The fish
  - The ecosystem
  - The biophysical environment
- *The human system*
  - The fishers
  - The post-harvest sector and consumers
  - Fishing households and communities
  - The social/economic/cultural environment
- *The fishery management system*
  - Fishery policy and planning

- Fishery management
- Fishery development
- Fishery research

These various components, and some of the interactions between them, can be depicted graphically as in Fig. 1.1 which, perhaps better than any number of words, contains the idea of the fishery system, and its structure and dynamics. Indeed, Fig. 1.1 forms the basis of how the details of each component will be examined sequentially in subsequent chapters. (It is worth noting a point about terminology here. While the natural, human and management components are referred to above as 'systems', they can also be referred to as 'sub-systems' of the fishery system as a whole. These terms are generally used interchangeably in this book.)

However, while Fig. 1.1 provides a useful and detailed graphical description, it is important to note that this is by no means the sole 'correct' system description – other approaches are

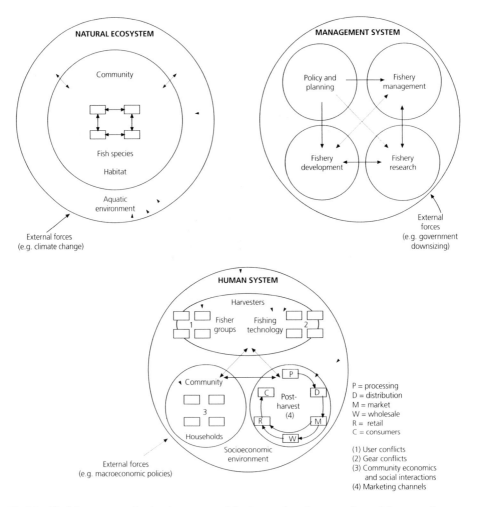

Fig. 1.1   The fishery system, showing the structure of the three major sub-systems (natural, human and management), the major components within each of these, and the key interactions between sub-systems and their components. Also indicated are the impacts of external forces on each part of the system.

equally reasonable. A selection of alternative approaches focusing on the use of flow charts and tables is described in this section. To begin, let us build up a view of a fishery system from the simple idea noted earlier: fish in the sea and a fleet of boats catching them. First, add to this the fact that the harvest is returned to land and sold in a market to give a system that can be depicted as in Fig. 1.2.

This simple figure displays a basic set of inputs (fish and fleet) combining to generate an output (harvest of fish) in the fishery system. Suppose we now recognise that both the fish stock and the fishing fleet are subject to their own inherent *dynamics,* i.e. processes of change over time. The fish are driven by natural population dynamics through the processes of reproduction (or 'recruitment') and mortality. The fishing fleet varies according to *capital dynamics* as fishers invest in new boats and fishing gear (physical capital), which then depreciates over time. Both the population dynamics and the capital dynamics are affected by the level of harvest. Clearly, the catch directly reduces the fish stock, but in addition, when that catch is sold in the market, the profits generated return to the fishers, who may adjust their investment (capital dynamics) as profits vary (depending on harvest and market conditions). These dynamic relationships can be incorporated by expanding the above system diagram as shown in Fig. 1.3.

There are still many aspects missing and/or oversimplified within this system diagram. For example, on the human side, it is useful to broaden the picture from simply a 'fishing fleet' to highlight the fleet and the fishers separately, as well as the dynamics of each. More fundamentally, it is crucial to obtain a more complete sense of the fishery system by looking beyond the internal 'core' of the fishery (fish and fishers) shown in Fig. 1.3, to incorporate interactions with the many other elements of the ecosystem and the human system. This more holistic, *integrated* approach helps to overcome past tendencies to analyse and attempt to manage the fishery as if it were merely 'fish in the sea, people in boats'. Instead, a broader perspective views fish as living in an ecosystem, within a biophysical environment, and fishers as living in households within communities, within a broader socioeconomic environment. Harvests move into the post-harvest sector to be transformed into products in the market-

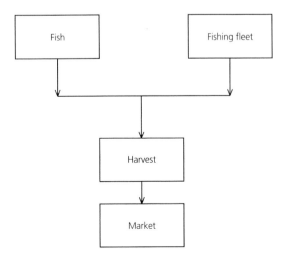

Fig. 1.2   A highly oversimplified view of a fishery system: boats catch fish, and the harvest is sold in the market.

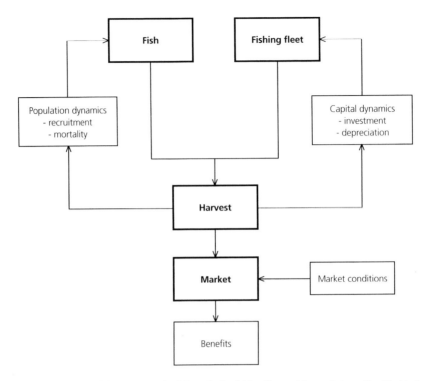

Fig. 1.3   A fuller view of a fishery system: the fish stock, the fishing fleet and the market are all subject to dynamic processes, which in turn are influenced by the harvest of fish. Economic benefits are produced from the sale of the harvest.

place, thereby increasing the total benefits produced. Finally, it is important to recognise the multiplicity of the benefits accruing from the fishery system, and how they feed back into other aspects of the fishery system. These various features can be added to Fig. 1.3 to develop a fuller picture (Fig. 1.4). This figure is not as comprehensive as Fig. 1.1, but it provides a base from which to engage in the task of examining the interactions amongst relevant components of the fishery system.

The diagrams in Figs 1.1–1.4 are orientated 'organisationally' as flow charts. An alternative approach to depicting the structure of fishery systems is a *strategic planning matrix*, which can be used to display the system in the form of a grid showing key components of the fishery system, major elements of the structure of these components, aspects of the associated ecological, social and economic environment, and relevant actions and/or impacts. Such a scheme is essentially a structured listing of representative or particularly relevant components of the system (since it is never possible to include *all* elements of the system). An example of a strategic planning matrix is shown in Fig. 1.5, where the fishery system is organised within seven principal categories, one per row, with category labels listed in the first column.

Of the seven categories in the matrix of Fig. 1.5, the first six are somewhat aggregated versions of the fishery system components used in this book. The first deals with the natural system (fish and ecosystem), the next three with the human system (harvesters, post-harvest and consumers, communities and socioeconomic environment) and the following two with

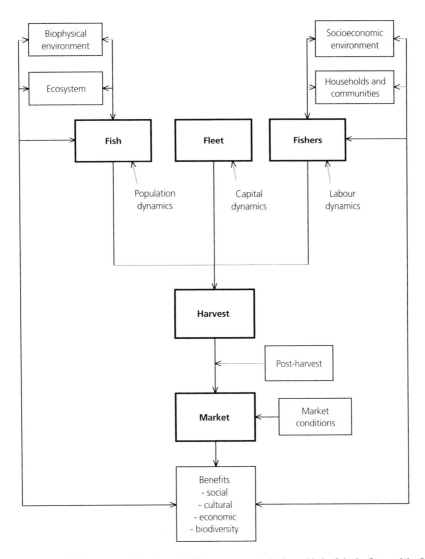

Fig. 1.4 A substantially more complete view of a fishery system, beginning with the fish, the fleet and the fishers, each subject to its own dynamics. The fish interact with the ecosystem and the biophysical environment, while the fishers interact with their households, communities and the socioeconomic environment. The post-harvest sector plays a role between the harvest and the market. The multidimensional benefits obtained from the fishery feed back to the natural and human components of the system.

the management system (fishery management, development and management needs). Each of these categories contains (in the corresponding row) a number of elements which have been roughly organised into those that are more local and/or internal to the fishery (on the left) and those of a larger spatial scale and/or greater external orientation (on the right). The seventh category (last row of the scheme) lists sources of external impacts on the fishery system which are not components of the system *per se*.

For example, the fish/ecosystem category covers the three main environments from in-shore (coastal) to offshore (deep water) to migratory and high seas situations, as well as

Fish and the ecosystem	Inshore (coastal)	Offshore (deep-water)	Migratory and high seas	Fishery habitat and env. quality	Ecological interactions
Harvesters	Subsistence fishers	Artisanal/small-scale fishers	Industrial fleets	Joint ventures	Foreign fleets
Post-harvest and consumers	Traditional processing	Industrial processing	Marketing and distribution	Domestic consumers	Export market
Communities and socioeconomic environment	Organisations and institutions	Communities and fishing households	Women in fisheries	Historical, cultural and legal aspects	Fishery-related activity; other econ. sectors
Fishery management	Objectives and policy	Management, surveillance, enforcement	Fishery development	Resource assessment and research	International policy (EEZ, law of sea)
Development and management needs	Institution building	Human resource development	Science and information systems	Appropriate management structures	Economic diversification
Major external impacts	Mariculture	Agriculture	Tourism	Industry	Transportation

Fig. 1.5 A strategic planning scheme highlighting several major groupings of fishery system components (shown in the first column) and key elements within each grouping (shown in each row). The elements in each row are arranged from those which are local-scale and fishery-focused on the left to those on a larger spatial scale and/or with less fishery orientation on the right. EEZ, exclusive economic zone.

aspects of fishery habitat and environmental quality, and ecological interactions. Running parallel to this, the harvesters category ranges (within a domestic fishery) from subsistence to artisanal/small-scale to industrial, as well as joint ventures and foreign fleets when other players are included. (Note that all of these terms are discussed in detail later.)

The strategic planning scheme, while undoubtedly an oversimplified approach to a complex system, is a simple tool that can complement more extensive assessment tools such as those to be described in Chapters 10 and 12. For example, to assess a given project or programme within a fishery, the scheme can serve as a template to give a visual indication of the components of the fishery system likely to be directly or indirectly affected, and to highlight the interactions which should be monitored as the fishery evolves. This allows an immediate assessment of a situation, as well as providing a graphical summary for use in *a posteriori* analysis, and shows which fishery components and interactions were seen as important in that particular situation.

Various other means are used to depict fishery systems.

- The fishery can be depicted as an image, picture or drawing (e.g. IIRR 1998), which can be particularly useful in community-level management and education efforts, or for publicity purposes.
- A fishery system is often depicted geographically in the form of a map, typically highlighting boundaries between zones related to fishing activity and fishery management, but also showing such matters as fish migrations and fishing vessel locations. A map has the advantage of highlighting spatial features (and spatial heterogeneity), although it is less useful for dynamic aspects. An example of a map depicting large-scale fishery systems is that in Fig. 1.6, showing fishing zones of the North Atlantic, in waters around Europe and off the coast of North America.
- A fishery system can be depicted with a focus on the underlying ecosystem, and in particular the food chain (food web).

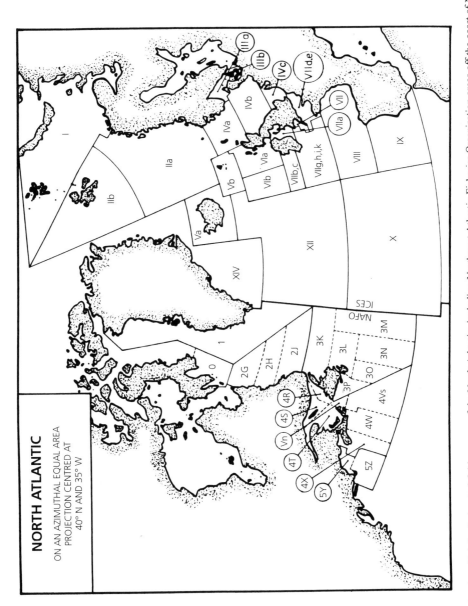

Fig. 1.6 A map showing the officially designated fishing zones of the North Atlantic, including Northwest Atlantic Fisheries Organisation areas off the coast of North America and International Council for the Exploration of the Sea areas in waters around Europe (courtesy of Ransom Myers).

- Fishery systems can be depicted using mathematical and computer models. In this approach, the most important features of the fishery are abstracted into mathematical (symbolic) language. This approach is discussed further in Chapter 12.
- Finally, a cultural perspective can be taken in depicting fishery systems. For example, Elisabeth Mann Borgese, in her ground-breaking book *The Oceanic Circle* (Borgese 1998), draws on the images of Mohandas Gandhi (1947) describing a desired future for his home country of India. Borgese uses this poetic representation of a system as a means to describe the complex systemic interactions of the oceanic world, as well as the embeddedness of individuals (such as fishers) in coastal communities, and of the latter in the larger society.

1.2   Characterising fishery systems

While all fisheries share some common features (people catching fish, for example), there is also an amazing diversity between and within fishery systems. How can we characterise and differentiate between the various types of fishery systems? This question is explored here, with particular attention to the matter of *scale* – something which arises in all aspects of the system and which may well impact on the success of fishery management.

1.2.1 Spatial scale

The spatial scale of a fishery system relates to its size, both geographically and administratively. Matters of spatial scale arise both in the practice of planning and management – such as delineating the appropriate boundaries of a particular fishery system, and designing an appropriate management system – and in the more conceptual considerations involving differentiation among the many possible fishery configurations. For example, one might consider the following fishery systems which have varying spatial scales.

- A coastal community, together with its local fishery resources (e.g. the fish in a tropical coral reef, or the sedentary lobster stock resident near a New England fishing community) and the corresponding small-scale management system.
- Fishery systems ranging from those at the state or provincial level to those at a national level, typically organised around formal jurisdictional boundaries.
- Regional multinational fishery organisations, such as (a) the FAO's Regional Fishery Bodies (e.g. the General Fisheries Commission for the Mediterranean (GFCM) and the Western Central Atlantic Fishery Commission (WECAFC)), and (b) the structures of the European Union revolving around the Common Fisheries Policy.

In a given situation, which of these spatial scales is to be preferred? When is it best to focus on the small scale, and when should we look at the fishery as a larger system? These are important questions, since the spatial scale at which the fishery system is examined will probably affect how interactions among fishery components (and the dynamics of those components, as discussed in Chapter 8) are addressed. Furthermore, given that a fishery system involves fish stocks and other *natural* components, fishers and other *human* components, and a variety of *management* components (including science, enforcement, policy, etc.), it is not surprising that the appropriate scale to view each of these components might differ considerably, depending on the specific circumstances. (See Fig. 1.7 for a depiction of the range of spatial scales in a fishery.)

Furthermore, there may be differences between the 'natural' ecological or physical boundaries for the fishery system, those most suitable from the perspective of the human system (involving economic and sociocultural factors) and the desirable boundaries for the management system (involving legal, institutional or political factors). How can the best balance be obtained between the 'natural' delineations of watersheds or coastal zones and the *de facto* boundaries of the system from the perspective of the human population? Does this require apportioning ocean currents or water flow between human-defined areas, or apportioning people between ecosystems? There seems to be no easy answer.

Such mismatches between the resource, the harvesting activity and the management system may well require the creation of new institutional arrangements for management. Indeed, persuasive arguments can be made for resolving the mismatch by decentralising some systems and taking a broader perspective for others.

Consider, for example, a fishery exploiting a highly migratory stock, such as tuna or herring, that ranges over the waters of many countries. Within a specific nation, the relevant fishers may operate under that nation's management framework, so that on the human side, the fishers and the national management system form a cohesive unit. However, that unit is focusing on a fishery resource that is transient within the nation's borders. In this case, the spatial scale of the natural system is quite different from that of both the human and the management systems. There is little hope of changing the way the fish behave, so we are obliged to adjust the human components of the system, typically by creating international management mechanisms in order to coordinate resource use. In this way, the overall management system comes into line with the 'natural' perspective, leading to the types of system which are in place for various large pelagic fisheries worldwide, notably those for tuna and swordfish, in which local management is 'nested' within a fully internationalised system.

The 'nesting' approach is also used by the European Union to deal with inconsistencies between the scales at which the natural, human and management sub-systems operate. In the EU case, the Common Fisheries Policy has led to a process by which fish stocks within the EU are assessed scientifically according to their biological nature and in the context of the relevant ecosystem, and the total allowable catch (TAC) set for the stock is then subdivided

Spatial scale

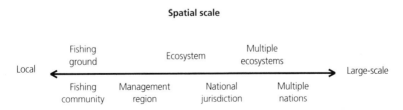

Fig. 1.7 The various spatial scales relevant to fishery systems, ranging from local to large-scale. Aspects of the natural world are shown above the axis, while those of the human and management systems are shown below the axis. Note that the positioning of these relative to one another reflects a common situation, but one which is by no means universal; for example, a *fishing ground* may in fact encompass an ecosystem (such as the North Sea) and may involve multiple nations.

among the relevant countries. Then each country is able to manage its own fishery system, which consists of that nation's allocated portion of the TAC, together with the corresponding set of fishers, their communities, post-harvest activities, the socioeconomic environment and the management sub-system (Karagiannakos 1996).

The above examples may be seen as incorporating an international coordinating layer within the management system. In other circumstances, however, the need may be more towards decentralising the existing scale of management, essentially creating a local layer within the management system. On the resource side, as we learn more about the fish, we find that in some cases, genetically distinct sub-stocks exist within what had been managed as a single fish stock (e.g. in Canada's Bay of Fundy), so that excessively homogenous large-scale management fails to recognise biodiversity, and can thus be detrimental to long-term conservation.

On the human side, fishers often note the desirability of decentralisation to resolve the mismatch between the scale of fishery management and what they view as a 'natural' system, thereby making more allowance for the specific local conditions in the ecosystem and the human system. As an illustration, consider the case noted earlier of a system in which fishers in a community harvest a largely sedentary stock. The relevant *system* might be seen as naturally comprising the fish stock and the environment in which the fish live, together with the fishers and their local community. However, modern management approaches have often failed to recognise systems on such a fine geographical scale, and instead function on a more extensive scale, i.e. provincial, state or national. Decisions impacting on the community and the local stock were made at a 'higher' level and then applied uniformly, even if local conditions differed dramatically. In such a case, there may be much merit in adjusting management arrangements to create a local component in the management system.

Even if the fish stocks are somewhat migratory, as is the case in the groundfish fisheries on Canada's Atlantic coast, there may be a case for partial decentralisation of management. In this fishery, the migratory behaviour of the fish, amongst other factors, led government fishery management to operate on a larger spatial scale, believing local management to be impossible. The management of groundfish was therefore very different from that of the lobster fisheries in the area (and in neighbouring New England), which tend to be managed on a more local level, both formally and informally. (Acheson (1975, 1989) noted that in some cases, each lobster fisher's fishing area is set locally and accepted locally, with community self-regulation and enforcement of these arrangements.) Using a large spatial scale, groundfish management units were designed to incorporate most of a stock's range in a single management unit, but this created a mismatch with the human system. It led to a view of fishers as pawns on a chessboard, covering vast areas of the ocean in search of the fish, rather than as a group of people living in a cohesive fishing community. The problem with this approach is that it tends to forego the very benefits of local-level management that are found in the lobster fishery, particularly the self-regulatory nature of local peer pressure and the ties that fishers have to the norms of their home community. The introduction of community management boards in recent years (discussed in Chapters 5 and 14) has helped to resolve this mismatch between the human system on the one hand and the natural and management systems on the other.

1.2.2 Small-scale versus large-scale fishery systems

The dichotomy between small-scale and large-scale fisheries is a fundamental distinction in fishery debates, and relates to a range of organisational and structural factors. This dichotomy can be defined in several ways:

- based on the size of a typical fisher's operation (e.g. vessel size), so that a small-scale fishery is one with a fleet of small boats;
- based on an 'inshore' versus 'offshore' split, related both to the distance from shore that the fishery can operate and the extent to which the fishery is tied to coastal communities;
- based (in developing countries, and in the social science literature) on what is referred to as 'mode of production' (artisanal versus industrial), which depends both on actual activity, and on fisher self-perception, i.e. whether fishers view themselves as 'artisans'.

For each of these factors, however, there are no clear-cut, universal boundaries between small and large scales. Thus, fisheries might be better classified as small-scale or large-scale on a case-by-case basis, depending on an assessment of a set of relevant characteristics. There is a substantial literature relating to this classification challenge (e.g. Charles 1991a; Durand *et al.* 1991, etc.) which has led to the identification of certain key characteristics. Some of these are shown in the first column of Table 1.1, with relevant features of small- and large-scale fisheries also indicated for each characteristic. In this table, the first row simply indicates the terminology used to describe small-scale and large-scale fisheries, while subsequent rows show characteristics that relate to five key areas:

(1)   the physical location of the fishing activity (row 2);
(2)   fishery objectives, e.g. nature of goals, developing region goals, labour goals (rows 3–5);
(3)   economic factors, e.g. mode of production, ownership, labour–capital mix (rows 6–8);
(4)   social factors, e.g. rural–urban mix, extent of local community ties (rows 9 and 10);
(5)   external perceptions, i.e. how the fishery is viewed from outside (row 11).

Note that the specific characteristics indicated above, and in the table, are those that seem to have been most often discussed in the literature on small-scale and large-scale fisheries, but this is by no means to be considered as a comprehensive listing.

    As an example, consider the case of the Atlantic Canada groundfish fishery, harvesting cod, haddock, redfish and other stocks. As is the case for the fisheries of most developed nations, the term *small-scale* (and even more so, *artisanal*) is rarely used in this fishery. Fishers, managers and scientists simply do not use such language. Instead, the dominant dichotomy historically has been made between *inshore* and *offshore* fisheries. To what extent can these be considered small-scale and large-scale? The inshore/offshore split is made traditionally on the basis of the size of the fishing vessels and the distance from shore at which fishing takes place, both of which are major criteria in Table 1.1. Furthermore, the inshore fishery also has more labour-intensive operations, and stronger connections with coastal communities. All of these indicators suggest that the Atlantic Canadian inshore fishery classifies as small-scale.

Table 1.1   The dichotomy between small-scale and large-scale fishery systems.

	Small-scale fisheries	Large-scale fisheries
Alternative terminology	Artisanal (developing areas); inshore/small-boat (developed areas)	Industrial (developing areas); corporate (developed areas)
Fishing location	Coastal, including tidal, inshore and near-shore areas	Offshore, operating relatively far from the coast
Nature of objectives	Multiple goals (social, cultural, economic, etc.)	Tendency to focus on a single goal (profit maximisation)
Specific objectives in developing regions	Food production and livelihood security	Export production and foreign exchange
Objectives relating to utilisation of labour	Focus on maximising employment opportunities	Focus on minimising labour costs (i.e. employment)
Mode of production	Subsistence fisheries as well as commercial ones, selling into appropriate markets	Market-driven commercial fisheries, often with a focus on export
Ownership	Typically individual/family; often a small business in developed nations	Typically corporate, often based on foreign fleets in developing nations
Mix of inputs	Labour intensive, relatively low technological level	Capital intensive, emphasis on applying new technology
Rural–urban mix	Predominantly rural; located typically outside mainstream social and economic centres	Often urban or urban-tied; owners within mainstream social and economic centres
Community connections	Closely tied to communities where fishers live; integral part of those communities	Relatively separate and independent of coastal communities
Common perceptions	'Traditional', romantic, technologically simple	Modern, impersonal, multinational corporations

On the other hand, this fishery has become quite heavily capitalised, and in most cases inshore fishers (as with their counterparts elsewhere in the 'developed' world) are more likely to define themselves as business people in an 'industrial' fishery than as 'artisans'. However, the present situation is even more complicated. What was formerly a relatively clear split between inshore and offshore has been muddied by the development of a *midshore* fleet, composed of intermediate-sized vessels that operate under inshore regulations but have the capability to fish in offshore areas. While this fishery may be small-scale relative to an *offshore* operation, its development (and similar technological advances in other fisheries) certainly confuses the small-scale/large-scale dichotomy.

Indeed, Panayotou (1985: p.11) has noted that 'it is not unusual to find that what is considered a small-scale fishery in one country would be classed as a large-scale fishery in another'. The 'inshore' groundfish fishery of Atlantic Canada (as with those elsewhere in Nortl. America and Europe) would be classified as large-scale if located in a country such as Costa Rica. At the same time, the so-called 'advanced artisanal' shrimp fishery of Costa Rica would be seen as large-scale if transplanted into many parts of Africa.

1.2.3 Other approaches to characterising fishery systems

In addition to classifications on the basis of scale, as described above, fisheries can be characterised in various other ways:

- by their geographical location, particularly the latitude (tropical, temperate, arctic);
- by the type of ecosystem in which they are located (upwelling, estuary, reef, etc.);
- by the physical environment in which they take place (rocky bottom, bay, lake, etc.);
- by the depth in the water column at which they take place (ocean bottom, surface);
- by the nature and behaviour of the harvesters (e.g. organised vs. unorganised, multi-occupational vs. specialised);
- by the socioeconomic environment (e.g. urban vs. rural, developed vs. undeveloped infrastructure, poor vs. wealthy, level of local community involvement).

1.3   Summary

This chapter has explored approaches to implementing an integrated 'systems' approach to looking at fisheries, in particular how fishery systems can be *depicted* (notably through graphical means such as flow charts and strategic planning matrices) and how they can be *characterised* (particularly in terms of spatial scale and the dichotomy between small-scale and large-scale systems). In Chapters 2–7, the components of the fishery system (as shown in Fig. 1.1) are discussed sequentially, while the dynamics of such systems are examined in Chapter 8 and case studies are discussed in Chapter 9.

Chapter 2
The Natural System*

It is not particularly profound to note that without fish, there can be no fishery system. Moreover, the fish are not isolated within their respective fisheries, but rather live together with other fished and unfished species, in complex ecosystems. These ecosystems, in turn, involve not only living creatures, but also the physical and chemical features affecting life in the ecosystem.

All this constitutes the fishery system's *natural* (i.e. non-human) *sub-system*, the subject of this chapter. A reasonable understanding of this natural sub-system is surely crucial to proper management of the fisheries that rely on harvests from that environment. Yet no attempt to cover the natural world in a single chapter can possibly do justice to the enormous diversity and complexity inherent in aquatic life and the aquatic environment. It is also an impossible task to compile, within a single chapter, the vast array of research undertaken on the natural system by biologists, oceanographers and many other natural scientists. While much remains to be explored and studied – and indeed, it is often suggested that the underwater world is less understood than the moon's surface – nevertheless, far more work has been carried out in this area than for the human and management sub-systems of the fishery.

Given all this, the goal of this chapter must be rather modest, i.e. to provide some sense of the structure of the natural system, and how this relates to aspects of fishery management. The discussion begins with an overview and classification at the level of the various species caught in fisheries around the world, and then proceeds to the levels of populations, communities and habitats, and finally to the biophysical environment (Fig. 2.1). The reader is urged to refer to the relevant references for a fuller discussion.

2.1 The fish

What is caught in fishery systems? Is it 'fish'? In reality, global production from capture fisheries is comprised not only of 'fish' as such, but also many other types of animals that inhabit aquatic ecosystems. Nevertheless, the term 'fish' is typically used as shorthand to refer to the range of species caught in fisheries, and this convenience is adopted here.

*I am particularly grateful to Dr. Bruce Hatcher (Biology Department, Dalhousie University, Canada) for the considerable advice and assistance he provided in the writing of this chapter.

NATURAL ECOSYSTEM

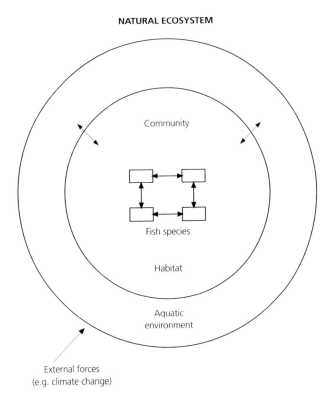

Fig. 2.1   The structure of the natural sub-system: fish species interact with the ecosystem, and in turn with the biophysical environment. External forces impact on the entire system.

Given this, what can be said about the nature of such species? What sense of order is there among the wide range of species exploited in fisheries around the world? King (1995: p. 1) points out that:

> 'In spite of the large diversity of marine species, most exploited species are contained in one of four large scientific groups. Species are included in a particular group on the basis of having similar [morphological and reproductive] characteristics and larval stages, as well as having what is believed to be a common ancestor, perhaps many millions of years ago.'

The four main groups of exploited species are:

- *Fishes* – with internal bony skeletons, fins and scales (e.g. cod, herring, salmon, tuna, carp)
- *Crustaceans* – invertebrates with external skeletons (e.g. prawns, shrimp, lobsters, crabs, krill)
- *Molluscs* – invertebrates with external or internal shells (e.g. clams, oysters, mussels, abalone, squid)

- *Echinoderms* – invertebrates with tube feet, with or without external plates (e.g. sea cucumbers, sea urchins).

To these major groups should be added two further groupings (taxa) that support less important fisheries:

- *Elasmobranchs* – cartilagenous fishes (e.g. sharks, rays)
- *Porifera* – sea sponges

The marine plants, notably seaweeds, constitute another major category of harvested aquatic species, Some of these are collected as they grow, but the vast majority of production arises in aquaculture.

The set of 'fished species' can be organised along taxonomic lines as in Fig. 2.2, and in more detail along market lines (e.g. for the European Union) as in Table 2.1. Note that fish and shellfish constitute the most important categories. These are given greater attention in the discussion that follows.

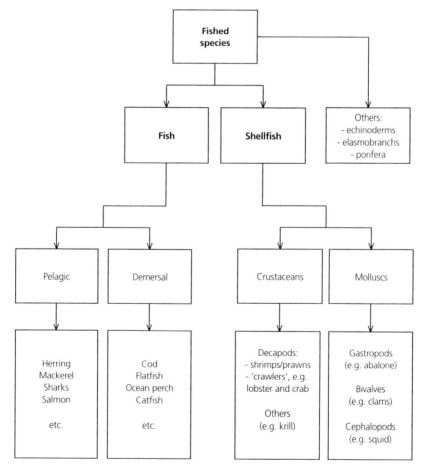

Fig. 2.2   A rough classification of those aquatic species of interest to fisheries.

Table 2.1   European Union catches by species group in 1995 (tonnes live weight) (Eurostat 1998).

Species group	Catch	%
Fin fish		
Marine		
Herring, sardine, anchovy	2 763 958	24.6
Cod, hake, haddock	2 616 972	23.3
Redfish, bass, conger	1 552 516	13.8
Jack, mullet, saury	1 383 418	12.3
Mackerel, snoek, cutlassfish	692 135	6.2
Tuna, bonito, billfish	526 820	4.7
Flounder, halibut, sole	343 620	3.1
Miscellaneous marine fishes	170 553	1.5
Sharks, rays, chimaeras	92 775	0.8
Freshwater fish	109 872	1.0
Diadromous fish	41 801	0.4
Invertebrates		
Crustaceans	230 762	2.1
Molluscs	597 318	5.3
Total	11 222 520	100

2.1.1 Fishes

Nine of the twelve common groupings of commercially important aquatic animals formulated by Sainsbury (1996) are fin fishes, which can be divided into two common categories according to the part of the ocean where they spend their adult lives: pelagic fish and demersal fish. The *pelagic* fish live principally in the upper layers of the ocean, near the surface, and the *demersal* fish live near the ocean bottom. Sainsbury's main groupings within these two categories are described below.

Pelagic fish

- *Herring and anchovy.* This group of relatively small pelagics includes herring, pilchards, tropical sardines, menhaden and anchovies. Sainsbury (1996: p. 8) notes that their spatial distribution typically 'matches the areas having greatest levels of primary production' (i.e. marine plant production). For example, major stocks of the tropical sardine 'are found in the Arabian Sea, off southeast Asia, Gulf of Guinea and northwest Africa', all areas in which upwelling occurs (see below). The species in this group often support major industrial fisheries.
- *Mackerel and tuna.* Sainsbury (1996: p. 9) notes that these species 'school in surface waters from tropical to temperate regions, are noted for rich, oily flesh and are the basis for major fisheries throughout the world.'
- *Sharks.* The many species of shark are caught throughout the world. As with bony fish, the meat of the elasmobranchs is used for food, but in addition, the livers and skin of sharks are utilised for various products. Shark fins are particularly valuable. The high demand, plus their characteristic low fecundity, has led to the over-exploitation of some shark stocks.

- *Salmon and trout.* These anadromous species breed in fresh water but spend most of their lives at sea. They are typically high-valued, the subject of large recreational fisheries, and culturally important to indigenous native groups in locations such as the northwest coast of North America and the Arctic (Fig. 2.3). Salmon is also farmed in many countries, supporting a major aquaculture industry.

Demersal fish

- *Cod and hake.* Species such as cod, haddock and pollock are found particularly in temperate and boreal regions such as the North Atlantic, where they have great historical importance and support many present-day fisheries. Hake species are widespread in many parts of the world, and while less valued, are often caught in abundance.
- *Flatfishes.* 'There are over 500 species of these bottom dwellers … Flatfishes live in close association with the sea bed on continental shelves from the tropics to the Arctic, and active fisheries exist worldwide …' (Sainsbury 1996: p. 11).
- *Spiny rayed fishes.* The most highly prized as food fish are the reef-dwelling snappers and groupers, which exist in tropical and subtropical regions worldwide (Sainsbury 1996).
- *Ocean perch.* 'Considerable commercial fishing operations are directed towards the ocean perch of the North Atlantic and rockfish of the North Pacific.' (Sainsbury 1996: p. 11)
- *Catfish.* 'These are bottom feeders able to tolerate poor water conditions found in estuaries, deltas, along coasts and in fresh water worldwide. They have few bones in the flesh and are farmed extensively in North America.' (Sainsbury 1996: p. 10)

Fig. 2.3   These Arctic charr, related to salmon, are caught in far northern ocean areas, where they are an important source of food for native people.

Note that the above list represents perhaps the most popular grouping of commercial fin fish species. Other taxonomic classifications are possible. For example, following Passino (1980) and Nelson (1994), three categories of fishes may be noted.

- Agnatha (e.g. lampreys, of little economic value and generally seen as destructive)
- Chondrichthyes (sharks, rays, skates and chimaeras)
- Actinopterygii ('bony fish', including most commercially important fish)

2.1.2 Shellfish

Crustaceans and molluscs together comprise Sainsbury's (1996) third grouping of aquatic species, the shellfish, which live largely on the sea bed of the continental shelf. The following list categorises the various shellfish, and gives some examples of each category (together with the scientific names of species).

- *Crustaceans.* Passino (1980) notes that most of the 26 000 species of Crustacea are aquatic, and that of the eight subclasses, seven comprise small animals that are often important in food webs (e.g. krill), but are not generally harvested. The eighth includes the economically important species within the biological order 'Decapoda', which are distributed widely around the world. The decapods include two groups of mostly high-valued species.
  - The shrimps and prawns, which swim just above the seabed, including the giant freshwater prawn (*Macrobrachium resenbergi*) and Sugpo prawn (*Penaeus monodon*).
  - The 'crawlers' such as the spiny lobster (*Panulirus japonicus*), American lobster (*Homarus americanus*), crayfish (*Astacas astacas*), Dungeness crab (*Cancer magister*) and snow crab (*Chionoecetes bairdi*).
- *Molluscs.* Many of these animals have highly valued muscle tissue. Passino (1980) divides the molluscs into three groupings.
  - *Gastropods* include those animals with a single shell that live on the seabed, including the abalone (*Haliotis rufescens*), conch (*Strombus gigas*) and escargot (*Helix aspera*).
  - *Bivalves* include clams, mussels, cockles, oysters and scallops. Only a few of the several thousand species are commercially important. Examples of high-value species are the Bay (or 'blue') mussel (*Mytilus edulis*), Japanese oyster (*Crassostrea gigas*) and giant clam (*Tridacna gigas*). These species 'have two symmetrical shells ... and live on the sea bed in shallow areas of the continental shelf, bays and estuaries' (Sainsbury 1996: p. 11).
  - *Cephalopods* include the swimming molluscs with an internal shell: species such as cuttlefish (*Sepia esculenta*), squid (*Loligo opalescens*) and octopus (*Octopus hongkongensis*). Sainsbury (1996: p. 12) compares the behaviour of the latter two species: the squid, caught in some major fisheries, 'swim in midwater schools from temperate to subtropical regions' while 'the octopus ... is a bottom dweller, living singly'.

Marine mammals

Marine mammals are not among the major fished species of the world, but they attract a disproportionate level of attention. As Passino (1980: p. 103) notes, these species 'have long been of interest to mankind, not only for meat, oil, fur and other commercial products, but also as objects of curiosity because of the similarity of their behaviour to humans.' Indeed, one of the most fundamental debates over human exploitation of the ocean revolves around whether or not marine mammals should be caught deliberately or even inadvertently in fisheries. Passino's categories of marine mammals (with some examples) are given below.

- *Carnivora* – California sea lion, walrus, harbor seal
- *Cetacea* – sperm whale, blue whale, killer whale, common dolphin, Amazon dolphin
- *Sirenia* – manatee, dugong

Some of these species (such as certain seals and whales) are harvested in directed, and often controversial, fisheries, while some, in contrast, have become the subject of major efforts within fisheries to avoid their capture (e.g. tuna fisheries avoiding dolphins). Still others, such as those in the third of these categories, are typically impacted only incidentally by industrial fisheries (although they are harvested in some small-scale fisheries in the tropics).

Fish stock versus fish population

In fishery systems, discussion of the fish is generally in terms of *stocks* and *populations*. Tyler & Gallucci (1980) note that there are 'distinctions between the fisheries concept of a stock and the biological concept of population', but there is some ambiguity in defining both of these terms, because they both refer to hierarchies of interrelated groups.

Consider first the latter idea of a *population*, which Royce (1996: p. 221) defines simply as 'a biological unit'. Iversen (1996: p. 106) suggests that this 'includes all individuals of a given species when there are no subspecies or, if there are subspecies, when their distributions are not discrete. It includes only all individuals of a subspecies when the distributions of the subspecies are discrete. Obviously, there is gene flow, or opportunity for such, throughout a population.' In a similar vein, Tyler & Gallucci (1980) refer to a population as 'the breeding unit of a species … Members of a population would have the opportunity to interbreed, and most of the larval fish in the population would develop in the same geographic area. Because of the relative lack of mixing between populations, and because populations are breeding units, gene frequencies within one population tend to be different from other populations.' Nonetheless, the physics of the ocean and extended pelagic stages of most fish larvae mean that small gene flows do occur between geographically distant populations, leading to the evolutionarily important concept of the *meta-population* (Tait & Dipper, 1998) – which has become central to studies of population biology.

Iversen (1996: p. 106) differentiates between the idea of a population, as defined above, and that of a sub-population: 'a fraction of a population that is genetically self-sustaining'. This, in turn, contrasts with the *stock* concept: 'a population or a portion of a population, all members of which are characterised by similarities that are not heritable, but are induced by the environment. A stock may or may not include members of several different subpopulations.'

Thus, a stock need not be a well-defined biological entity. Rather, as Gulland (1983: p. 21) puts it, 'the choice and definition of a unit stock can be considered as essentially an operational matter ... a group of fish can be treated as a unit stock if possible differences within the group and interchanges with other groups can be ignored without making the conclusions reached depart from reality to an unacceptable extent.' The latter conditions may be very difficult to verify, but if such an approach is used, Gulland notes that the fish defined as being in a certain stock may not even be all of the same species! Tyler & Gallucci (1980) put the approach similarly: 'Essentially a stock is defined so that fisheries yield models will work when they are applied ... In this way, a basic fish sampling unit is established that is characterised by homogeneity of natural production parameters. A stock may be a portion of a population, or include more than one population.'

There are, however, different perspectives on this point. For example, Cushing (1975: p. 19) has stated that a stock 'is a large population of fishes distinct from its neighbours and differences between stocks should be detectable genetically ...' Royce (1996: p. 221) agrees that 'the ideal definition of *stock* is that of a single interbreeding population', but notes that 'this condition is so rarely demonstrable ... that *stock* must be more or less arbitrarily defined.' The overall point, then, is that, in a manner analogous to the statistical method *analysis of variance* (ANOVA), stocks need to be defined so that 'between' variation (across different stocks) clearly exceeds 'within' variation (within a stock).

In summary, while fish populations are certainly of interest to scientists, it is fish stocks that are typically the focus of assessment and regulatory management efforts. As Iversen (1996: p.105) emphasises: 'Rational fisheries management requires knowledge of the extent to which exploited populations comprise a discrete, uniform and self-sustaining stock.'

There is clearly great diversity amongst the many aquatic species, only a small fraction of which is exploited in fisheries (but most of which are affected in some fashion by fishing). The range of variability in key characteristics of fished species is outlined in Table 2.2, beginning with where the species are typically found (the broad 'biogeographic province' and specific 'obligate habitat'), turning to biological features of the individual animals (longevity, extent of migratory movement and allometry, i.e. relative size) and finally trophic status (position in the food chain).

Table 2.2   Range of variability in key characteristics of fished species.

Characteristic	Range of cases	Examples
Biogeographic province	Cold waters	Antarctic krill
	Temperate waters	Cod, American lobster
	Tropical waters	Tiger shrimp, croaker
Obligate habitat	Tropical coral reefs	Parrotfish, groupers
	Tropical mangroves	Bivalves, juvenile fishes
	Tropical deep waters	Tuna
	Temperate demersal	Cod, flounder
	Temperate pelagic	Herring, salmon
Longevity	Short-lived	Shrimp (~1 year)
	Intermediate	Cod (~15–20 years)
	Long-lived	Redfish (~40 + years)
Extent of movement	Sessile	Mussels, sponges
	Sedentary	Clams, sea urchins
	Site-attached	Crabs, reef fishes
	Migratory	Cod, herring, mackerel
Allometry	Small	Krill (<) shrimp
	Medium	Herring (<) cod
	Large	Tuna (<) whale
Trophic status	Herbivore	Krill, abalone
	Omnivore	Lobster
	Detritivore	Prawn, sea cucumber
	1st-order predator	Herring baleen whale
	Top predator	Shark, killer whale

Australia's fisheries resources

'Fishing takes place throughout the large Australian Fishing Zone ... an area of about 9 million square kilometres that extends 200 nautical miles from the shore ... The Australian Fishing Zone is third largest in the world after those of France (because of its external territories) and the USA. The waters fished comprise many different habitats, including inland rivers and farm dams, mangrove-lined creeks and estuaries, shallow coastal bays, coral reefs, the continental shelf and continental slope to over 1.5 km in depth. Because of its long geographical isolation from other continents and its wide range of habitat types, one of the most diverse marine faunas in the world has evolved in Australia's waters. In contrast, by world standards, Australian inland waters have comparatively few freshwater fish species.

Though they are diverse and occupy one of the largest fishing zones, Australia's fisheries resources are not as abundant or productive as those in many other parts of the world. This is thought to be because, on average, Australian waters are low in nutrients due to little run-off from the dry Australian continent, a narrow continental shelf, the predominantly southwards flow of the main Australian coastal currents and the lack of permanent upwellings.'

Government of Australia (1999)

2.2   Spatial distribution of fished resources

A number of factors influence the spatial arrangement of fished species. Classification of fishery-related aquatic life with respect to their position in the water column (demersal versus pelagic) was noted above. Laevastu *et al.* (1996: p. 12) put it in a related way, suggesting that the three major components of marine ecosystem life are those on or near the ocean bottom (benthos) and those higher in the water column, i.e. the plankton (drifting animals, usually small) and the nekton (swimming animals).

The plankton are particularly important members of the marine communities that support fisheries. Sainsbury (1996: p. 4) notes that phytoplankton, which 'occurs near the surface where sunlight penetrates', is 'food for a wide range of sea life including some fish species, but mainly small animals known as zooplankton, which in turn are food for larger fish'. The point is that 'the abundance of phytoplankton is a useful guide to the ability of various regions to produce fish'. Thus, from a spatial perspective, organisms low in the food web (and the biophysical conditions that affect them) can strongly influence where fished species are concentrated.

Another major factor that explains the spatial distribution of aquatic species is latitude. In broad terms, species diversity tends to decrease with increasing latitude, although the explanation for this is not entirely clear. Furthermore, there is a greater tendency to observe 'dominant' species at higher latitudes as opposed to a more even mix of species in the tropics. As Laevastu *et al.* (1996: p. 58) note: 'The biodiversity of tropical marine environments is great … and the number of species in tropical communities is high, but few if any species dominate, whereas in higher latitudes strong dominance by a few species is a rule.' (These authors (p. 37) qualify their comment by noting that in tropical upwelling regions, it is usual to find that 'one or two pelagic species dominate'.)

Of course, the physical and chemical factors that vary with latitude (e.g. temperature, light, nutrients) also influence which species are caught where in the world's fisheries. For example, in a summary of the mix of fished species found in each of several key regions of the globe, Coull (1993) contrasts those found in two different parts of the Atlantic Ocean, the cold northern and the warm temperate areas.

- 'In the assemblage of stocks in the northern North Atlantic, the most fully researched of the oceans, demersal species like the gadoids (mainly cod, haddock, saithe) and flatfish like the plaice are a prominent part of the resource base, but it includes major pelagic species like the herring. It also includes much sought after crustaceans like the lobster and edible crab.'
- 'By contrast, in the warm temperate Atlantic … species like the octopus and cuttlefish become prominent, but the most important species are pelagic – the sardine, anchovy and the bluefin tuna (the tuna species best adapted to lower temperatures) for example. There are also species with a wider temperature tolerance: the demersal hake and the pelagic mackerel, for example, are both found in both cool and warm temperate waters.'

Natural capital

A traditional economic analysis of automobile manufacturing might focus on how labour (workers) and capital (machines and materials) combine together to produce goods (cars). In this sense, the term *capital* refers to a physical *stock* of machines and material that is needed to generate a *flow* of output (automobiles). In this case, it is manufactured human-made capital that we are discussing – the machines were made by people. If we now think of *capital* as being *stocks* of any inputs to a process, then we can view labour as *human capital* (including in that term not just a simple measure of person-hours of work, but also human ingenuity and other human attributes). Now note that in a fishery, in addition to human capital and human-made capital (which together provide a *flow* of fishing activity), there is another key *stock,* the *natural capital* of fish, water, ecosystems and physical–chemical environment, that provides flows not only of fish (through natural population dynamics), but also of ecological services such as maintenance of the water cycle, biological diversity and scenic beauty. This natural capital is essentially what keeps the planet functioning, just as capital produced by humans, i.e. roads and buildings and factories, forms the basis of the 'built environment'. Natural capital also combines with fishers and boats to generate a *flow* of seafood products.

Clearly, fish stocks are only part of the *natural capital* in the aquatic environment. Nevertheless, they are obviously of immediate importance from the fishery perspective! It is impossible to monitor change over time in the entire natural capital stock of the sea, but considerable effort is expended in many cases to measure changes in the biomass of key fish stocks. This just measures change in one component of natural capital, which is important in itself, although increases or decreases in the biomass of fished stocks do not necessarily imply that the overall natural capital of the ocean is moving in the same direction.

How valuable is natural capital? We regularly put a monetary value on human-made capital by measuring investment costs, i.e. the value of roads and bridges, office blocks and fish processing plants. With natural capital, however, it is not so easy. There are many roles played by elements of the ecosystem that are truly invaluable, so we can never measure the full value of natural capital in money terms. (Actually, it is not so different with human-made capital: do we value the pyramids of Egypt merely by what it cost to build them? Or to replace them? Some things have a cultural value that transcends monetary accounting.) Yet it has proved useful to look at natural capital partly in money terms (Costanza *et al.* 1997a) for two reasons.

First, this helps to highlight that it is not just dead fish on boats that have value (as in a typical gross domestic product accounting); so do fish stocks and other natural capital in the sea, which serve as a 'savings account' from which interest may be harvested year after year. The money value of natural capital serves as a lower bound on the true value.

Second, a natural capital analysis shows precisely how large harvest levels, which look good in standard economic terms, lead to *depreciation* (reduction) of the natural capital. This is like spending more than we earn, so our savings account declines to zero, or not maintaining our machines, until eventually the factory falls apart. In this way, we can explain what underlies the all too common fishery collapses around the world.

2.3   The ecosystem

The preceding discussion highlights the fact that the populations of fished species on the long list presented above do not live in isolation from one another. They interact in a complex manner with each other, with populations of unfished plants and animals, with humans (the top predator), and with the physical–chemical environment in which they live. Predators eat prey. Species compete for common food sources. (Furthermore, with humans involved, by-catch species are caught in the fishing gear targeted on other species.) Individuals, local populations, meta-populations, species, local assemblages of species, the communities they form, and the habitats, environments, and, ultimately, the global oceans they inhabit form a hierarchy of the sort fundamental to the science of ecology (Odum 1974). O'Neill *et al.* (1986) recognise a dual hierarchy of organisation, biological and functional, the essential concept being that of multi-scale linkages and interactions among the living and non-living components within what are called ecosystems.

Ecologists are among those most accustomed to thinking about *systems*. A systems approach is, after all, at the heart of ecology. As Liss & Warren (1980: p.45) note: 'A major objective of ecology is explanation of the structure and organisation of systems of populations.' Countless textbooks and research papers have been written about ecological systems, what they are, how to study them and how to protect them. The discussion here provides merely a cursory treatment of this important subject. Consider first the matter of defining what is meant by an *ecosystem*. Here are three possible definitions among many.

(1)   The Food and Agriculture Organisation defines an ecosystem as a 'functioning, interacting system composed of living organisms and their environment' and notes that 'The concept is applicable at any scale, from the planet as an ecosystem to a microscopic colony of organisms and its immediate surroundings.' (FAO 1999a)

(2)   The Convention on Biological Diversity defines an ecosystem as 'a dynamic complex of plant, animal and micro-organism communities and their non-living environment interacting as a functional unit.' (Arico 1998: p. 41)

(3)   Schramm & Hubert (1996: pp. 6–7) state that 'an ecosystem is a system formed by the interaction of a community of organisms with their environment. By this definition, an ecosystem may be as small as the array of microbes on the underside of a leaf on a stream bottom or as large as all of the interacting biotic and abiotic components in the Pacific Ocean.'

Some commentators suggest that the term *ecosystem* is bound to be ill-defined ('fuzzy'), and that the definition of an ecosystem is very dependent on the context – in particular on precisely what aspects are of interest and what boundaries we choose to draw. Others differ, pointing to the ability to define and operationalise ecosystems in practical circumstances (such as coral reefs), and to make use of certain characteristics of ecosystems, such as their hierarchical organisation. A 'systems' perspective on ecosystems, and on ecosystem–human interactions, is shown in the box on p. 34.

A systems view of ecosystems and human connections

Berkes & Folke (1992) present a systems view of the environment and of the human–environment relationship.

The systems view of the environment

'The structure and function of the ecosystem is sustained by synergistic feedbacks between organisms and their environment. The physical environment puts constraints on the growth and development of the biological subsystem which, in turn, actively modifies its physical environment to enhance its chance of survival … The ecological system as a whole is seen to be in a dynamic process of self-organisation and self-maintenance (homeostasis). Solar energy drives the use of matter for self-organisation, and complex, interdependent hierarchical structures evolve. It is this self-organising ability, the resilience, organisation, and vigour of the ecosystem that generates and sustains the goods and services which form the necessary material basis for human societies.'

The systems view of the human–environment relationship

'The structure and function of the ecosystem is sustained by synergistic feedbacks between human societies and their environment. The physical and biological environment place basic physical constraints on the growth and development of the human subsystem … The human subsystem, in turn, actively modifies its physical and biological environment … the self-organising ability and homeostasis of the ecosystem is paralleled by the self-organising ability and homeostasis of the human subsystem.'

2.4   Aquatic/fishery ecosystems

In discussing ecosystems of relevance to fisheries, we can speak broadly of *aquatic ecosystems*, or we may focus on *marine ecosystems* if our interest is in an oceanic fishery, or we may highlight the connection between fisheries and aquatic ecosystems by referring to *fishery ecosystems,* which Royce (1996: p. 200) defines simply as 'those that support a fishery activity'.

Despite their obvious importance to fisheries, aquatic ecosystems, at least in the sea, are by no means well understood. It is often noted that humans know more about the surface of the moon than they do about the bottom of the ocean. Many remarkable findings deep in the ocean have only recently been discovered, and undoubtedly many more are yet to be discovered. Biodiversity in the marine environment is not limited to the ocean floor, and aquatic biodiversity is, of course, also found in freshwater lakes and rivers. What is common to many aquatic environments is our lack of understanding of the detailed nature and functioning of the relevant ecosystems.

Interactions are complex, poorly understood and potentially bi-directional. For example, suppose that in a particular ecosystem we focus on a specific predator and prey pair. If a fishery targeting the prey reduces its population through harvesting, this may imply a reduction in the food available for the predator. Should the allowable harvest of the prey be reduced below what would otherwise be the case to provide greater availability for the predator?

Conversely, what would be the effect of a reduced catch by the predator? This would seem to imply greater mortality for the prey, and thus a reduced prey population. At the same time, however, a larger predator stock may well produce more juveniles in the future, and it is sometimes the case that those juveniles are themselves a food source for the *prey* (as with juvenile cod being eaten by herring). In such a case, the overall impact on the prey is uncertain – a balance between increased predation and increased food availability. This is potentially

---

Large marine ecosystems (LMEs)

Large marine ecosystems are areas of ocean with impressive characteristics in terms of spatial extent, ecological functions and economic importance. In particular, large marine ecosystems (LMEs):

- 'are relatively large regions, on the order of 200 000 km^2 or larger, characterised by distinct bathymetry, hydrography, productivity, and trophically dependent populations.' (Sherman 1993: p. 3)
- 'represent distinct geographic regions characterised by ecological criteria rather than by political expediency. The LMEs represent a natural ecological unit for resource and environmental assessments and management. On a global scale, 49 LMEs have been described. They account for 95% of the annual global marine fishery yields ...' (Sherman 1998: p. 46)

Examples of LMEs currently defined are (Sherman 1993):

- Yellow Sea ecosystem (eastern Asia)
- Sea of Japan (eastern Asia)
- Bay of Bengal (southern Asia)
- Great Barrier Reef (Australia)
- North Sea ecosystem (northern Europe)
- Baltic Sea (Europe)
- Mediterranean–Adriatic Sea (Europe/Africa)
- Gulf of Guinea (west Africa)
- Benguela (southern Africa)
- California current (western North America)
- Scotian shelf (eastern North America)
- Gulf of Mexico (Caribbean)
- Humboldt current (western South America)
- Patagonian shelf (eastern South America)

Fig. 2.4   This tuna, caught off the Atlantic coast of North America, was highly migratory and passed through a variety of ecosystems before being caught. By considering such species in the context of ecosystems, we can take into account, as best possible, the multi-species and ecosystemic interactions affecting them.

compounded by aspects of competition. An increased abundance of the predator may imply greater competition with other predator species that have the same prey as a food source. The impact of this on the prey will depend on the strength of competition between the predators, and the preferences of each species for the specific prey as a food source. The interactions can be complex!

Traditionally, assessments of fishery resources have been of a *single-species* nature. In other words, the multi-species fishery system is studied one species at a time, typically ignoring the non-commercial species, and generally not considering interactions among species, except in an aggregated form, as a *natural mortality* coefficient which is typically assumed to be constant. This is clearly a great oversimplification of the range of factors known to affect the survival of fish (Gulland 1982; Hilborn 1987; Laevastu *et al.* 1996). Similarly, effects of the physical–chemical environment on the dynamics of fished populations are rarely incorporated in the models used to predict fish stock status, despite strong scientific reasons to do so (Longhurst 1981; Hilborn & Walters 1992; Mann & Lazier 1996). In short, the fact that fish are small and integral parts of large marine ecosystems has rarely been incorporated in stock assessment and prediction for management decision making (Fig. 2.4).

While fishery scientists have always recognised the desirability of broadening to an *ecosystem approach* (e.g. Gulland 1982), this has rarely taken place owing to the difficulties in doing so, the lack of resources to do so, and an inertia that developed in some cases to perpetuate well-accepted, if flawed, single-species assessment methods. However, a new trend is developing to find feasible methods for multi-species and/or ecosystem-level assessment within an *ecosystem approach*. This is discussed in detail in Chapter 12.

2.5   A typology of fishery ecosystems

How can fishery-related ecosystems be classified? Royce (1996: p. 200) suggests that whereas 'many terrestrial systems are identified by the dominant biological complex, such as tundra or tropical rain forest', 'most aquatic systems are identified by the dominant physical factors; examples are streams, lakes, continental shelves, or tropical seas.' This generalisation clearly has exceptions, i.e. cases where it is in fact the biological features that define the system, as with mangals, seagrass meadows, coral reefs and kelp beds – although the latter biological communities, combined, occupy a very small fraction of the global ocean area (Mann 1982).

The following typology of fishery ecosystems follows Royce's (1996) approach, which begins with marine systems, followed by freshwater systems and finally aquaculture systems (although the latter are not part of the present discussion). Most ecosystem descriptions are from Royce, but the quotations are taken from the Committee on Biological Diversity in Marine Systems (1995).

- *Estuaries* – major mollusc fisheries, spawning/nursery areas for crustaceans and fishes (Fig. 2.5).
    'Estuaries have long been associated with … some of the world's greatest fisheries for oysters, clams, shrimp, crabs and fishes [and aquatic communities such as] marshes, seagrass beds, and mangroves … The majority of people in the world live within 100 kilometres of bays and estuaries … and such environments command enormous attention and use.' (p. 37)
- *Continental shelves* – The major demersal fisheries of the world including cod and flounder in temperate areas and multiple species of groundfish in subtropical/tropical areas.
    'Continental shelves represent the great interface between the continents and open oceans … sites of most of the major world fisheries, located on fishing banks, in upwelling zones, and on broad shallow platforms.' (p. 41)
- *Continental slopes* – rockfish and hake.
    'The seemingly distant and more immune slope waters are no longer far away and no longer immune. Deep-water fisheries – and their attendant physical effects (e.g. habitat alteration due to dredging and trawling) – have entered slope waters …' (pp. 41–42)
- *Oceanic surface waters* – The sites of the world's most extensive driftnet, midwater trawl and purse seine fisheries for pelagic species such as herring, anchovy and tuna, as well as for whales.

Fig. 2.5   Mangroves are found in many bays, estuaries and other coastal regions of the tropics. Along with coral reefs, they are of critical importance to the coastal environment; mangroves provide key habitats for juvenile fishes and other aquatic life.

- *Oceanic near-surface waters* – substantial potential for fishery resources.
  'The public's view of the open sea as a vast homogeneous body of water is belied by the complexity of oceanic subsystems: fronts and eddies, upwelling and downwelling regions, boundary currents, and large ocean gyres' (p. 42)
- *Deep sea* – low probability of substantial fishery resources.
  '… the previous notion of a global deep-sea bottom that is uniformly featureless has been shattered over the last two decades by countless discoveries of unique, sometimes bizarre, and highly diverse deep-sea communities … Each major ocean basin has a distinctive fauna, and bottom assemblages vary according to latitudinal gradients and topographical features such as basins, canyons, and areas of strong currents …' (p. 44)
- *Tropical reefs* – The protein source for many of the world's artisanal fishers is provided by the many species of fish and invertebrates that live in these topographically complex structures.
  'Reefs provide good examples of the importance of linkages between habitats – reef biodiversity is dependent on adjacent ecosystems for feeding areas and nursery grounds and as buffers against land runoff of sediments and nutrients …' (p. 39)
- *Oligotrophic lakes* – located in cold environments, at high altitude or high latitude, often dominated by salmonids.
- *Meso and eutrophic lakes* – located in warm, temperate to tropical areas, having high productivity, and often dominated by cichlids.
- *Cold-water streams* – located typically in higher altitude, forested areas, and often dominated by salmonids.

The Barents Sea

'The Barents Sea covers an area of 1 405 000 km^2. It is a high-latitude, shallow-water ecosystem and one of the world's most biologically productive oceans. The high level of biological production is due to the sea's shallow waters (5–400 m), a high level of vertical mixing, and the confluence of the warm North Atlantic current and cold Arctic waters. Living resources of the Barents Sea include fish, marine mammals and seabirds. The sea serves as a nursery area for commercially important stocks of herring, cod and haddock, while other species such as capelin and polar cod spend their entire life cycles in the area. Several species of cetaceans (minke and humpback whales, white-sided dolphin and white-beaked dolphin) visit the area during their feeding migrations. The sea supports some of the world's largest seabird populations … The Arctic ecosystem of the Barents Sea region is extremely vulnerable to impacts from human activities and is slow to recover from disruptions or damage. The high degree of vulnerability is due to slow biological and chemical processes caused by low water temperatures, high concentrations of biological resources, and the relatively high level of body fat in organisms where organic contaminants readily accumulate.'

Eglington *et al.* (1998)

- *Warm-water rivers* – located in lower altitude, agricultural areas, typically with perciform and cyprinid fishes.
- *Impoundments* – algae, prawns and fish for aquaculture.
- Raceways, fish pens and cages – aquaculture.

2.6 The physical–chemical environment

The discussion in this chapter began with an overview and classification of the various species caught in fisheries around the world. The obvious point was then made that these individual species, far from sitting in isolation within their respective fisheries, live together with other fished and unfished species within complex ecosystems. The discussion proceeded to examine some characteristics and classifications of these aquatic ecosystems. This logic leads us to want to understand the fish and their ecosystem in order to manage the corresponding fishery.

In discussing ecosystems, it is important to take note not only of the living creatures in the system, but also of the physical features affecting life in the ecosystem. Such features operate on various spatial scales. These range from currents flowing across an ocean basin, to upwellings that affect specific segments of the ocean, to more modest (albeit locally important) effects driven by freshwater/saltwater interfaces (e.g. river outflows, as for the St Lawrence River on Canada's Atlantic coast, variations in which can have a major impact on fish stocks in the Gulf of St Lawrence). Some types of forcing, such as the tides, can affect fishery systems on large spatial scales as well as very locally, within a given bay, for example.

This section focuses on aquatic systems in the ocean environment, providing a brief overview of some of the most significant biophysical interactions, beginning at a large spatial scale and then turning to more localised phenomena. The discussion here draws on the work of Mann and Lazier (1996) and Tait and Dipper (1998), to which the reader is referred for a more comprehensive treatment.

2.6.1 The winds

A discussion of large-scale physical impacts on fisheries logically begins outside the water. While fish stocks can be strongly affected by ocean currents, these in turn are driven by the winds, as Mann & Lazier (1996: p. 242) note:

> 'All the major surface currents in the oceans are created by the drag of the wind on the surface of the water. The winds, in turn, are created because the earth's surface is heated unevenly by the sun, making the tropical regions warm and the polar regions cold.'

The planet's major wind systems are of two principal types (Mann & Lazier 1996: pp. 242–243):

- The northeast and southeast *trade winds* occur roughly between 30° north and the equator, and 30° south and the equator, respectively. 'The trade winds arise because the warm air in the equatorial regions rises and is replaced by air flowing towards the equator in both the northern and southern hemispheres. The Coriolis force deflects the equatorward flows to the west giving rise to the northeast and southeast trades …'
- The *westerlies* are found in two bands, between 30° and 60° north, and between 30° and 60° south latitude. These 'continuous bands of flow right around the earth' are strong westerly winds arising from a horizontal pressure gradient caused because in the relevant latitudes (>30°), 'the temperature at all levels in the troposphere [the lower 10–20 km of the atmosphere] decreases rapidly'.

In addition, there are important monsoon wind systems in the northern Indo-Pacific, Indian Ocean and Arabian Seas. While somewhat more localised, these have dramatic impacts on the affected fisheries, and indeed are perhaps the dominant driving forces on both resource and human behaviour in these regions.

The most significant direct effect of winds on fisheries is to limit fishing effort (and hence fishing mortality) through foul weather. The impact is most severe in artisanal fisheries because of the vulnerability of their small, often poorly outfitted vessels. The most important indirect effect of wind on fisheries lies in the major upwellings on the eastern margins of the ocean basins, where trade winds blow offshore (see below).

2.6.2 Ocean currents

Elisabeth Mann Borgese, a political scientist best known for her work on the Law of the Sea and her advocacy of conservationist use of the oceans, nicely summed up the essence of the world's ocean currents (Borgese 1998: p. 28):

'Ocean space is traversed by a system of currents, driven by the temperature differences between equator and poles, by winds, and by the rotation of the earth which deflects their course (Coriolis force). Compared to the earth's rivers, the quantity of water transported by these currents through ocean space is imposing … The volume of water that flows in the Gulf Stream is about 100 times that of all rivers on earth combined.'

Tait & Dipper (1998: p. 9) describe ocean currents in a similar vein. 'The major currents of the oceans are caused by the combined effects of wind action and barometric pressures on the surface, and density differences between different parts of the sea.' They note (p. 11) that:

'the North-East and South-East Trade winds blow fairly consistently throughout the year, setting in motion the surface water to form the great North and South Equatorial Currents which flow from east to west in the Atlantic, Indian and Pacific Oceans. Across the path of these currents lie continents which deflect the waters north or south.'

This behaviour leads to the existence of *gyres*, i.e. large-scale circular flows of water around the various ocean basins (Mann & Lazier 1996). There are subtropical gyres found between latitudes 15° and 45°, and sub-polar gyres in the north Pacific and Atlantic. The subtropical gyres include, on their western sides, strong poleward-flowing boundary currents such as the Gulf Stream off North America and the Kuroshio current off Asia, which 'transport large quantities of heat away from equatorial regions' (p. 280). The sub-polar gyres produce 'some of the most productive waters in the world' (p. 241), such as the Labrador current off Newfoundland, where the northern cod stock resides.

Mann & Lazier (1996) describe the importance that gyres and other aspects of the ocean's currents (meanders, rings and eddies) play in the ocean's biological production. They note (p. 261) that 'organisms have adapted to the various physical regimes'; for example, some species 'use the boundary currents for long-range transport between breeding grounds and feeding areas', while others 'make several circuits of an ocean gyre while growing to maturity'.

The most profound effect of ocean surface currents on fisheries is to connect populations of fish across space from spawning grounds to nursery grounds, and among sub-populations in a meta-population. The transport of larvae blurs stock boundaries (confounding fishery models and management), but also replenishes locally depleted stocks (providing a major biological argument in favour of marine protected areas, as discussed in Chapter 12).

Water moves not only around the oceans, but also up and down in the water column. As Tait & Dipper (1998: p. 14) note, 'oceanic circulation should be visualised in three dimensions.' Indeed, the surface currents described above cause movements of water that must be replaced, either with influx of surface water from elsewhere, or with deep water rising to the surface (these are known as upwellings; see below). In the latter case, vast and poorly understood sub-surface currents are created: the bottom current, the deep current and the intermediate current (Tait & Dipper 1998).

2.6.3 Upwellings

As noted above, upwelling occurs when surface water is swept by the wind away from the coast, and this is replaced by deeper water rising to the surface close to shore. In more localised settings, seabed topography (e.g. steep-sided reefs) may deflect bottom currents towards the surface. In either case, such water, typically nutrient-rich, supplies essential plant food to the lighted (euphotic) zone of the area, creating high fertility and consequently important fishing grounds. As Mann & Lazier (1996: p. 139) note:

> 'the key to high biological productivity is the upwelling of "new" nutrients from deep waters into the euphotic zone and the retention of phytoplankton in well-lighted waters by stratification of the water column.'

They note (p. 140) five major coastal currents associated with upwelling areas, in all of which fish that feed low in the food web, such as anchovy, sardine, pilchard and mackerel, dominate:

- the Canary current (off northwest Africa);
- the Benguela current (west coast of southern Africa);
- the Peru current (west coast of South America);
- the California current (west coast of North America);
- the Somali current (western Indian Ocean).

2.6.4 Other relatively localised phenomena

In addition to upwellings, various other biophysical features may be found at a local to intermediate spatial scales (i.e. smaller than that of continental margins). Among these are fronts, tidal currents and various freshwater/saltwater interactions (Mann & Lazier 1996).

- *Fronts* lie at the boundary between waters of differing characteristics, e.g. strongly mixed versus stratified, or lower-salinity run-off versus fully saline. Such fronts can occur between mixed and stratified waters on continental shelves (tidal fronts, due to tidal forces), or between shelf and slope waters (shelf-break fronts), or on boundaries due to upwelling or freshwater run-off.
- *Tidal currents* can have several impacts. First, they can create vertical mixing in the water column, potentially allowing a year-round flow of nutrients and thereby particularly rich fishing grounds. Second, they may generate flows that affect spawning and larvae dispersal. Third, 'internal waves' can be generated, increasing nutrient levels in certain parts of the water column.
- *Freshwater/saltwater interactions* occur within estuaries, where specific local water circulation patterns are found, in 'plumes' where fresh water from rivers and estuaries is carried onto the continental shelf, and more generally due to tidal mixing in coastal areas of the ocean.

2.6.5 Physical features

The discussion above has focused on the specific biophysical aspects of the ocean system that relate to flows, i.e. winds, ocean currents, freshwater outflows and tidal waters. These flows are certainly of fundamental importance to fisheries, but also worthy of note are the specific physical aspects of the coast and the ocean bottom.

- The physical shape of a coastline, its curves and indentations, in the form of bays, inlets, estuaries and the like, can be a dominant factor determining the behaviour of currents, tides and fish migrations. For example, in Canada's Bay of Fundy, known for the highest tides in the world, fishers operate 'in tune with the tides' by timing their fishing activity to the semi-diurnal rhythm of the fishes' lives.
- The physical nature of the coast and of the ocean bottom also influences the type of fish habitat available. As fishers are well aware, this largely determines which species are to be found where. Whether a tropical coastline supports coral reefs or seagrass meadows, or whether a temperate coastal area has sandy or rocky ocean bottoms, clearly plays a major role in determining the benthic/demersal life that can be supported in a particular location. Some habitats appear to be critical to the survival of certain (usually juvenile) stages of life cycles, and hence are candidates for special protection.

2.7 Summary

This chapter has presented a brief overview of the *natural sub-systems* within fishery systems, focusing on the structure of these systems, and approaches to classifying the relevant fish species, ecosystems and physical environments. As noted at the outset, the chapter does not attempt to do justice to the wealth of research on the subject, nor can it reflect the subtleties of the complex nature of aquatic systems. The focus has been on some major structural aspects, as listed below.

- The vast array of exploited species in fisheries, which fall into a few categories (for example, fish vs. shellfish, demersal vs. pelagic species).
- The diversity among species, which exists in terms of such characteristics as morphology, life cycle and habitat.
- The types of aquatic ecosystems of which fished species are a part.
- The various physical and biophysical phenomena which play important roles as driving forces behind change in the fishery system.

Later in the book, we return to some key issues that have arisen in this chapter. One of these is the need for integrated approaches to management and research on fishery systems, notably focusing on links between the human and natural systems. Related to this is the need for an 'ecosystem approach' to fishery management (Chapter 12), one that integrates the ecosystem into management, considering the impacts of human activity on the ecosystem, and the role of the ecosystem in affecting fishery outcomes.

Chapter 3
The Human System

'Any system is a set of interrelated components. In a fishery system, one of the primary, most dynamic components of the system are people and their behaviors.'

Orbach (1980: p. 149)

This chapter examines the range of human elements in the fishery system, beginning with the fishers (harvesters) and then moving to the post-harvest components, fishing households and communities, and the broader socioeconomic environment. Figure 3.1 depicts the human

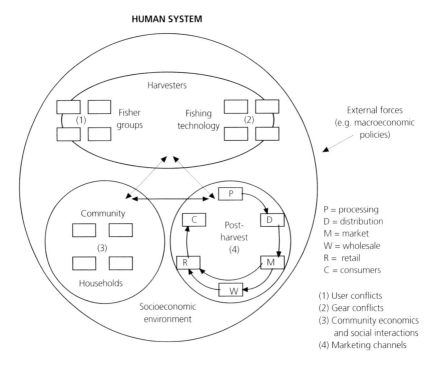

Fig. 3.1 The structure of the human sub-system. Harvesters interact with one another through fisher groups (sectors or organisations) and through their fishing technology. User and gear conflicts can arise as a result (Chapter 13). The fishers also interact with their households and communities, where economic and social interactions are important. The post-harvest sector involves a flow of activities from processing to consumers. Finally, these various components interact with the socioeconomic environment, and external forces impact on the entire system.

sub-system, with emphasis on the internal structure within fisher, technology, community and post-harvest elements, and the interactions among the various elements.

3.1   The fishers

Like the fisheries in which they operate, fishers are diverse in their variety and widespread in their distribution around the world. Fishers are at the heart of the human sub-system of the fishery, although discussion about (and statistics on) participants in the fishery often focus as much on boats as on people. The following discussion focuses on two typologies, one of fishers and the other of fishing methods. Later in the chapter, the discussion moves 'beyond the fishers' to look at the many components of the post-harvest sector, fishing households and communities, and the broader socioeconomic environment in which the fishery operates.

3.2   A typology of fishers

Fishers around the world seem to fit within four principal categories (Fig. 3.2), based on the nature of, and background to, their particular fishing activities.

- *Subsistence fishers:* those catching fish as their own source of food.
- *Native/indigenous/aboriginal fishers:* those belonging to aboriginal groups, often fishing for subsistence.
- *Recreational fishers:* those catching fish principally for their own enjoyment.
- *Commercial fishers:* those catching fish for sale in domestic or export markets; these fishers are traditionally viewed as falling into *artisanal* and *industrial* categories (see below).

These are not mutually exclusive groupings, however, with overlap possible between groups.

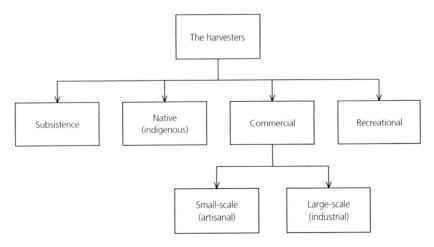

Fig. 3.2   A simple classification of the harvesters (fishers) into four main groupings, with the commercial sector being further subdivided. Note, however, that there is considerable diversity within any single group, there are gradations between groups, and in many cases, some fishers may fit into more than one group.

The relative presence of each of these fisher categories varies from fishery to fishery. For example, in developing regions, subsistence fishers and artisanal commercial fishers are typically prevalent in the near-shore coastal fisheries, while industrial (notably foreign 'distant water') fleets may dominate fisheries further from shore, exploiting offshore/deep-sea resources. The relative presence of recreational fishers varies widely, and is particularly notable in developed regions (where disposable incomes are on average higher) and in developing regions where both the natural conditions and tourist infrastructure are suitable (e.g. the tourism-oriented islands of the Caribbean). Native/indigenous/aboriginal fishers may play a major role in specific locations, typically where a segment of a nation's citizenry is of much longer historical standing than the now-dominant population groups (most often of European decent). In some cases, these fishers have played a large role in the local fishery over long periods of time (e.g. in Panama), while in other cases, recent agreements with governments have greatly increased their role: a notable example is that of the Maori in New Zealand.

In some settings, all four of these forms of fishers are found. For example, in Atlantic Canada, where a range of fisher groupings operate, there is also considerable mixing of fishing activities and modes.

- The indigenous Mi'Kmaq and Malisseet peoples fish for subsistence, but also have the right (recently recognised by the Supreme Court) to fish commercially, on a small-scale basis.
- Along coastal Newfoundland, fishing is primarily seen as a small-scale commercial activity, but coastal residents also have the long-standing use of fish resources as a subsistence food source.
- Throughout the region, cod is caught in industrial and small-scale fishing, operating side by side, as well as in recreational fishing.

3.2.1 A focus on commercial fisheries

Looking specifically at the commercial fishery category globally, the principal, dominant dichotomy (as noted in Chapter 1) lies between artisanal/small-scale and industrial/large-scale fishers:

- artisanal/small-scale fishers are those catching commercially but at low levels;
- industrial/large-scale fishers are the corporate fleets of capital-intensive vessels.

Panayotou (1985: p. 11) gives a broad depiction of small- and large-scale fishers:

- *large-scale fishers* are 'those who have a broad spectrum of options both in terms of fishing grounds and non-fishing investment opportunities';
- *small-scale fishers* are 'those who, by virtue of their limited fishing range and a host of related socioeconomic characteristics, are confined to a narrow strip of land and sea around their community, are faced with a limited set of options, if any, and are intrinsically dependent on the local resources'.

Small-scale fishers are characterised by various attributes, such as:

(1)   a high level of dependence on the fishery for their livelihood, with few other job op-
       portunities, and often with relatively low net incomes;
(2)   utilisation of vessels that are relatively small and individually owned;
(3)   a tendency towards use of a 'share' system to divide fishing income among boat owner,
       captain and crew, rather than a wage system (as is common in industrial fisheries);
(4)   traditionally being outside the centres of economic and political power, on the periph-
       ery of the larger society, whether owing to location (e.g. in rural or remote areas) or
       membership in particular minority groups (e.g. indigenous native peoples), although it
       should be noted that in some industrialised countries, the political influence of small-
       scale fishers has been considerable;
(5)   often being viewed by analysts in one of two very different ways: as participants in an
       activity 'ripe for modernisation and rationalisation', or as people (and communities)
       threatened by external economic forces and in need of protection.

Small-scale fishers seek out a livelihood in most countries of the world. Recently, a worldwide network of small-scale fisher organisations has been formed (Fig. 3.3).

The characterisation of fishers into small-scale and large-scale groupings obviously provides only an indication of the reality. It seems better to envision a spectrum from one extreme
to the other. Consider, for example, the data in Table 3.1 for European Union fishing fleets,
which indicate that in some nations (such as Sweden and Greece), a small-boat fleet dominates, while in others (e.g. Belgium and the Netherlands), the fleet is clearly of a larger scale,
on average.

Fig. 3.3   These artisanal fishers operate on the outskirts of Lima, Peru. In the background is their union office; an
international association of fisher organisations has been formed to support small-scale fishers world-wide.

Table 3.1   The European Union fishing fleet in 1995 (European Commission 1998).

	Number of boats	Power (kW)	kW/boat	GRT/boat	<10 m (%)	10–24 m (%)	>24 m (%)
Finland	2 959	174 608	59	9			
Sweden	4 349	328 686	76	14	88.5	8.8	2.7
Denmark	4 993	412 723	83	20	67.7	26.3	4.4
UK	9 983	1 104 406	111	24	70.0	24.9	5.0
Netherlands	508	436 197	859	301	1.7	39.6	58.7
Belgium	157	65 889	418	146	0	52.8	47.1
Germany	2 452	167 692	68	31	76.9	20.6	0
Ireland	1 421	190 501	134	39	44.8	49.2	5.7
France	6 650	997 548	150	27	59.9	35.9	4.2
Portugal	12 317	416 010	34	11	87.4	10.4	0
Spain	19 103	1 849 993	97	32	71.8	21.7	6.4
Italy	16 434	1 513 871	92	16	53.2	42.0	4.8
Greece	20 354	662 768	33	6	88.1	10.9	0
Total	101 680	8 320 892					

3.2.2 Diversity of fishers

Whatever the particular class of fishers we may examine, there is bound to be some degree of internal heterogeneity. This will arise in a number of respects, some of which are listed below.

- Within any given group of fishers, there are variations in many social and demographic aspects, such as age, education, social status and religion. Between fisher groups, there may be differences in internal social cohesion (how attached the fishers feel to their group) and in community connections (attachment to their local community).
- In commercial fisheries, there is also variation by occupational commitment (e.g. full-time versus part-time) and the level of occupational pluralism (with some fishers specialised entirely in fishing for a single species, some utilising a range of resources, and others drawing income from outside the fishery as well as from fishing).
- Fishers vary in their motivation and behaviour. For example, some may be profit-maximisers (behaving as ideal capitalist 'firms'), while others may be satisficers (fishing to obtain 'enough' income). The latter behaviour has been noted in the Costa Rican community of Puerto Thiel, and contrasts with other communities also fishing in the same region, the Gulf of Nicoya (Gonzalez *et al.* 1993; Charles & Herrera 1994).

3.2.3 Women in fishing

In the above discussion of fishers, there is no mention of what may be globally the most basic classification of humans, namely that by gender: males and females. In the past, such a distinction may have been seen as irrelevant, since attention was focused on the fishing activity, and women, it was assumed, played no role in this. As Nadel-Klein & Davis (1988: p. 1) put it:

'Images of fishing tend to be male … It is safe to say, in fact, that Western tradition has stereotyped fishing as an exclusively male occupation. Unfortunately, this perspective robs us of the ability to think about women in fishing communities, to ask what women do and to contemplate the possibility that "fishing" as an economic enterprise might require, and even value, women's labor.'

Indeed, the reality in much of the world seems to be that women are either involved in fishing itself, or play a major role in the on-shore components of the fishery system (such as processing, in industrial contexts, or marketing, in artisanal settings). A good example of direct involvement in fishing is the activity of women in the South Pacific (Oceania). Chapman (1987) concludes that this activity, while looked down upon by the men in these communities, in fact provides a major and especially regular source of food protein. She also quotes evidence that women are often more knowledgeable than men on fisheries and ecological matters in this region, while noting that 'consultation of women as fishing experts would be a complete reversal in policy for most modern fisheries development officers in Oceania' (Chapman 1987: p. 283).

The role of women in the fishery and in fishing communities has received a certain degree of attention, particularly in the social science literature. For example, the contributions in Nadel-Klein & Davis (1988) explore a range of such activities across most continents of the world. Davis (1988: p. 214) suggests that there are two major forms of participation by women:

'Studies of women and the fishery tend to regard women's contributions in two different ways. The first mode is tangible and functional and entails women's fishery-related

---

The Norwegian Fishermen's Association

'Norges Fiskarlag (the Norwegian Fishermen's Association) is a politically independent national organisation based on voluntary membership of fishermen via their county associations and group organisations. The members are owners of boats and/or gear, and fishermen working on a share or percentage basis. It includes fishermen on small fishing boats as well as those on board the largest deep sea fishing vessels. The highest governing body of Norges Fiskarlag is the Congress, which consists of some 70 delegates elected by the county associations and group organisations. Ordinarily the Congress meets every second year. In the intermediate periods, authority is exercised by the national Committee ... Organisational activities embrace economic, social and cultural fields, as well as matters directly connected with fishing. Norges Fiskarlag works in close co-operation with the government authorities. For instance, the Main Agreement for the Fishing Industry, entered into on June 3, 1964, provides that the Association is responsible for negotiations with the authorities on matters that concern economic conditions in the industry as a whole. Because of the great importance of the fishing industry in the Norwegian coastal districts, Norges Fiskarlag has an important responsibility regarding the development of the communities in these areas.'

Norwegian Fishermen's Association (1996)

work roles. The second contributory mode is less tangible and involves the emotional, ideological, more rarefied contributions women make to the fishing enterprises – in terms of their roles of wife, mother, and sister of fishermen and as carriers of the family tradition.'

An especially important role of women in the fishery lies in organising the community to respond to threats to the livelihood of the local fishery. In addition, Ruddle (1994) highlights the important role of women in the building-up and holding of fishery and marine environmental knowledge within the community.

3.3   A typology of fishing methods

Fisheries vary according to the technology used, both in terms of the vessel (size and construction) and the gear type. With respect to the latter, Cushing (1975: p. 5) describes four main types of fishing gear: seines, trawls, gill nets and lines. Sainsbury (1996) provides a somewhat different set of key gear types, noting that 'it is convenient to group the fishing methods according to the demands they place on the operating vessel' (p. 15). He notes that worldwide, 'the most commercially important methods are purse seining and trawling, when including the many variations of each' (p. 25). The following listing of fishing methods draws on a combination of Cushing's and Sainsbury's categorisations.

3.3.1 Seines/encircling gear

Seines are encircling nets that can be used for pelagic species such as salmon and herring, as well as for bottom-dwelling fish such as flounder. In the former case, the fishers set a purse seine to encircle a school of fish, and the net is then closed ('pursed') from below to trap the fish, which are then moved into the vessel. Another form of seine net, the Danish seine, can be used for groundfish, particularly flounder. It is placed in a circular arrangement on the ocean floor, and is gradually closed by towing. The fish therein are then brought to the surface.

3.3.2 Trawls and other towed/dragged gear

Trawls are (Royce 1996: p. 292) 'cone-shaped nets that are towed through the water'. These may be beam trawls (using a beam to hold the net open) or otter trawls (using 'otter boards'). The gear may be pulled over the ocean bottom to harvest benthic/demersal species (such as cod), or used as a pelagic trawl towed in midwater (e.g. for redfish/perch). Sainsbury (1996) also includes dredging (e.g. for scallops) in this category as a 'dragged gear'.

3.3.3 Gill nets: drift and static gear

The essential idea of a gill net is to catch fish by the gills when they swim into the net; such nets may be in the form of *set nets* or *drift nets*. The former are attached to posts or other fixed objects, while the latter 'drifts or drives with the tide' (Cushing 1975). Royce (1996: p. 293) notes that gill nets 'are used extensively by small-boat fishermen in sheltered waters because they can be hauled by hand or by simple hauling machinery.' On the other hand, gill nets in the form of large-scale drift nets, several kilometres in length, are also used in industrial settings in deep-sea locations.

3.3.4 Traps and pots

Along with the set net within the category of 'static' gear, Sainsbury (1996) includes traps and pots, particularly baited traps placed on the ocean floor (e.g. metal or wooden traps for lobster and crab), as well as barrier traps placed at varying depths (e.g. traps for cod).

3.3.5 Lines

Use of a 'hook and line' is a particularly ancient method of fishing and remains perhaps the most common fishing method today. As Royce (1996: p. 294) puts it, 'The fish are enticed to bite the hook by edible bait or by one of an endless variety of artificial attractants … Commercial practices include holding the lines in the hand while fishing for bottom fish [i.e. *hand-lining*] or attaching them to a moving boat while trolling for fish closer to the surface [e.g. for salmon] … But the most advanced commercial method is an adaptation of long-line or set-line fishing.' In particular, long-lining is a major form of commercial fishing, involving the use of lines of baited hooks, either placed along the ocean bottom (e.g. for groundfish) or suspended in the water column (e.g. for swordfish).

3.3.6 Other methods

As well as those mentioned above, an amazing range of other fishing methods are used around the world. These include some of the most selective approaches, such as harpooning (swordfish, whales) and diving/spears (conch, reef fishes, sea urchins), as well as particularly non-selective and destructive methods (notably poisons, dynamite and small-mesh nets in tropical waters). The principal fishing methods described above are shown graphically in Fig. 3.4.

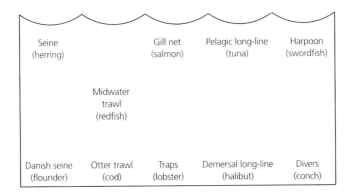

Fig. 3.4 A schematic view of some fishing methods, showing where in the water column they apply. Each pair (shown vertically) has a similarity in methodology; for example, gill nets and traps are both 'fixed gear', while divers often use spears, not unlike harpooners.

3.3.7 The choice of fishing method

Clearly, the choice of which fishing method to use in a given circumstance will depend on a wide range of factors. Indeed, in any fishery system, the fishers will probably have experimented over the years to find suitable gear to catch the desired mix of species within the constraints of available financial and technological resources. The choice will represent a balancing of factors related to the *biological nature* of the fish being harvested, the *economic nature* of the technologies, and the *social* considerations underlying the fishery.

With respect to biological characteristics, a sedentary, bottom-dwelling species such as lobster, on a seabed that is sufficiently flat, might be caught with traps. On the other hand, flatfish such as flounder are also bottom-dwelling, but being more mobile, they are caught not with traps, but with gear such as otter trawls and Danish seines. Meanwhile, a migratory species might be caught using gill nets laid along the migration paths (as with salmon, for which the migration routes are relatively clear) or by using highly mobile vessels to chase the fish (as with tuna or swordfish that roam widely). Finally, some species seem to be amenable to capture in a wide variety of ways; for example, cod is caught using fixed traps, gill nets, Danish seines, hand-lines, long-lines, bottom trawling, midwater trawling and various other methods.

Economic and social factors can also be expected to affect the choice of fishing methods. For example, economic realities will prevent low-valued species from being harvested using high-cost methods, unless there are *economies of scale* (with harvesting done in large quantities, as with some small pelagics). The 'bottom line' is that the net social and economic benefits from harvesting must be positive. However, several perspectives on this are possible. A focus on short-term versus long-term benefits will affect the level of concern for conserving fish and habitat (e.g. destructive methods can be very profitable in the short term). An emphasis on private profit (market value of the catch minus the cost of the fishing activity) versus a balance of multiple objectives (benefits of income and food production minus the time, energy and cost expended in fishing) may also influence the choices made. Finally, social and governance factors can also be important. For example, bottom-dragging as a method to harvest lobsters is used in the United States but is banned in Canada. These differing choices relate to social dynamics and the perception of ecological considerations in the two jurisdictions.

Of increasing importance is the selectivity of the fishing method, i.e. its capability to catch only target species and sizes of fish. The world is becoming more concerned about the by-catch issue, in terms of both the undesirable mortality of certain species (such as dolphins caught in tuna fishing) and the waste of natural resources (and food for humans) when low-valued by-catches are dumped overboard. This issue has both biological and economic implications, since increased selectivity can have biological and conservation benefits, while possibly increasing the cost involved in catching a certain quantity of fish.

3.4   Beyond the fishers

'The human component of a fishery involves more than fishermen themselves, that is, others besides those who harvest the resource from its habitat. These harvesters are only a small part of the total set of people involved in fisheries. For every commercial fisherman, for example, there are three sets of people who are equally a part of the human dimension of his activity: his family and "community" in the social or political sense, the people in the boat yards, supply stores, and service facilities who are both integral to and dependent upon the harvesting activity; and the distributors, marketers, and consumers who create the demand for his product.'

Orbach (1980: p. 150)

The remainder of this chapter explores the nature of these additional components of the human system. This begins with the various stages within the post-harvest sector, including aspects of processing, marketing, distribution, markets and consumers. Next is an examination of the structure and functioning of fishing households and communities, and finally some aspects of the broader socioeconomic environment are discussed. Figure 3.5 provides one depiction of the human sub-system, emphasising the central interrelationship between the individual fishers, the fishing households, the communities and the broader region. Related considerations arising at each of these stages, such as the post-harvest components, are also shown.

3.5   The post-harvest sector and consumers

Historically, in looking at fishery systems, the dominant focus of managers, planners, re-

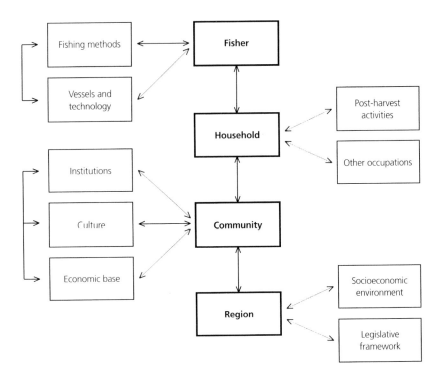

Fig. 3.5   The fundamental linkages between the fisher, the household, the community and the region more broadly. Related aspects of each of these are also shown.

searchers and others has been on the harvesting sector. However, the post-harvest sector is taking on a larger significance with the recognition that the world's available wild fishery resources are now generally unable to sustain significant increases in harvest, and that it is therefore crucial to focus on a goal of maximising the benefits to society provided by each fish that can be caught sustainably. This *sustainable development* approach ensures that the limited quantities of fish available are used as efficiently as possible to meet the many nutritional, employment, social and economic development goals. This point has particular relevance to the post-harvest sector, implying the need for attention to:

- reducing waste and post-harvest losses;
- maximising the *value added* through appropriate processing;
- developing and/or improving distribution and marketing systems;
- integrating the fishery into overall rural development efforts.

So let us examine what happens to the fish once they are brought to shore by the fishers. This section provides a brief overview of activities in the post-harvest sector, organised into three steps:

(1)   marketing and distribution;
(2)   processing and markets;
(3)   consumers.

Note that the critical element at the first step is to ensure that a mechanism exists by which fish leave the fisher's vessel and move either to the processing level, or directly to market. This involves (a) appropriate attention to the marketing of the fish, 'the channel of communication between the producer and the consumer' (Lawson 1984), (b) defining any role for intermediary *fish dealers*, i.e. middlemen who buy fish from the fisher for resale, and (c) consideration of the means available for the physical distribution of the product. At the second step, a varying degree of processing will be likely to occur, combined with a process of buying and selling in a market-place. The third step is sale to the various consumers. These may be individuals who are buying fish for household consumption, large institutions and corporations, such as hospitals and airlines, or other industries buying fish or fish products for specific uses (e.g. feed for livestock or in aquaculture).

The process by which fish move from the sea to the dinner table can be complex, and there is no single route, but Fig. 3.6 gives an idea of the steps involved.

3.6   Marketing and distribution

3.6.1 *Marketing*

Marketing is a crucial activity in the fishery; in a commercial context, a good catch is only of benefit if it can be sold. Marketing is the act of locating and arranging a market (specifically a buyer) for the catch obtained by a specific fisher, cooperative, company or community. Clearly, such efforts can make the difference between a reasonable income for fishers and others, versus a sad situation in which large quantities of unsold fish sit rotting on the shore.

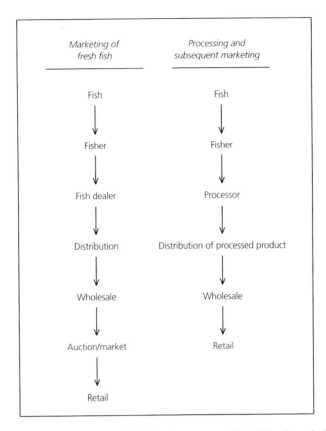

Fig. 3.6   The flow of fish from the sea to the retail level, for the two cases of fresh fish sales and of marketing a processed product.

In the corporate world of fishing, companies therefore pay due attention to the development and maintenance of marketing arrangements. At the small-scale level, fishers may well hand the catch over to the women of the household or community for marketing locally, or may need to rely on intermediaries (fish dealers) to take the fish to market. In the latter case, there is typically an ongoing tension between the fisher and the dealer; on the one hand, the fisher is often paid a very low sum for the fish, but on the other hand, the dealer is taking on the risk of locating an eventual buyer. Whether the arrangement can be viewed as 'fair' depends on the extent to which the low price paid to the fisher reflects a risk premium, and to what extent it reflects a case of monopsony, in which the fisher has no bargaining power because no other dealers are available. Certainly, many past development efforts have involved seeking to increase fisher incomes by reducing the role of middlemen. However, such efforts must be based on a good understanding of the complexities of the coastal system, to avoid creating unexpected problems, for example, by reducing the role played by women, or reducing the stability and cohesion of the fishing communities.

While marketing – to ensure that fish can leave the shore and move on to subsequent stages – is crucial at the 'micro' level of fishers, cooperatives, companies and communities, there is also a wider sense in which marketing is important. At the 'macro' level, marketing involves promoting fish consumption in general (as in broad national campaigns by fishery organisations and/or marketing boards), and promoting consumption specifically of fish from a particular region or nation (e.g. through advertising in target countries). Efforts at this scale to improve marketing and distribution can play important roles in economic development, as indicated in Table 3.2.

3.6.2 Distribution

Finally, attention to mechanisms for distribution of the fish is also important. After all, a sale is only complete if the fish can be delivered to the buyer. This is an obvious point, yet the reality in some developing countries is that a large fraction of the population, sometimes reaching 50% or more, do not eat marine fish simply because it is unavailable. (Of course, in other countries, fish consumption per capita is low because it is expensive, while in still other nations, fish is not a traditional food and thus there is an intrinsically low demand.) In any case, an assessment of the post-harvest 'marketing channels' must take into account the level of transportation capability within and external to the fishery system.

3.7   Processing

The processing of a harvest includes anything done to the fish before eventual sale to wholesalers and/or retailers. This could include the following common methods:

● heading, gutting and icing (in preparation for selling in fresh form);
● freezing (a common form for industrially caught fish);
● smoking and salting (traditional forms, lengthening shelf life at low cost);
● canning (a common form for many species, such as tuna, sardine and salmon);
● reduction (fish meal production, as with Peruvian anchovy).

Note that of these, all tend to be relatively labour-intensive except the last two, which are generally more capital-intensive. Processing can be a critical step in the flow of fish through the fishery system (Fig. 3.7). Clearly, a variety of benefits are obtained from fish processing. Some of the principal ones are the following:

Table 3.2   Interaction of fish marketing and economic development.

Marketing impact: ↓	Increase consumer demand	Improve distribution system	Improve market access	Increase alternative employment	Increase fisher empowerment
Intermediate impact: ↓	More production of under-utilised fish	Better marketing channels	Increased exports, foreign exchange	Less dependency among fishers	Fewer middlemen, more fisher income
Developmental impact:	More employment and food available	More protein available	Improved balance of trade	Decreased need for high-interest credit	Fishing community development

- Processing represents a *secondary industry* within the fishery system, one which usually creates additional employment in fishery-based regions, and which provides *added value* to the fish landed by harvesters. Indeed, it is not uncommon to find that, as a 'rule of thumb', the *market value* of fish products after processing and marketing may be around twice the *landed value* received by fishers for their catch.
- Processing provides a means to transform fish into more manageable forms. For example, fish meal (for feed in aquaculture, and for poultry and livestock) is typically made from fish with a high oil content, that needs rapid processing and thus is not easy to sell fresh. Processing into canned, salted or frozen products makes distribution easier, and reduces spoilage.
- Processing transforms fish into more marketable forms, with consumer preferences being met through a variety of product forms and packaging options.
- Improved processing can result in better utilisation of by-catch and the development of new resources, often leading to economic development in marginal areas.

The scale of processing activity can vary widely. Perhaps the ultimate capital-intensive version is that carried out on factory freezer trawlers, where the catch obtained on the vessel moves through an on-board processing facility, so that frozen product is brought to shore and into the distribution system. A similar level of processing is involved in fish meal production, carried out at large plants. At the other extreme are relatively minimal levels of processing, perhaps carried out by fishers and their families prior to going to market. This may involve a simple process of heading and gutting, or other similarly labour-intensive activities. In

Fig. 3.7   Processing, distribution and marketing are linked together in the fishery system. The truck bearing this sign transports fish from a processing company for distribution to exporters, wholesalers and the retail market. The sign provides some moving advertising at the same time.

Fish processing in Iceland

- Most fish processed in Iceland is exported, with the total value of seafood exports in 1995 being ISK 90 billion (US$1.3 billion). Given the importance of exports, the relative role of the various processing activities can be seen in terms of their percentage contribution to gross export value: land-based freezing operations are the most significant (44%), followed by on-board freezing (20%), salted products (16%), iced fish (8%), and fish meal and oil (10%). (The remaining 2% of gross export value arises from 'sundry processing categories'.)
- 'Considerable changes have taken place in the disposition of catches during the past decade. In 1981, just over half the demersal catch was frozen at land-based facilities; salting fish came next and around 100 000 tonnes of landed fish were dried. Drying of fish has virtually disappeared as a processing category, and the shares of both salting and land-based freezing have declined somewhat. However, the combined share of land-based and on-board freezing is growing.'
- 'Value-added processing for retail-packaged products has become an important area within both the freezing and salting sectors. Almost 140 freezing plants were authorised to operate in Iceland in 1995, along with 210 salting plants, 30 herring processing factories, 13 scallop plants and 13 canning factories. In addition, a large number of small-boat fishermen are authorised to process lumpfish roe from their own catches.'

Source: Government of Iceland (1999)

---

between are various levels of processing, carried out at processing plants ranging from community centres to industrial/corporate facilities.

While most attention within the processing sector is typically paid to the material being processed, it is also important to look at those doing the work. These individuals are very often relatives of the fishers, and are usually residents of the same communities in which the fishers live (see below). Particularly notable is the major role of women in onshore fish plant work (see, for example, discussions on Atlantic Canada (Lamson 1986a) and Iceland (Gunnlaugsdottir 1984)). Also of importance is the relationship between the work environment and corresponding productivity levels in fish plants (Chaumel 1984). For example, Baldursson (1984) examined interactions between health, work stress and the effects of salary systems on processing workers in Danish fish plants. Such analyses highlight the perceptions about fish-plant work relative to other employment.

3.8   Markets

The process by which fish is actually bought and sold is known as the *market*. This term is used conceptually (e.g. 'let the market decide the price') and also to refer to physical entities where the buying and selling takes place. These may be located in a community setting or in a major urban centre, such as the famous Billingsgate fish market in London (England) or the

---

'Value-added' in European groundfish processing

'The European processors have always been the key developer of value-added ground-fish products. Some 40 years ago, the success story of fish fingers started in European fish processing plants, soon followed by "Schlemmerfillets" (breaded fish steaks with various sauces and marinades). The processing of these products is still mainly carried out in the EU countries. In the tradition of this innovative market bracket are the following recent developments: breaded, smoked, and wet fish products with improved packaging, which incorporates more information on the type of fish and how to cook it. The wet fish is now presented on a traditional tray with an overwrapping film, and the breaded fish lines are packaged on coloured trays. These new "fish in sauces" products include salmon in watercress sauce, cod in mushroom sauce, smoked haddock in cheese sauce and "moules marinière". Coating is another important form of adding value to groundfish products for the European market: skinless, boneless cod, haddock or plaice fillet in a light oven-crisp crumb.'

Josupeit (1998)

Boston (USA) fish market. Local-level markets may involve independent intermediaries or family members (especially women of the household) in the selling process. Large markets serve regional, national and even international roles, e.g. fish from across New England and Atlantic Canada finds its way to the Boston market for sale and subsequent export around the world.

The theory of supply and demand is the cornerstone in discussing fish markets (see, for example, Cunningham *et al.* 1985). The 'benchmark' system is typically one that is *perfectly competitive*, satisfying a set of assumptions which include those listed below.

- The number of both buyers and sellers is large, no individual controls enough of the quantities supplied or demanded to be able to influence the price, and there is no collusion among buyers or sellers.
- For a given product, the factors determining price are supply and demand (with the price level, in turn, influencing harvesters and consumers); an equilibrium price is arrived at, at which supply and demand are balanced.
- There is 'full knowledge' about the information that is available to, and the subsequent actions of, all players in the fishery.

However, markets for fish are never *perfect*. It is important to be aware of possible market imperfections, which arise for a variety of reasons; two of these are described here.

3.8.1 Market power

The distribution of the total retail value obtained from the sale of fish between the fishers and the other intermediary stages can be very variable. For example, the price received by fishers as a percentage of the final retail price has been assessed at 20–25% in developed countries versus 30–40% in developing countries (Lawson 1984: p. 104), This fraction varies with the level of processing of the products, which in turns varies with the buying power of consumers.

Market power in a given fishery system will depend on the internal social structure, such as the role played by producer organisations and cooperatives on the fisher side, and by vertical integration and food wholesaling on the processor side. In addition, the level of foreign participation in the fishery may also have a role in the operation of the market. (The specific impacts of economic globalisation, the increasing level of interconnectedness in the world economy, will be discussed in Chapter 8.)

3.8.2 Contractual constraints

A common market imperfection arises due to the role of middlemen, not only as fish buyers but also as financiers, lending money to fishers, who agree to sell fish to the middlemen in return. Often, the fisher becomes beholden to the trader/financier, who may exert monopolistic selling power for capital and equipment, and monopolistic buying power for fish. This means that subsequent market interactions are not based solely on supply and demand, but rather on links between these individuals, links that may be seen as exploitative or symbiotic depending on one's perspective. Examples of this situation are the 'fish merchants' that in the past dominated the marketing of fish in Newfoundland, and the local fish buyers in many developing countries today.

3.9   Consumers

After passing from fisher to intermediary to processor and to wholesaler, most fish (apart from that used for animal feed or industrial purposes) will eventually end up in the retail sector, for sale to consumers. In examining the consumer sector of the fishery system, two key determinants must be considered: consumer preferences and consumer demand.

3.9.1 Consumer preferences

Preferences are the inherent desires that people have for certain products, and these are typically influenced by local traditions and cultures. Such consumer preferences must be understood in order to predict the impacts of developmental and management policies. Three key levels of consumer preferences should be considered.

- The inherent preference for fish versus other meats and protein sources is a fundamental issue. Consider the case of proposed policies to increase fishery production, perhaps through infrastructure development or subsidies on fuel and/or vessel construction. If fish markets are restricted to the local population, and if that population prefers other food sources, as is the case in parts of Africa, then such fishery development efforts are unlikely to be successful.

- Consumer preferences can also vary across fish species being harvested, and indeed across the strains/breeds/varieties within a given species. For example, on the Atlantic coast of Canada, the traditional 'fish and chips' dish in Newfoundland uses cod, while in Nova Scotia, just a few hundred kilometres away, the same dish is made almost universally with a different groundfish, haddock. In Java, Indonesia, common carp is a popular food, and the consumers typically prefer green-coloured fish of that species, not for price or taste reasons, but simply because that is the cultural preference.
- Finally, consumer preferences also vary with respect to the mode of preparation of a given fish. Lawson (1984: p. 117) notes that '… fish in many parts of the developing world, particularly the tropics, is preferred smoked or dried and also this gives a good flavour and texture in spicy soups and stews. To introduce frozen fish, even though it may be hygienically preferable, may not be appreciated by the consumer …' Consider the case of Ghana in the early 1960s, where well-intentioned measures were taken to improve the distribution of fish. This was based on a high-cost capital-intensive distribution of frozen fish. However, in reality, consumers actually wanted smoked fish, and thus the fish was eventually smoked anyway; the freezing process was both contrary to consumer preferences and a waste of money.

A recent development in fisheries is the increase in efforts at *eco-labelling*. This process parallels equivalent ones in, for example, forest products production. It involves seafood products being assessed according to the extent to which their production meets certain desired conservation and ecosystem-protection features. Products passing the test are labelled (on the package, say) as being ecologically friendly. The best known example of this is the 'dolphin safe' label placed on cans of tuna, when that fish was caught in a manner that protects dolphins from being killed in the process. A more comprehensive eco-labelling initiative, initiated by the Marine Stewardship Council, seeks to certify fish caught in 'sustainable fisheries' (Sutton 1998). A particularly difficult aspect of this process lies in finding a generally accepted definition of the latter term; for example, is it sufficient that the fish was caught in a fishery managed according to a total allowable catch? Despite such challenges, however, the move to eco-labelling is likely to gain momentum. This is driven by the recognition that consumer preferences, traditionally focused on the choice between product options, or more generally between protein sources, also involves choices that relate to our concerns about environmental protection.

3.9.2 Consumer demand

Discussion of consumer demand focuses on the relationship between the price of the product and the amount of the product that consumers are willing to purchase, qualified by the *ability to pay* on the part of those consumers (i.e. purchasing power). As with consumer preferences, it is important to understand consumer demand in order to analyse the impacts of actions in other parts of the fishery system. For example, efforts to improve quality control in fish processing may lead to healthier fish products, but the resulting price may be higher. Thus, depending on the availability of substitutes in the market-place (e.g. fish from other sources, or other forms of protein), what appeared to be an obviously beneficial move to improve the desirability of a product could also lead to drastically reduced demand, and therefore lower incomes for fishers and processors.

It is important to emphasise that an assessment of the market for a particular seafood product obviously depends on who are the relevant consumers. This, in turn, has changed over time as a result of economic globalisation (Chapter 8). For example, in developing countries, the drive to maximise the value of the fish caught has two major impacts. Fish is being diverted (a) from local markets to those in Northern countries, and (b) from use as food fish to use as fish meal in salmon and shrimp farms (again serving the demand of Northern consumers). Both of these impacts result in lower availability for local nutritional needs.

3.10 Fishing households and communities

One of the most noticeable, and regrettable, manifestations of a failure to examine and understand the fishery system as a whole has been a preoccupation in fishery analyses with fish and fishing 'firms' as the elements of study, rather than the broader context of where the fish live (the ecosystem) and where the fishers live (coastal communities). This is not to say that there are not many social science studies of fishing communities; the literature is full of those. Rather, there is often a lack of linkage between what goes on in the fishery itself, and how communities operate socially, economically and in terms of the functioning of community institutions.

What can an understanding of these linkages, between the fishery on the one hand and fishing households and communities on the other, contribute to the pursuit of sustainable fisheries, and the successful practice of fishery management in particular? This question is explored briefly here.

3.10.1 Households

A fishing household is one in which at least one member is involved in the fishery. Most discussions focus on those households with at least one member being a harvester, although more broadly, we could also include those households where the only fishery connection is with post-harvest aspects, particularly processing. Considering the former perspective here, we can focus on the specific matter of how household structure and operation influence harvesting behaviour at sea. Several such influences can be noted.

First, several household members may be involved in harvesting. Often these 'kin relationships' involve children (often sons) assisting the parent, who is the captain and/or vessel owner, but there may well be the involvement of less immediate relatives as well. The impact of such practices can be complex. For example, there may be impacts on productivity. On the one hand, Ullah (1985), in an analysis of fishers on the River Jamuna in Bangladesh and the relative use of outside hired labour versus family labour, concluded that (for a given amount of labour hired) productivity increases with the proportion of family labour used. On the other hand, if the vessel owner is under implicit or explicit social obligations to hire kin, the level of such hiring may be uneconomical, thereby reducing direct family income. The behaviour of the enterprise may be affected as well. If labour is based on household participation, this implies that more of the gross income received by the enterprise is kept internally by the household. This could lead to *greater* harvesting intensity by a *profit-maximiser* (since costs of fishing are lower), or conversely, *lower* intensity in the case of a *satisficer* (since a sufficient household income will have been obtained more rapidly). Finally, in the long term, the availability of household labour may provide greater income security, since internalising labour costs allows the enterprise to survive better during fishery downturns.

There are also other contributors to income security. First, in many cases household members not involved in harvesting may be highly involved on the post-harvest side, perhaps working in processing plants (in an industrial setting) or marketing and distributing the catch within the community and beyond (for those in an artisanal context). Depending on the motivations of the household, this may reduce pressure on the resource. Second, the harvester and others in the household may hold jobs entirely outside the fishery system (Table 3.3). This could have the effect of stabilising family income and reducing the risk of major loss if a disaster in the fishery system were to occur (such as an unexpected stock collapse).

Finally, family members may be involved in organisational aspects of the fishery system. For example, in some cases the spouses of fishers will have responsibility for the financial and book-keeping aspects of their family 'enterprises', as well as involvement in various aspects of fishery management. The latter may take the form of direct involvement in fisher organisations and/or a role in support organisations, such as women's groups within fishing communities that take part in campaigns to protect the livelihood of the fishers. (Whatever the specific activities of fisher spouses, it is of interest to note the results of Kearney (1992), who found that higher net incomes of fishing enterprises significantly correlated with the number of fishery-related tasks performed by those spouses.)

3.10.2 *Communities*

With the concept of *community-based management* (discussed particularly in Chapters 4 and 13) rapidly emerging as a major focus in many of the world's fishery systems, the need for careful study of fishing communities, something often seen as the social scientist's purview, must now be more broadly recognised. There are two principal approaches to the discussion. On the one hand are the many debates about the merits of community-based management, often taking place without much attention to the diverse nature of such communities. On the other hand are debates dwelling on a question that has occupied social scientists for centuries, namely 'what is a community?' In the fishery context, this debate focuses principally on two concepts of fishing communities:

Table 3.3   Fishery households in Japan (Ministry of Agriculture, Forestry and Fisheries 1999).

| | | Independent fishery household | | | |
| | | | Part-time | | |
	Total	Total	Full-time	Mainly independent operation	Secondarily independent operation	Fishery worker household
Households (thousands)	198.3	145.6	45.1	59.1	41.4	52.8
Employment (thousands)	277.7	220.5	69.5	97.1	53.9	57.2
% of employment	100.0	79.4	25.0	35.0	19.4	20.6

- *geographically based communities*, which are those referred to in common usage, such as villages or towns located along the coast;
- *communities of interest*, which are groups of fishers sharing some attribute in common, such as similar vessels, a common ethnic background, or a common target species.

While the latter has some relevance in fishery systems, and will be explored again in Chapter 13, the focus here is on geographically based fishing communities (Pinkerton 1987). These can be defined simply as 'an association of people living in a given area or sharing some general commonality in addition to geographic proximity' (IIRR 1998: p. 63). Such communities (and analogies in forest environments, urban centres and elsewhere) have been the subject of many ethnographic and socioeconomic case studies over the years (e.g. Panayotou 1985; Doeringer *et al.* 1986). In recent years, however, research attention seems to have shifted from the nature of fishing communities themselves, more to the specific matter of the role communities can play in management of fishery systems.

In any case, the present brief discussion cannot hope to summarise the body of knowledge on communities in general and fishing communities in particular, but rather attempts to highlight some of the key features in the communities, and the relevant factors that may need to be examined in relation to an understanding of fishery systems. In such discussions of the structure and operation of fishing communities, it must first be noted that there is great diversity among such communities. This diversity implies that it would be foolhardy to speak of a 'typical' community. Nevertheless, there are certain major components of fishing communities that tend to be universal, or at least very common.

Some key components of fishing communities

- Facilities at which fishers obtain provisions when preparing to go to sea.
- Facilities at which fishers land their catch (beaches, wharves, etc.).
- Facilities at which fish are marketed, processed and/or distributed elsewhere.
- Other fishery-related economic activities, such as boat repair facilities.
- Other non-fishery economic activities, such as agriculture, tourism and industry.
- Community facilities, such as schools, churches and meeting places.
- Community institutions, such as municipal government and legal systems.
- General community infrastructure, such as roads, electricity, water and sewers.
- Social and cultural facilities, such as central squares, bars, theatres, festivals, etc.
- Facilities provided by upper levels of government (e.g. post offices).

While the above list indicates some features of fishing communities that need to be examined in seeking to understand the broad fishery system, it is also useful to envision the factors that may be relevant in classifying and differentiating among fishing communities, i.e. in examining the diversity among fishing communities. Table 3.4, while by no means exhaustive, identifies some of the elements which are relevant to looking at fishing communities.

Also important in assessing the state of fishing communities is the matter of how fishers perceive and value the communities in which they live. Apostle *et al.* (1985: p. 256) note that:

Table 3.4 Some relevant factors in fishing communities.

Demographic	• Community population • Population trends • Levels of migration • Age and gender structure • Education levels
Sociocultural	• Identified community objectives • Religious stratification • Gender roles • Social stratification and power structure • Level of social cohesion • Local traditions and norms
Economic	• Income levels and distribution • Wealth levels and distribution • Degree of dependence on the fishery • Degree of fishing-related activity • Diversity in livelihood opportunities • Household economic structure • Types and location of markets
Institutional	• Pattern of community organisation • Pattern of local resource management • Pattern of resource ownership and tenure • Level of community infrastructure • Regulatory and enforcement approaches • Interaction with upper levels of government • Use of traditional ecological knowledge • Involvement of women in local institutions
Environmental	• Availability and condition of fish stocks • Quality of aquatic and coastal habitat • Oceanographic/environmental conditions

'it is essential to understand how inhabitants perceive their present-day existence ... Do people continue to live in these small villages by choice, or from lack of alternatives? Is work satisfaction a prime reason for wishing to remain within the community, or is the work secondary to other factors related to place?'

These authors, in their synthesis of community attachment and job satisfaction among fishers in Nova Scotia and New England, conclude that fishers in their study areas fit within three broad groupings:

- those with a high level of job satisfaction and strong community attachment;
- those with a high level of job satisfaction but weak community ties;
- those with a weak commitment to the fishery as an occupation, but a strong community attachment and/or interest in the maintenance of kin and status positions.

This classification is compatible with the results obtained by Panayotou and Panayotou (1986) in a very different context, i.e. the fisheries of Thailand, where fishers and fishing communities could be categorised according to their degree of geographical mobility and occupational mobility, and these factors were likely to have been determined by community ties and job satisfaction, respectively.

3.11 The socioeconomic environment

On the human side of the fishery system, there is clearly great variety inherent in the components discussed so far: the fishers, processors, fishing households and fishing communities. However, the story does not stop there. Just as the fish are part of an ecosystem, so too are the fishers and post-harvest sectors situated within a broad socioeconomic environment. This environment incorporates human, social and institutional elements at the community, regional, national and global levels, all of which can influence the objectives pursued in the fishery, as well as fishing activity itself (Fig. 3.8). Many questions might be asked about the links between the fishery system and the socioeconomic environment, and some possibilities are given in Table 3.5.

Fig. 3.8   All those in the fishery sector, such as these processing plant workers, are affected by the fishery's socioeconomic environment – in their local community, as well as in the broader economic, social and cultural systems. Individual decisions about entry into and exit from the fishery, and about the fishery activities to pursue, will depend on such diverse factors as the state of labour markets, the social support system and the strength of community culture.

Table 3.5   The fishery system and its socioeconomic environment.

Demographic	• How do demographic aspects of the fishery system, such as participation by age and gender, interact with external influences, such as national population and migration trends?
Sociocultural	• What are the broad aspects of society, culture, history and tradition that impact on decision making in the fishery system? To what extent do those outside the fishery system have power over internal choices?
Economic	• How does the fishery economy interact with the economic structure and dynamics at the regional and/or national levels? How are the economic inputs in the fishery, notably labour and capital, affected by the broad economic environment?
Institutional	• How do local fishery objectives relate to broader regional and national policy goals? • How does the local institutional structure interact with institutions, legal arrangements, legislation and policy frameworks at national and/or sub-national levels?

3.11.1   *Individual decisions and the socioeconomic environment*

Incorporating the broader socioeconomic environment into our analysis of fishery systems allows a clearer recognition of the difference in perspective between 'private' decisions made by individuals in the fishery and the pursuit of broader community and/or societal objectives. For example, consider the reality in many fisheries, located in a region of isolated fishing communities, where few alternative employment possibilities are available. Often the maintenance of sustainable livelihoods, i.e. stable employment with reasonable incomes, is a priority among society's fishery objectives. This is not just a matter of providing jobs in the fishery, but also of maintaining a strong 'engine' of the coastal economy, given the extent of spin-off benefits from the fishery into coastal communities.

From an economic point of view, this relates to the idea of the *social cost of labour*. This is the true cost to society of having a fisher working in the fishery rather than doing something else in the economy – in other words, a measure of the benefits that society *could have obtained* if the crew member had instead worked at the 'best alternative' job outside the fishery. However, if fishers in such a socioeconomic environment actually have *no* job alternatives, then there is no social cost to keeping those individuals in the fishery. Indeed, the loss of fishers from the fishery may lead, through a multiplier effect, to an economic *loss* to the regional economy. Social costs may rise, due, for example, to increased crime and/or decreased health and welfare levels. In such circumstances, the social cost of labour may even be negative; far from being a 'cost' to be minimised, employment of fishers might be seen as a positive 'good'.

Given all this at the societal level, let us now look at individual decisions. When an owner of a fishing vessel employs an individual crew member, it is often necessary to follow an agreed-upon crew wage level and/or minimum wage laws set at a national or state level. From the owner's private perspective, therefore, a labour cost is incurred, given by that set wage level. This is referred to as the *private cost of labour*. From the vessel owner's perspective, this is a positive value, and creates an incentive for the owner to minimise the use of labour. However, in the scenario described above, private incentives, although 'rational', do not reflect society's objectives. Thus, it is important to understand these different levels of decision making, and to design management and policy measures to make the various levels compatible.

3.11.2   The fishery system, labour markets and the socioeconomic environment

Just as harvests from a given fishery system interact with a broader market-place, so too do the harvesters in the fishery interact with their socioeconomic environment through labour markets, in which people choose their form of employment and employers choose their employees. Wage rates or crew shares on fishing vessels will depend on the balance of this labour supply and demand process. It is crucial to note that what goes on outside the fishery system *per se*, but within the broad socioeconomic environment, can operate through the labour market to influence the fishery system.

One form of this impact is caused by the balance between profits and wages in the fishery, arising either from the ongoing economic climate in a coastal area, or changes in the economy. Consider, for example, the case of a high-unemployment or isolated fishing region. In such a case, wages are likely to be relatively low (since workers in the economy will have little bargaining power), so that other things being equal, higher profits may be available for boat owners. These excess profits could (if unregulated) lead to greater investment in vessels, resulting in excess catching power and threatening the sustainability of the resource base (Chapter 8).

3.12 Summary

This chapter has given an overview of the human side of the fishery system. The topics considered range from the fishers and their fishing methods, to post-harvest aspects of marketing, distribution, processing, product markets and eventual consumers, to fishing households and communities, to the broader socioeconomic environment in which the fishery operates. In many ways, the discussion has paralleled that of the natural system. Fishers, like the fish, are at the core of the fishery system. The households and communities, as well as onshore post-harvest activities, may be seen as analogous to the ecosystem in which the fish live. Finally, the socioeconomic environment is the broader setting for human activity in the fishery, not unlike the role played by the biophysical environment within the natural system. In the next four chapters, attention turns to the management sub-system, and mechanisms to study and control behaviour emanating from the human sub-system, in part to maintain the integrity of the natural system.

Chapter 4
The Management System: Policy and Planning

The components of the fishery system described so far are those directly involved in fishing and post-harvest activities (including the fishers and the fish) and those surrounding such activities, notably the ecosystem, the fishing communities, and the biophysical and socio-economic environment. We now turn to the management aspects of the fishery system. While these do not generate direct fishery outputs, since they are not what economists refer to as 'productive' activities, they are crucial to ensuring that such productive benefits are obtained. Indeed, in most fisheries, the need to control exploitation through management has become clearer over time for the reasons given below.

- Fish stocks are not only depletable, but have the potential to be driven to extinction if exploitation is uncontrolled. This is notably the case within complex, uncertain aquatic ecosystems, where poorly understood inter-species effects make a lack of control (*laissez-faire*) all too likely to produce over-exploitation of one or more interacting species.
- Conflicting biological, social, economic and cultural goals inherent in most fisheries must be balanced through management if total fishery benefits are to be maximised.
- Controls are needed over the rate of fish stock exploitation in order to balance present-day needs with maintenance of the resource at suitable levels for future use.

The extent of these concerns depends in part on the type of fishery in question. Historically, management systems became relatively well established in localised (and therefore vulnerable) fisheries, as in those for salmon stocks (within or near their natal streams) or sedentary species such as lobster. For example, it is well understood that in the case of salmon (Parsons 1993), which are harvested primarily on returning from the open sea to reproduce in their native rivers, the simple act of placing a net across a river mouth could result in the capture of *all* returning adults, with extinction the eventual undesirable result.

The fishery management components form the 'guidance mechanism' for the complex fishery system, with its inherent natural and human dynamics. It is therefore crucial that the management system takes into account the many components of the fishery and its environment, in order to avoid the failures that can result if important interconnections among these are missed. Furthermore, since it is the management system that directs the fishery towards achieving society's objectives, it is also crucial that these objectives are clearly understood. This point is highlighted within the discussions of management and development in the following chapters.

Much has been written in recent years about 'priorities' and 'new directions' for fisheries management. We leave the discussion of what fishery management *should* look like for later in the book (particularly Chapter 15). Here, the focus is on the structure and appearance of the existing management components in fishery systems, and on the various sets of regulations that may be placed on commercial, subsistence, recreational and indigenous fisheries, as well as on subsets of any of these fisheries. The fishery management system can be viewed as being made up of four components (Fig. 4.1).

- *Fishery policy and planning (strategic management)*
  - Overall objectives to be pursued in the fishery system.
  - Policy directions to meet the declared objectives.
  - Legislation related to fishery management and regulation.
  - Decisions regarding the structure of the management system.
- *Fishery management (tactical and operational management)*
  - A portfolio (a suitable 'mix') of management measures to control the impact of harvesting on the fish stocks and the ecosystem.
  - Annual levels for each management measure (e.g. allowable catches).
  - Day-to-day decisions to achieve operational plan (e.g. fishery openings).
  - Research and data collection to provide the necessary knowledge base.

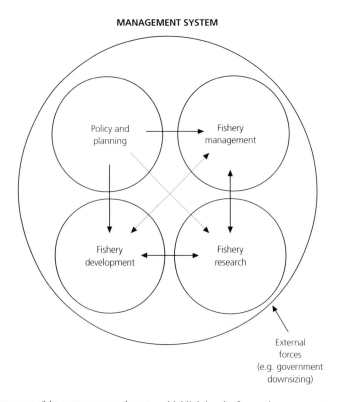

Fig. 4.1   The structure of the management sub-system, highlighting the four major components: policy and planning, management, development and research. Interactions among these components, and the role of external forces, are also indicated.

- *Fishery development*
  - Measures to improve the physical infrastructure, technological capabilities, institutions and/or human productivity in the fishery system (including the management system as well as the human system of fishers, processors, etc.).
  - Measures, other than through fishery management *per se*, to improve the flow of sustainable benefits from the fishery, including market development, quality control and improvements to distribution processes.
  - Development of new fisheries, where feasible and desirable.
- *Fishery research*
  - Measures to collect, analyse and disseminate relevant data on the various components of the fishery system, to support the fishery management and development activities.
  - Measures to assess and conserve fish stocks.
  - Measures to understand the natural and human systems in the fishery.

The first of these, fishery policy and planning, is discussed in this chapter, while subsequent themes are presented in Chapters 5–7.

4.1   Fishery policy and planning

The policy and planning stage is arguably the most critical element of the fishery management system. The focus in the following discussion is on policy and planning for fishery management specifically, as opposed to the consideration of issues that arise in fishery development, which are treated separately in Chapter 6.

4.2   Fishery objectives

The effective management of fisheries (or any economic activity) is based on the achievement of societal goals and objectives through the use of appropriate policy and regulatory instruments. This implies a natural structure for discussion of the management sub-system, first focusing on the variety of objectives pursued in fisheries management.

   As noted throughout this book, fisheries are very much multi-objective activities, serving a variety of social, cultural, political, economic and ecological goals. In any given situation, the multiplicity of objectives to be pursued will depend on societal policy decisions, and in turn, the choice of fishery institutions and management approaches will depend on those objectives and the priorities attached to each one (FAO 1997b). Without such a structure and process, the fishery management system cannot proceed intelligently. There is no reason to use valuable human and physical resources in developing and managing fishery systems if this is not to achieve some desired outcomes, i.e. the objectives. Furthermore, the set of objectives must be clearly understood at all levels, from fisher through to government. Attempts to manage fisheries in the absence of clear and explicit objectives are bound to be less than successful.

   Finally, it is important to note that not only do the priorities placed on fishery objectives help to determine a *desired structure* of the fishery (fleet composition, preferred gears, etc.)

and a *level* at which fishing should take place (e.g. optimal harvest rate or total allowable catch (TAC)), but also the choice of a 'best' regulatory framework. For example, Regier & Grima (1985: p. 855) point out that, in a study of fisheries management for Canada's Pacific coast (Pearse 1982), market mechanisms were advocated for regulating fisheries (such as herring) that tend to be dominated by economic objectives, but were not seen to be appropriate for fisheries serving significant social goals (such as salmonids and small coastal stocks).

4.3   A portfolio of fishery objectives

The first step in considering fishery objectives is to examine the 'big picture': in the context of a particular society, what do we want the fishery system to provide? This is a matter of determining *strategic* goals. Theoretical discussions often assume that there is but one single objective in the fishery (for example, fishery economics typically focuses on a rent maximisation goal), but in reality there are usually multiple strategic objectives being pursued simultaneously in a given fishery system, although these are quite likely *de facto* (undeclared) objectives, rather than explicitly declared.

Fishery objectives are often viewed as falling into three principal categories (e.g. FAO 1983: p. 20):

- biological/resource conservation;
- social/equity;
- economic/productivity.

In fact, the first of these, the conservation requirement, is more often treated not as an objective, but rather as a *constraint* to be respected as we go about pursuing our other goals in the fishery. For example, two biological considerations that are often stated as the most fundamental of fishery objectives, but could alternatively be viewed as firm constraints in the fishery system, are:

- conservation of the fish stocks/avoiding stock collapse;
- maintaining or enhancing biodiversity and ecosystem health.

Most specific objectives seem to fall within the social and economic categories listed above. Indeed, the most frequently discussed objectives may well be:

- production of fish, whether for food, livelihood (income) or profit;
- economic efficiency, economic viability and rent generation;
- employment;
- export promotion and generation of foreign exchange.

These are elaborated in sequence below.

4.3.1 Production of fish

In cases where the fishery is an important contributor to the food supply (whether nationally or locally), a major objective may be harvest maximisation: producing as much fish as possible, presumably constrained by the need for fish stock conservation. Indeed, since greater catches also imply greater fishing opportunities (and greater incomes, up to a point), the pursuit of this objective will tend to please fishers. This objective also tends to be compatible with that of maximising total employment (since, at least in the processing and distribution sectors, this is likely to be proportional to production).

4.3.2 Economic efficiency

The goal of economic efficiency is a common one, and has undoubtedly played a central role in economic studies of fisheries. Indeed, everyone wants their fishery to run 'efficiently', but it is often unclear what exactly this implies; the meaning often depends on the philosophy and ideology of the fishery players. How should economic efficiency be measured? Focusing on the individual boat owner, an efficient operation generates maximum profits by that individual. This implies, for example, keeping the costs of employment and of other community-level spending as low as possible for a given level of production. Focusing on the harvesting sector as a whole, efficiency may imply the maximisation of resource rents. Finally, efficiency measured at a broader scale of the community, or regional economy, involves maximising a multifaceted mix of overall benefits produced at that level (e.g. for the community) relative to the *net* costs incurred. The concept of efficiency is the same in each case; it is the interpretation that differs according to the scale at which we are looking. In each case, it is possible to measure the effects of 'inefficiency' by comparing an 'economic optimum' with a current approach.

4.3.3 Employment

Fishery employment is often considered an important objective, not only for its own sake, but also as a common means of supporting rural communities and thereby providing a measure of social stability. Indeed, in many social studies of fisheries, employment is considered as the principal objective to be pursued, entirely replacing the efficiency criterion which prevails in economic analyses.

4.3.4 Foreign exchange/balance of payments

In developing countries, exports provide the means to obtain foreign currency; something which is important at both the individual level (allowing the purchase of desired consumer products) and at the national level (bringing increased wealth into the country, and allowing greater imports without upsetting the balance of the economy). Nations subjected to structural adjustment programmes, whether imposed internally or by the International Monetary Fund, may be obliged to focus on export production (at the expense of other objectives above) as a means to pay debts and/or improve a balance of payments.

Other social and economic goals may be pursued, particularly those that apply not just to the fishery, but to other economic sectors as well. For example, some possible objectives arising at the 'macro' policy level are:

- industry diversification;
- sociopolitical stability;
- decreasing rural–urban drift;
- maintaining a regional balance of development.

4.4   Objectives, priorities and conflict

The formulation of fishery objectives is a crucial, but by no means simple, task which will probably involve conflict among the many players, notably the fishery participants and the various levels of government. Many of the debates over desired directions in fishery systems have their origins in the underlying systematic differences in objectives between these fishery players (Chapter 13). The real challenge is not to list all the possible objectives in a fishery system, but rather to prioritise the list. Some objectives will be considered more important than others. In balancing a set of multiple goals, which are the priorities? For example, the rent (profit) maximisation objective noted above (and the related goal of export earnings) may be top priorities in an *industrial* fishery, while maintaining stability and employment in fishing communities may be at the top in a small-scale coastal fishery (Fig. 4.2). A major

Fig. 4.2   A major source of conflict in many fisheries worldwide results from the balancing act between two major sets of objectives: increasing production, profits and export revenues through fishery industrialisation, versus maintaining rural stability, employment and incomes through small-scale fisheries and coastal communities (as shown here).

aim of fishery planning, something that is properly a task for policy makers, is to determine a suitable balance or blend of legitimate objectives by formulating a priority listing of the objectives to be pursued.

Naturally, this process leads to considerable disagreement, usually between those with conflicting philosophies or ideologies. As but one example, Smith (1981: p. 21) discusses potential conflicts between community/equity objectives and national goals, noting that programmes to encourage investment in the least capitalised components of the fishery 'may distribute incomes more equitably among individual fishermen and communities, but to the extent that they decrease the sustainable yield they make the pie to be divided that much smaller, thus conflicting with national goals of resource conservation and management'.

The conflict raised by discussions of objectives is a key reason why policy makers often completely avoid the subject. However, a lack of clear objectives has led to so many problems, and indeed crises, in fishery systems that the need for explicit objectives, and for means to resolve the conflict inherent in them, is now apparent. This process can draw on a variety of analytical tools to examine the implications of placing 'weights' on the various objectives, and making trade-offs among conflicting goals (e.g. Healey 1984).

4.4.1 A key source of conflict: allocation

In a totally unmanaged fishery, anyone who wishes to go fishing has a chance of catching fish. Even in a situation where the number of legal fishers is limited and the total catch is limited by a TAC, each fisher, in theory, has an equal opportunity to catch fish. However, suppose that the fish migrate along a coastline, so some fishers have access to the fish before others. It is seen as unfair that those with first access might catch the entire TAC, leaving nothing for the others. A similar situation could arise if fish move from offshore to coastal waters, in which case an industrial fleet may be able to exploit the stock before the inshore artisanal fishers. Or suppose multiple gear types are involved in a fishery, and a recognition emerges that it is physically unsafe to have these sectors competing on the fishing grounds. In all these cases, measures are needed to *allocate* access to the fish and/or the available catch of fish.

At the heart of the allocation issue are debates over how much access or fishing time or harvest each group of fishers is to receive. This topic frequently dominates fishery discourse (and is discussed in detail in Chapter 13). Indeed, many fishery managers note that arguments over allocation seem never-ending, and may overwhelm other fishery discussions.

How can allocation issues best be approached? The most logical path would seem to lie in making conscious decisions about allocations based on the declared objectives in the fishery – those determined within the policy and planning process. This approach may have occurred, implicitly if not explicitly, in some fisheries in which allocations have shifted. For example, moves toward more capital-intensive fishing (through large-scale boat-building) presumably reflect a focus on the goals of export promotion and/or fishery 'modernisation', while a move in the opposite direction (i.e. Indonesia's decision to ban trawling in its coastal waters) served the policy goals of reducing conflict and favouring artisanal fishers (Bailey 1997). Of course, there are reasons why this logical approach is not always (perhaps even rarely) followed; one likely reason is a reluctance of governments to antagonise one sector of the fishery by favouring another. In any case, most of those engaged in fishery management and planning will need to deal somehow with the allocation issue. We return to this theme in Chapters 13 and 14.

Setting Management Priorities: Agenda 21, Chapter 17, 'Protection of the Oceans, All Kinds of Seas, Including Enclosed and Semi-Enclosed Seas, and Coastal Areas and the Protection, Rational Use and Development of Their Living Resources'

Chapter 17 of Agenda 21, the key document emanating from the United Nations Conference on Environment and Development, deals broadly with conservation and use of the oceans, but provides in particular an important reference point for fishery management. Discussion is divided into six programme areas.

(a) Integrated management and sustainable development of coastal areas, including exclusive economic zones.

This theme area focuses on environmental management and policy, including implementation of management plans, impact assessments, improvements to coastal living conditions, etc.

(b) Marine environmental protection.

This is a major area of discussion in Agenda 21, focusing principally on technological aspects such as sewage treatment facilities, emission control policies, pollution monitoring mechanisms and international and/or regional cooperation.

(c) Sustainable use and conservation of marine living resources of the high seas.

This theme has a management and policy focus, with the two key aspects being treatment of straddling stocks, and the monitoring, control and surveillance ('MCS') needed to manage high seas fishing.

(d) Sustainable use and conservation of marine living resources under national jurisdiction.

This theme receives substantial coverage in Agenda 21. The discussion is wide ranging: better resource assessments, strengthening legal and regulatory frameworks, implementing strategies for sustainable resource use, appropriate development of mariculture and fisheries, promoting the use of environmentally sound technologies, reducing post-harvest wastage, expanding recreational and tourist activities based on marine living resources, and so on. Much of the discussion emphasises small-scale fishery management and development, involving fishers and fishing communities. There is also a focus on capacity building, for example through 'support to local fishing communities, in particular those that rely on fishing for subsistence, indigenous people and women, including as appropriate, the technical and financial assistance to organise, maintain, exchange and improve traditional knowledge of marine living resources and fishing techniques, and upgrade knowledge on marine ecosystems'.

(e) Addressing critical uncertainties for the management of the marine environment and climate change.

This theme has a significant research component, in terms of such needs as: methods for improved forecasting of marine conditions, development of standard methodologies for assessing and modelling marine and coastal environments, examining the effects of increased UV on marine ecosystems, systematic observation of coastal habitats, sea level changes and fishery statistics, and other research on the impacts of climate change on the marine system.

(f) Strengthening international cooperation and coordination.

This theme deals with promoting information exchanges and intergovernmental/inter-agency cooperation, as well as issues relating to links between trade and environmental conservation.

(g) Sustainable development of small islands.

This area focuses on the 'special environmental and developmental characteristics of small islands', calling for suitable coastal area management and environmentally sound technology, all to be done while 'taking into account the traditional and cultural values of indigenous people'.

4.5   Fishery management institutions

The concept of an *institution*, which has been used at various points previously in this book, is seen typically in one of two ways. In common usage, we might think of an institution as an organisational arrangement of some sort by which people interact, pursue society's goals and manage themselves, for example, a Department of Fisheries, an association or cooperative of fishers, or a public education system. Social scientists, on the other hand, tend to define an *institution* not as an organisation *per se*, but more broadly as a set of rules or 'norms' that govern the behaviour of individuals in the system. As North (1990) put it:

> 'Institutions are the rules of the game in a society or, more formally, are the humanly devised constraints that shape human interaction. In consequence, they structure incentives in human exchange, whether political, social, or economic.'

Examples of such institutions include the market-place, the legal system, a municipal council, and so on. It is important to recognise that the structure and operation of a fishery system can be affected greatly by the evolution, or choice, of these various institutions, within the fishery or external to it, and at various governmental levels. Successful management requires the 'right' institutions – ones that are structured properly, with widespread support, and which are seen as fair and just. It is clear that in the past, poor institutional arrangements in many fisheries led to disastrous conservation failures.

Institutions play many roles. Many focus usefully on facilitating interaction among fishery participants, in order to improve mutual understanding, group dynamics, marketing efforts, etc. Another major role is the creation and reinforcement of incentives that shift the behaviour of those in the fishery in desired directions. For example, the market as an institution can create economic incentives inducing changes in the levels of fishing effort or of capital investment in vessels, in at-sea practices such as the dumping of unwanted fish overboard, or in other actions, positive or negative. Similarly, community institutions can create social incentives for the sharing of fishery resources, or for responsible behaviour in fishing. Having the 'right' institution is a prerequisite for creating the 'right' incentives, since no matter how suitable the incentives, they will probably be unsuccessful if imposed from within an otherwise dysfunctional institution.

What factors make a fishery management institution work effectively? Fully addressing this question is a crucial component of the policy and planning process, and is discussed further in Chapter 13 (where the rapidly emerging co-management approach is examined) and in Chapter 14 (where the focus is on ensuring appropriate *rights* in fishery systems). Whatever the form of the institution, it is important to undertake a regular assessment of its performance in policy making and management. A key point is to ensure that the institution truly pursues its objectives, and that due attention is paid not only to fishing activities as such, but also to the links between fishery management and the broader fishery system, including fishing communities (Chapter 13).

4.5.1 Legislative framework

While the first institution that may come to mind in terms of those impacting on the fishery is the governmental agency charged with fishery management, in fact equally important, although less visible on a day-to-day basis, is the legislative system underlying governance of the fishery. Legislation sets the tone of fishery management and lays out the overall directions to be pursued.

Legal frameworks are of critical importance at national, regional and global levels. At the national level, not only do such frameworks lay out overall policy directions, but the legislation also allots jurisdiction over fisheries and their management between federal, state/provincial and local levels. In Canada, for example, the country's constitution provides the national government with responsibility for the management of ocean fisheries, while the provinces have responsibility for inland fisheries, aquaculture, and any land-based activity (including fish processing) relating to ocean fisheries. In the Philippines, the 'Philippine Fisheries Code of 1998' (Republic Act No. 8550) has produced a remarkable decentralisation of management authority over inshore 'municipal' fisheries (Congress of the Philippines 1998: p. 17).

A National Fishery Management Institution: The Ministry of Fisheries in Norway

'The Ministry of Fisheries is the secretariat of the Minister of Fisheries and introduces laws and regulations. The general objective of the Ministry is to create sustainable and profitable fishing and aquaculture industries, secure sea traffic and ensure efficient and competitive sea transport. The Ministry of Fisheries is responsible for:

- the fishing industry;
- port, lighthouse and pilot services;
- the aquaculture industry;
- electronic navigation devices.

The four departments of the Ministry of Fisheries are:

- The Administration Department;
- Department of Coast Resource Management;
- Department of Aquaculture, Industry and Exports;
- Department of Resource and Research.

Under the Ministry are the Coast Directorate, the Directorate of Fisheries, the Guarantee Fund for Fishermen and the National Fishery Bank of Norway. These institutions are the Ministry's executive bodies within their respective fields of competence ... The Norwegian Fishery bank grants loans and manages a number of subsidy schemes for the fishing fleet. The bank has four branches. The Guarantee Fund for Fishermen manages social benefit schemes for fishermen.

The Norwegian Seafood Export Council is a publicly appointed body with its own administration. The Council performs administrative duties, and is also an advisory body to the Ministry. In addition, the council coordinates marketing strategies for the industry and has officials posted to the chief export markets.'

Norwegian Fishermen's Association (1996)

'The municipal/city government shall have jurisdiction over municipal waters as defined in this Code. The municipal/city government, in consultation with the [Fisheries and Aquatic Resources Management Council] shall be responsible for the management, conservation, development, protection, utilization, and disposition of all fish and fishery/aquatic resources within their respective municipal waters.'

At a regional level, legislative frameworks may be in place to govern multinational fishery systems. A major example of this is the European Union's Common Fisheries Policy (CFP), which evolved from the Treaty of Rome in 1957 (which established the European Union) to adoption in 1976 of a set of measures on fishery structures and markets, through to the full implementation of the CFP in 1983 as a mechanism for the conservation and management of

A Multinational Fishery Management Institution: The European Union's Directorate-General XIV

'DG XIV is the Directorate-General responsible for the Common Fisheries Policy (CFP), which covers all fishing activities, the farming of living aquatic resources, and their processing and marketing, on the legal basis of Article 39 of the Treaty of Rome. Since the first decisions adopted in 1970, the CFP has been through many changes, and its current form centres on four main areas: the conservation and management of marine resources, relations and agreements with non-member countries and international organisations, structural measures, and the common market organisation for fishery products … In line with the basic principles of the CFP, DG XIV prepares legislation, implements management policy and monitors compliance with Community law in this field, by means of discussions, analyses and studies. The DG is also responsible for running a scientific research programme for fisheries as part of the Community's framework research programme.'

Organisational structure of DG XIV

Director-General
Units reporting directly to the Director-General
(1)   Legal coordination of internal and horizontal matters, legal issues.
(2)   Budget and human resources, evaluation.
Directorate A: Horizontal measures and markets
(A/1)   Relations with other institutions, the Advisory Committee, and non-governmental and trade organisations.
(A/2)   Communication, information and studies.
(A/3)   Databases and fisheries economics.
(A/4)   Common organisation of markets, trade policy and health matters.
Directorate B: International fisheries organisations and fisheries agreements
(B/1)   International organisations and enlargements.
(B/2)   Baltic, North Atlantic and North Pacific.
(B/3)   South Atlantic, Indian Ocean, South Pacific and Antarctic.
(B/4)   Eastern Central Atlantic and Mediterranean.
Directorate C: Conservation policy and monitoring
(C/1)   Conservation and environment issues.
(C/2)   Research and scientific analysis.
(C/3)   Monitoring and licences.
(C/4)   Inspection.
Directorate D: Structures and areas dependent on fisheries
(D/1)   Coordination and general matters relating to structural measures.
(D/2)   Finland, France, Greece, Ireland, Italy and Portugal.
(D/3)   Austria, Belgium, Denmark, Germany, Luxembourg, the Netherlands, Spain, Sweden and the United Kingdom.

European Commission (1999)

fishery resources (Karagiannakos 1995; Symes 1997). The CFP has remained controversial over the years, owing in part to its allocation of substantial decision-making power away from individual EU members to governance at a more central, Europe-wide level. (See box on p. 80 for additional details on the CFP and related governance aspects.)

Legal frameworks relevant to fisheries have become particularly important (and indeed high-profile) at the global level. The 'cod wars' in waters around Iceland in the 1970s led to one of the world's major pieces of international law, the United Nations' Convention on the Law of the Sea, which now provides overall guidance regarding behaviour in the oceans, and specifically in fisheries on coastal stocks. Subsequently, the 'turbot war' off the coast of Canada in the 1990s led to the 1995 United Nations Conference on Straddling Fish Stocks and Highly Migratory Fish Stocks, and the formal 'Agreement for the Implementation of the Provisions of the United Nations Convention on the Law of the Sea of 10 December 1982 relating to the Conservation and Management of Straddling Fish Stocks and Highly Migratory Fish Stocks'. These developments are likely to ensure that international law relating to fisheries remains an important consideration in many fishery systems for years to come.

4.6   Time scales of management

The division of fishery management into a strategic level (discussed in this chapter), as well as tactical and operational stages (Chapter 5), tends to reflect the temporal scales at which corresponding actions take place, as illustrated in Fig. 4.3.

A well-planned fishery management system will seek to operate on a time scale compatible with changes in the fish stocks, the ecosystem, the fishery itself and the information available about all these. For example, if fish stocks (or indeed our level of *understanding* about these stocks) vary greatly from month to month, management decisions may be best made on a comparable scale. Otherwise, decisions are likely to become quickly out of date. It also helps if management decisions are made on a time scale compatible with the market for the fishery products.

A lack of fit in this regard can have significant consequences. For example, if marketing agreements are made annually to sell certain amounts of product to exporters or wholesalers, but management decisions are made on a shorter time scale, one could envision a situation where a fishery is closed at such a time as to leave harvesters and processors unable to meet market commitments. This could lead both to large immediate losses and loss of

Fig. 4.3   The time scales relevant to fishery management (top), along with the 'location' on this spectrum of the strategic, tactical and operational levels of management. Also shown are some examples of management actions and where they may fit within this range of time scales. TAC, total allowable catch.

future markets, owing to an inability to guarantee delivery. So differing time scales can have significant consequences for all players in the fishery system.

4.7   Spatial scales of management

An important strategic issue in fishery planning concerns the appropriate spatial scale at which to manage a particular fishery resource. Indeed, what exactly is the fishery system to be managed? Should we manage small genetically distinct units or aggregations of stocks? Should the fish and the ecosystem define the boundaries, or the human institutions and political divisions? These questions were addressed in Chapter 1 with respect to *characterisations* of the fishery system, but it is essential that they are also considered in a management context. Often there is no latitude in changing the definition of *the fishery*; for example, it may be based on historical evolution and not open to debate. Nevertheless, it is important to be aware of spatial considerations and the possibility of modifying the spatial nature of the fishery as it evolves over time (for example, by shifting to more local-level management where useful).

In existing fishery systems, we can observe management operating at a variety of spatial scales (Fig. 4.4). This depends on factors such as the fish stock involved, the organisational arrangements of the fishers, the responsibilities and powers of the various levels of government, and the traditions and customs involved. As but one example, management on a finer spatial scale is more likely to be feasible for a basically sedentary species than for a relatively migratory species, for a case in which fishers are restricted to relatively clear geographical boundaries in their local area than for one in which participants range widely, and for fisheries with a tradition of local-level self-regulation.

On a multinational scale, consider the case of European Community fisheries operating under a Common Fisheries Policy, as described above. TACs are set and allocated amongst member states, and common measures relating to fleet structure and size are determined where desired. However, once a TAC is set and subdivided among nations, each nation is able to manage its share of the harvest in a manner designed to achieve national objectives. Thus, the spatial scale of management is large when TACs are being determined, but potentially smaller when fleets are being managed. This is one reflection of the principle of subsidiarity – making decisions at the most local level possible. Such a philosophy fits closely with the trend in management away from a totally centralised format towards devolving responsibili-

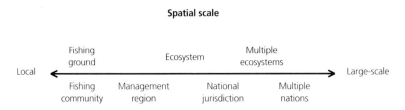

Fig. 4.4   As discussed in Chapter 1, fishery systems (and thus fishery management) function at a range of spatial scales from local to large-scale. Aspects of the natural world are shown above the axis, while those of the human and management systems are shown below the axis. Note that the positioning of these relative to one another, while common in practice, is by no means universal. In some cases, for example, a *fishing ground* may encompass an ecosystem (such as the North Sea) and may involve multinational management.

ties to a more local level, as is evident in co-management and community-based management initiatives.

Variability in spatial scale may also occur between the biological, harvesting and management aspects of the system. In particular, spatial scales can differ considerably for decisions concerning the fish, the fishers or overall fishery policy.

- Fish stocks are managed based on geographical units delineated into internationally agreed statistical areas.
- Fishers may be managed (or manage themselves) according to the same statistical areas, or using smaller geographical units reflecting 'community' demarcations.
- Broad, fundamental *strategic* decisions about the structure of the fishery and policy directions may be made on a larger-scale basis, e.g. for an entire region of a nation.

Indeed, a key strategic decision concerns the *design* of the fishery management system, and in particular how administrative arrangements allocate roles and responsibilities across the fishery, from the perspectives both of geography and of the structure of the natural and human sub-systems involved (Fig. 4.5).

4.8 Summary

This chapter has focused on the nature and structure of the policy and planning process in fisheries. The emphasis has been on three themes: (a) setting overall strategic objectives,

Fig. 4.5 Some fisheries operate on a large spatial scale, while others are more locally based, as is the case with this small-scale fishery in southern Thailand. The spatial scale of the natural and human components of the fishery will influence how management can and should operate.

(b) developing suitable management institutions (including matters relating to the appropriateness and acceptability of the management system), and (c) consideration of temporal and spatial scales in the management system. Successfully dealing with all of these lies at the foundation of fishery management, leading to matters of implementation to be discussed in the next chapter. Most of these themes will also be considered later in this book with regard to measures that move fishery systems in the direction of sustainability and resilience – particularly measures to deal with uncertainty, complexity and conflict in fisheries.

Chapter 5
Fishery Management

As explained earlier, the fishery management system consists of a wide range of activities: the strategic management involved in the formulation of legislation, policy and plans (Chapter 4); the development of new fisheries, infrastructure and fishing methods (Chapter 6); the acquisition of knowledge about the natural and human systems, and use of that knowledge to provide advice (Chapter 7); and the activities most closely associated with the term 'fishery management' – the *operational* aspects and 'tactics' of management that are the focus of this chapter. These involve developing a set of suitable tactics and appropriate operational plans to guide the fishery, in keeping with overall strategic fishery goals and policy directions. Three major aspects are involved in this:

- determining the level of fishing effort and/or catch corresponding to the objectives set out in the strategic management process;
- determining the management measures that can feasibly achieve the above effort or catch levels, and which are compatible with the strategic management policy choices;
- implementing the set of chosen management measures, including enforcement of the corresponding regulations.

The first half of this chapter discusses each of these aspects in turn. The second half focuses on a more detailed examination of each major management measure, organised under the headings: input/effort controls, output/catch controls, technical measures, ecologically based measures and economic instruments.

5.1   Appropriate effort and catch levels

Whatever decisions are made at the strategic level of management, the renewable nature of fishery resources will lead naturally to the fundamental question: how much fishing can take place, and how much catch can be harvested from the sea without being detrimental to fishing in future years? This question, and its equivalents in forestry and wildlife management, addresses the balancing act, inherent in all renewable resource harvesting, between present-day benefits and future rewards. It has been addressed to a varying degree over the past several centuries, and within this century has become a fully fledged *science of sustainability*, sometimes referred to as resource management science (Gulland 1977).

In seeking to answer this key operational question of management, the idea emerged of a *sustainable yield*, i.e. an allowable annual harvest which, even if repeated indefinitely into the future, would not lead to excessive depletion of the fish stock. The concept of sustainable yield has long dominated the analysis of renewable resources (e.g. for the fishery case, see Schaefer 1954; Beverton & Holt 1957; Gulland 1977; FAO 1983). A variety of tools are applied to determine sustainable yields and the corresponding level of fishing effort that can be safely allowed without over-harvesting the stocks.

For example, one of the most common tools is a graph (Schaefer 1954) showing how a fishery operating with a constant annual level of fishing effort, repeated indefinitely into the future, will generate a certain sustainable yield from the fishery (Fig. 5.1). This graph is highly simplified. It assumes a single fish species, an aggregated fishing effort, no uncertainty and a static equilibrium, but has nevertheless proved useful for purposes of illustration. The nature of this graph will be discussed in detail in Chapter 8.

The point here is that Fig. 5.1 shows how, given certain assumptions, any combination of effort and yield lying on the curve is sustainable biologically. This illustrates that the idea of sustainable yield is compatible with a spectrum of harvesting options, the choice of which will depend on the strategic goals being pursued (Chapter 4; Charles 1988). Indeed, strategic fishery goals are ultimately implemented in the fishery largely through the choice of a specific level of harvesting. This will reflect a balancing of multiple objectives, but to determine that balance it is useful to understand the implications of pursuing any particular strategic objective individually. Operational levels of fishing corresponding to five possible strategic objectives (cf. Salz 1986: p. 23) are described below.

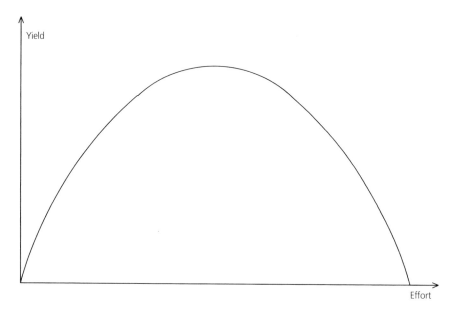

Fig. 5.1   Schaefer (1954) showed how a fishery operating with a constant annual level of constant fishing effort (a point on the horizontal axis), applied indefinitely into the future, will generate a certain sustainable yield (the height of the so-called yield curve). See Chapter 8 for a detailed treatment of this graph.

(1) *Maximum biomass.* An annual effort level of $E=0$ (zero) will provide no catch and no revenue, but will maintain the biomass at its greatest possible level. This may be desired, for example, if a particular society views the catching of a particular species as immoral (e.g. disapproving of the killing of a certain marine mammal). Note that while this may be seen as the ultimate conservation measure, it would be misleading to refer to such a situation as 'maximum conservation' since in general the term 'conservation' implies sustainable *use*, which allows for some level of harvesting.

(2) *Maximum fishing employment.* An annual effort level at the *bionomic* or *open access* equilibrium, E_{OAE} (where revenues just match costs, including opportunity costs), will maximise the fishing effort, while still producing normal profits for the fishers, but zero resource rents. (This contrasts with the maximisation of post-harvest employment, for which one would more likely want the catch level to be at maximum sustainable yield (MSY) – see below.)

(3) *Maximum sustainable yield* (MSY). Applying an annual effort level E_{MSY} will produce the maximum *harvest* of fish that can, in theory, be caught year after year indefinitely into the future. This may be desirable if society's goal is to maximise food production, or possibly employment in post-harvest activities.

(4) *Maximum economic yield* (MEY). An annual effort level E_{MEY} generates a maximum level of sustainable *economic rents* from the fishery, obtainable each year indefinitely into the future.

(5) *Maximum social yield* or *optimum sustainable yield* (OSY). An annual effort level E_{MScY} maximises a multi-objective blend of socioeconomic values, perhaps including equity, employment and rents, with appropriate weighting of each goal.

The implications of these fishery objectives can be shown by building on Fig. 5.1. Suppose we assume for simplicity a constant price for fish, so the revenue received by fishers is given by the yield multiplied by the price, and assume that fishing costs are proportional to the amount of fishing effort (the more you fish, the more it costs). With these assumptions, we obtain the so-called Gordon–Schaefer diagram in Fig. 5.2, where the U-shaped curve is now the total fishery revenue, the straight line shows the fishing costs, and the declining curve is the equilibrium biomass level for any given constant fishing effort level. The latter reaches zero at the extinction point, marked E (extinction); an effort level higher than this, imposed year after year, is assumed to lead the stock to extinction. (See Chapter 8 for a more detailed derivation of the Gordon–Schaefer approach.)

Figure 5.2 shows the long-term constant fishing effort levels corresponding to the first four of the objectives described above. Note that $0 < E_{MEY} < E_{MSY} < E_{OAE} < E$ (extinction) and that the sustainable biomass declines as we move from one to the next of these effort levels. The effort level E_{MScY} is not shown, since its location is arbitrary, and represents a balancing of objectives determined on a case-by-case basis. In some cases, a zero effort may be considered optimal, in other cases it could be any point below E (extinction). For example, suppose we wish to focus on optimising economic benefits from the fishery. Although fishing at E_{MEY} may maximise rents, if there is a very high unemployment rate in the economy, and thus a lack of non-fishery jobs for fishers (a zero *social opportunity cost of labour*), then fishing at a higher

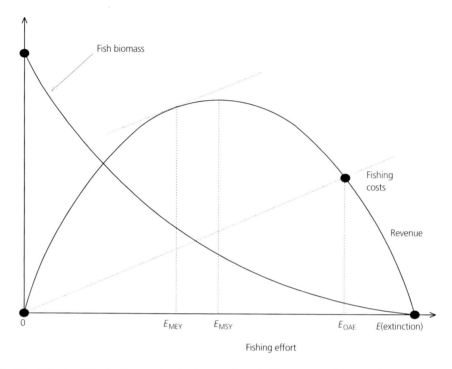

Fig. 5.2   A Gordon–Schaefer diagram showing the annual sustainable revenue and the operating costs, as well as the equilibrium fish biomass, produced if any given constant annual fishing effort (a point on the horizontal axis) is applied indefinitely into the future. The effort levels labelled as 0, E_{MEY}, E_{MSY} and E_{OAE} correspond to four of the objectives discussed in the text (E_{MScY} is not shown here) while E(extinction) is the effort level above which extinction will result.

level may be preferable. Thus, the determination of the 'best' means to meet fishery objectives can be a complex task.

5.2   Developing a portfolio of fishery management measures

Once the desired levels of exploitation have been determined, the next major prong in management involves determining the measures ('tactics') that can best be expected to achieve target fishing levels, in keeping with policy directions defined at the strategic management level. This may focus on the harvesting sector, but it is important to keep in mind that *integrated* management takes into account the full fishery system – not only the harvesting sector, but also the handling, processing, distribution and marketing, and consumption of fish harvests.

5.2.1 Evaluating broad classes of management measures

The first step in choosing from among possible management measures lies in assessing the conservation, socioeconomic and manageability implications of each broad category of management control.

- Input (effort) controls: i.e. regulating what fishers bring *into* the fishing process.
- Output (catch) controls: i.e. regulating what comes *out* of the fishing process.
- Technical measures (regulating technology): e.g. size of mesh in nets or size of hooks.
- Ecologically based management: e.g. marine protected areas, multi-species approaches.
- Indirect economic instruments: e.g. taxes on catches (*royalties*) or on fishing effort.

These measures are examined in detail later in the chapter. Note that these broad categories are not mutually independent; there is likely to be some overlap. In particular, technical, ecological and economic elements may be included to some extent within any of the other categories. For example, input/effort and output/catch controls may incorporate technical measures, may seek to take an 'ecological' approach, and certainly will have economic implications. Nevertheless, these categories are useful in ensuring that we consider all aspects of a management regime.

5.2.2 Evaluating specific management measures

The second step lies in evaluating specific control measures. For the first three of the broad management categories listed above, key specific measures might be described as in Table 5.1 (with entries detailed later in this chapter).

5.2.3 Choosing a portfolio of management measures

A suitable set of management measures must be chosen, based on an understanding of (a) the extent to which each category of management control, and each specific measure, achieves the stated objectives, (b) the extent to which each of these is compatible with the desired policies, and (c) the extent to which each is feasible from a manageability perspective.

5.3   Implementation at the operational level

At the 'lowest' level of management, the focus is on specific, detailed plans for implementing the various management tools selected at the tactical stage. This involves:

- determining precise (perhaps quantitative) levels of each management measure, such as the annual levels of the total allowable catch in a cod fishery, the carapace size in a lobster fishery, or the escapement target in a salmon fishery;

Table 5.1   Management tools: effort, catch and technical measures.

Input controls (effort)	Outputs controls (catch)	Technical measures (where, when, how)
Fleet size (no. of boats)	Total allowable catch (TAC)	Gear limitations (mesh or hook size)
Vessel capacity (catching power)	Individual quotas	Closed areas (nursery grounds)
Fishing effort (use intensity and time)	Community quotas	Closed seasons (spawning closure)
Fishing area/boat (e.g. TURFs)	Escapement targets	

TURFs, territorial use rights in fishing.

- determining management plans for each sector of the fishery, and detailing the operating requirements for a particular class of vessel or for participants located in a specific geographical area;
- implementing a mechanism for adjusting management measures over the course of the fishing season (*adaptive management*) when required for conservation purposes, to meet agreed sharing arrangements between sectors, or for other reasons;
- implementing a mechanism for enforcing management plans, both at the level of the individual fisher and fishery-wide;
- implementing specific measures required to support the management of the fishery system, such as research needed to improve our understanding of the marine fish resources, and fishers and fishing communities, as well as the broader ecological and human environment.

Example: choosing gear configuration (mesh size)

A common decision needed in fisheries concerns the allowable dimensions of the fishing gear to be used. For example, what is the 'optimal' hook size? If too large, the fisher will tend to catch only the very largest fish, and will thus earn a low income. If too small, primarily juvenile fish may be caught, again providing less income, as well as harming the stock. Presumably, an intermediate size will be best.

A similar argument applies to gill nets. A net with a very small mesh size (small 'holes' in the net) will entrap very small fish but will act more as a barrier to large fish, which will tend to swim away from the net rather than being caught. Thus, with a small-mesh net, the total biomass caught will be small unless so many small fish are caught that it decimates the stock. On the other hand, a gill net with too large a mesh size will catch nothing, since all fish will be able to swim through the mesh. Some intermediate mesh size presumably represents an 'optimal' design.

The situation is somewhat different with trawling. Trawl nets with a small mesh size will catch and retain much of the intercepted fish as the nets are pulled through the water. As the mesh size is increased, fewer small fish will remain in the net. Ultimately, at a very large mesh size, most fish will escape, and catches will be very low, as in the case of gill nets.

Figure 5.3 depicts the situation for the trawling case. The *average net economic benefits* (e.g. per day of fishing) obtained in the fishery are plotted against the mesh size utilised for each of a short-term and a long-term scenario. The short-term curve focuses on how much is caught in the *current* day of fishing (or tow of the trawl), without concern for the impacts of this harvest on the fish stock. On the other hand, the long-term analysis is in terms of average daily benefits obtained over time, taking into account the response of the fish stock to different levels of fishing pressure as given by different mesh sizes.

The idea is as follows. A hypothetical net with zero mesh size is like a large bucket; all fish encountering it will be caught. In the short term, this implies very large catches

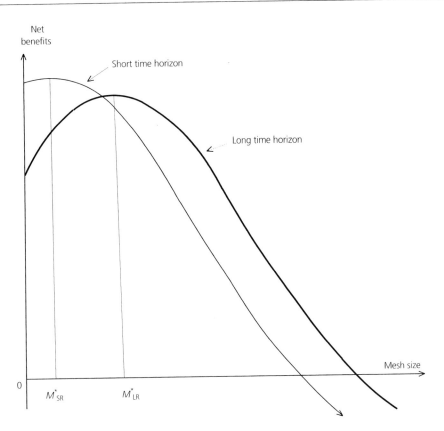

Fig. 5.3   The *net economic benefits* (e.g. per day of fishing) in a trawling fishery plotted against the mesh size utilised, for both a short-term and a long-term scenario. The short-run curve focuses on the *current* day of fishing (or tow of the trawl), while the long-run analysis is in terms of *average* daily benefits obtained over an extended time frame. The benefit-maximising mesh size is greater in the long-run analysis than that used in a short-run perspective ($M^*_{LR} > M^*_{SR}$).

per tow and large net benefits (assuming that the cost of pulling such a net is not too high). At small but positive mesh sizes, essentially the same argument applies. This is why trawlers often add liners to their nets, to reduce the mesh size, whether legally or illegally, increasing short-term profitability. However, if all vessels were to operate in this way, the average benefit in the long term will be low because all the fish will be caught in the first few years, so that future harvests will be negligible. In other words, there is a short-term economic incentive to use gear that is efficient in catching the most fish, but which is harmful in the long term. This is illustrated in Fig. 5.3, where the optimal mesh size (where the peak of the curve occurs) is greater in the long-term case than in the short-term analysis. This is a graphical way of highlighting the need for regulation and enforcement of fishing-gear configurations.

5.4   Fishery enforcement

Fishery enforcement is a crucial step in the management system. Its rationale lies in the realisation that some degree of illegal fishing can be anticipated as a response to a regulatory framework designed to limit fishing activities, and in the light of economic incentives that make such illegal fishing profitable in the absence of potential penalties. Indeed, illegal fishing and misreporting of catch levels, in an environment of insufficient enforcement, have caused serious overfishing as well as errors in stock assessment. (See Angel *et al.* 1994 for an analysis of the impacts of enforcement shortfalls in the Atlantic Canadian groundfishery.)

One can observe, particularly in poorer developing nations, that the good intentions inherent in fishery legislation can be thwarted by a lack of policy attention to (or financial capability for) enforcement of that legislation. For example, legislation to regulate the operation of foreign fishing vessels within a nation's territorial waters is only useful if the will and the resources are available for its enforcement. Capacity-limiting regulations designed to restrict harvesting power can easily be thwarted by fishers unless the regulations are designed cooperatively and enforcement is made *endogenous* through self-regulation.

The importance of understanding the enforcement problem is apparent throughout the real world of fisheries. As with fishery management broadly, there are strategic, tactical and operational aspects of fishery enforcement.

- At the strategic level, the challenge is to design an effective framework linking management and enforcement, to minimise the incentives for illegal activity, and to maximise the incentives for self-monitoring (self-regulation) by fishers. There is also the key question of how much to spend on enforcement. Since a major rationale of enforcement is to prevent overfishing, its benefits come in the future (as increased harvests), but it results in increased present-day costs.
- At the tactical level, the goal is to determine the most effective set of monitoring, control and surveillance mechanisms, given a certain budget constraint; measures that best minimise illegal behaviour, or the impact of that behaviour, given the enforcement resources available.
- At the operational level, the enforcement agency uses the resource allocations made in the tactical decisions above to carry out its day-to-day activities as well as possible. This involves assigning enforcement officers and vehicles (boats or aircraft) to specific routes and tasks at sea, as well as accompanying enforcement efforts on land (Fig. 5.4).

There is also a research component to fishery enforcement. Indeed, finding the answers at each of the strategic, tactical and operational levels requires interdisciplinary approaches incorporating the study of such areas as economic incentives, criminology and social behaviour, appropriate institutions, legal mechanisms and operational management. Some of the questions that must be addressed are listed below.

- How should regulations be developed? How can consensus in the formulation of regulations be achieved in order to maximise efficiency of enforcement?

Fig. 5.4   Enforcement of fishery regulations takes place both at sea and on land. A crucial aspect in both cases is to maximise the likelihood of compliance – at this landing site in Barbados, the regulation is clearly stated, and the process for complying is straightforward.

- How can fishery participants collectively and individually have a stake in enforcing agreed regulations?
- To what extent can enforcement be made more feasible and efficient through cooperative (government/industry) or community-based fishery management?
- How should enforcement effort be expended? What allocations of surveillance and search activities, and what routing of enforcement vehicles, will be most effective?
- How are fishers likely to respond to specific forms of enforcement?
- What is the link between the level of penalties (fines) and the likelihood of apprehension?
- How might biological and technological considerations interact with regulatory enforcement?

Unfortunately, enforcement is among the least-studied and least-understood components of the fishery system. A wide array of fishery management and regulatory measures have been implemented in fisheries around the world, from gear limitations to limited entry to individual quotas, but the critical question of enforceability is rarely addressed in any depth. In short, there is a gap in fishery analysis and a need for a rigorous approach, a 'theory of fishery enforcement' (cf. Sutinen & Andersen 1985; Sutinen & Hennessey 1986; Sutinen *et al.* 1990; Furlong 1991; Charles 1993; Charles *et al.* 1999).

Coastal states and regional cooperation: shared stocks and foreign fleets

Troadec (1982) describes the management process involved in achieving regional co-operation on the management of fish stocks shared between nations, such as along a stretch of coastline that includes a number of countries. The following list is adapted from Troadec's work (pp. 54–57). Note that strategic, tactical and operational aspects are intertwined here.

Evaluation of fisheries

- Collection and distribution of data among participating parties.
- Stock assessment, study of distribution and migration patterns.
- Assessment of changes in the current state of the fishery and in fishing intensity.
- Development of technical strategies for fishery management.
- Determination of a suitable programme of fisheries research.

Negotiations of resource-sharing agreements

- Development of an overall fishing strategy, acceptable to all parties; this must be based on settling on harmonised objectives and determining an acceptable common fishing rate.
- Allocation of resource access (fishing effort) and/or allowable catches (measured as portions of the TAC or in terms of catch capacity) between participating nations.

Formulation of national fishing plans

- Establishment of national objectives for the development and operation of national fisheries
- Allocation of domestic fishery resources among the various domestic fleets, and foreign fleets where relevant and desirable.

Exploitation

- Implementation of monitoring and enforcement systems (beach patrols, coast guard, aircraft).
- Implementation of reciprocal fishing agreements between nations, authorising fleets to operate throughout the zones within the jurisdiction of countries sharing the same resource.

'Follow-up'

- Follow-up on the state of fisheries.
- Follow-up on the implementation of regulations.
- Implementation of any legal action required.

5.5   A survey of fishery management measures

The remainder of this chapter provides brief overviews of the principal items in the management tool-kit, organised according to the five categories described above: input/effort controls, output/catch controls (quota), technical measures, ecologically based management, and indirect economic instruments. Note that several input and output control measures described below (particularly individual gear and catch limits) represent *use rights* approaches and are described in greater detail in Chapter 14.

5.6   Input (effort) controls

The basic idea behind input controls is to regulate 'fishing effort', one of the classic, if ill-defined, concepts in fisheries. 'Effort' measures *how much* fishing takes place; it is 'effort' that impacts on the fish stocks. Although effort *per se* does not measure the specific impacts of that fishing, it is fair to say that 'no effort implies no fishery'. While fishing effort is an amorphous amalgam of inputs, there are four major constituent elements that can be identified for any given component of the fishing fleet:

- the number of fishing vessels;
- the average *potential* catching power of a vessel in the fleet (taking into account the typical size, fishing gear, electronic gear and other physical 'inputs', as well as the vessel's crew);
- the average intensity of operation of a vessel per unit time at sea, measuring the fraction of the potential catching power that is actually realised;
- the average time at sea for a vessel in the fleet.

Thus, the total effort for the particular component of the fishing fleet is given as the product of these four figures:

Fishing effort = (Number of vessels) × (Catching power) × (Intensity) × (Days at sea)

Note that if a fishery system had no vessels, or no gear (implying no catching power), or no fishing time, we would expect that there would be no catch. In other words, if any one of the four ingredients above were zero, there would effectively be no fishery. Note also that both real-world experience and research results have indicated that a major factor in determining the outcome of fishing is the skill and experience of those doing the fishing. This factor could be incorporated in measures of fishing effort, but in most analyses of fishing activities this has not been done.

5.6.1 Limiting entry

Among fishing effort controls, one of the most widespread is the limited entry approach, by which the number of participants in a fishery is directly limited through providing fishing licences to a limited number of individual vessel owners. This approach essentially institutes a form of access rights, as discussed in Chapter 14.

5.6.2 Limiting the capacity per vessel

While limited entry is the principal means used to limit access to the fishery, it must be seen not as the stand-alone 'solution' to managing the people side of the fishery, but rather as just one tool needed in a portfolio of management methods in any given fishery. This is because the impact each vessel makes on the fishery resource, i.e. its capacity, or catching power, is not constant, but rather depends on the dimensions of the vessel, its physical capacity to hold fish, its fishing gear and electronic gear, and so on. Thus, in any form of input or effort control programme, it is important to place limitations on the key components of effort on a typical vessel. Common approaches to this are (a) limiting the dimensions of the vessel, notably the length or the hold capacity, and (b) limiting the amount of gear (e.g. number of traps or nets) that can be utilised from each vessel.

5.6.3 Limiting the intensity of operation

This input component is very difficult to control. While the total number of vessels (fleet size) and the potential catching power (based on the characteristics of the vessel, such as size and gear availability) can be fairly clearly regulated (if not always easily enforced), intensity of use relates to more nebulous matters of how hard the crew work, as well as uncontrollable matters such as weather conditions during fishing times.

5.6.4 Limiting time fishing

Any fleet, with whatever level of average catching power, and whatever level of motivation among the crew, cannot catch fish unless the vessels are at sea. Thus, controls over fishing time, often expressed as *days at sea,* are increasingly being examined as tools for fishery management. A challenge in setting an allowable fishing time lies in the uncertain nature of the 'intensity' noted above. If we assume that 100% of the catching power will be utilised, the reality may be that actual fishing effort ends up being far lower than expected. On the other hand, if effort is reduced by a scaling factor to account for what is anticipated to be unutilised catching power, the manager may be surprised by more intense fishing than expected, resulting in excessive effort levels. (There is also a major issue of changes over time in catching power per vessel, due to investment behaviour among fishers. This issue is discussed further in Chapter 8.)

5.6.5 Limiting the location of fishing

An important input into the fishing process is the location where the fishing takes place. Indeed, fishers are often highly secretive about their fishing locations and convinced that they know the best place to place their traps, nets or lines. Given this importance of location, it is not surprising that one of the major traditional management methods is that of territorial allocations, through which specific areas of water bodies are set aside for the exclusive use of designated individuals, families, fishery sectors or communities. This may be done on an individual level, in terms of each fisher's traditional fishing grounds, or on a governmental basis through some form of area licensing. The former is sometimes referred to as *territorial use rights in fishing* (TURFs) and is discussed further in Chapter 14, while Table 5.2 provides an example of government-designated spatial restrictions. In either case, such an approach can reduce conflicts between fishers and limit the intensity of exploitation of any single geographical component of the resource.

5.6.6 Challenges with input controls

If inputs are controlled, and if those controls are binding, a natural incentive is created among fishers to find ways to expand the use of effort. This response of fishers to input restrictions can affect the success of input controls. For example, restrictions on vessel length, designed to limit capacity expansion, can lead logically to the construction of *wider* vessels, so that capacity expands nonetheless. Similarly, increases in the minimum legal mesh size on trawlers can lead to increases in the use of other unregulated inputs, such as time spent fishing. Limitations on the number of traps allowed per fisher in a lobster or crab fishery can induce changes in other inputs, such as increased use of labour (perhaps hauling traps more frequently). Thus, it is necessary to place controls on a wide range of inputs. One cannot expect that a single control only on the number of days fishing, or only on the amount of gear, etc., will allow the fishery's objectives to be fully met.

5.7   Output (catch) controls

While *input* controls focus on limiting the various components of fishing effort, *output* controls focus almost entirely on what is taken from the fish stock, the catches (although in some cases, attention is also placed on what is left behind after the catch is taken).

5.7.1 Total allowable catch (TAC)

By far the most commonly discussed output control is that of regulating the total harvest of each fish stock in the fishery system. Such overall control in this situation is called the total

Table 5.2   Spatial zoning regulations in Malaysia (Abdullah & Kuperan 1997).

Zone	Distance from shore	Description
A	Within 5 miles (8 km) of the shoreline	Reserved solely for artisanal, owner-operated vessels
B	5–12 miles (8–19 km)	Reserved for owner-operated trawlers and purse-seiners of less than 40 GRT
C	12–30 miles (19–48 km)	Reserved for trawlers and purse-seiners greater than 40 GRT, wholly owned and operated by Malaysian fishermen
D	Beyond 30 miles (48 km)	Reserved for deep-sea fishing vessels of 70 GRT and above. Foreign fishing through joint ventures or charter are restricted in this zone

---

Quota management in Iceland

Danielsson (1997) provides a review of the initial introduction of quota management in the fisheries of Iceland. 'The regulation of fishing gear and the closure of fishing grounds have long been used as management tools in Iceland, and they still are. More direct fisheries management, in the form of restrictions on catches or effort, dates back to 1965 when fishing for inshore shrimp and scallop were subjected to licences, effort restrictions and catch quotas. In 1969, after the collapse of the herring stocks, TACs were set for catches of the Icelandic summer-spawning herring … In [1976], the spawning cod stock was estimated at a very low level, thus precipitating the introduction of TACs for cod …'

---

allowable catch (TAC), i.e. the quantity of biomass that is permitted to be caught. The TAC is usually based on biological grounds, i.e. the TAC is the amount of fish that can reasonably be killed in the fishing process. While this should properly be measured as the 'catch', in reality the limitation is generally implemented in terms of how much fish is brought to shore. There can be a major difference between these amounts, in particular because of the possibility that fish may be caught but dumped overboard before landing if that fish is of a lesser value than other fish, or if it is of a prohibited species or size. This situation is discussed below.

In many cases, the TAC is then subdivided into quotas for specific fishery sectors. This might be done through allocations by country, as in the case of TACs set by the European Union, which are then subdivided among countries, or TACs for stocks managed internationally, as with Canada's northern cod stock. The subdivision of a single nation's TAC might be by fishing gear (e.g. bottom trawlers, long-liners, etc.) or by vessel size (e.g. dividing the TAC into separate pieces for small, medium and large vessels). In some countries, TACs are subdivided down to the individual level (community, enterprise, vessel or fisher) (see below and Chapter 14). Note that a management system based on setting and subdividing a TAC is often referred to as *quota management*.

5.7.2 Individual quotas

Individual quotas (IQs) are quantitative output rights defining the amount each fisher can catch within a certain time period. These can take the form of 'trip limits' that restrict how much can be caught on each fishing trip, or they can be set on an annual basis, with the allowable harvest (*individual quota*) typically defined as a fraction of the total allowable catch. In the latter case, there are two principal options to choose between: individual transferable quotas (ITQs) or individual non-transferable quotas (INTQs). Both can allow transfers of quota within a given year (if, for example, a fisher has extra quota of a particular species); the difference lies in what happens between years. With ITQs, permanent transfers (buying and selling) of quotas can occur, while with INTQs, permanent sales of quota are not allowed, so that each fisher's share remains the same from year to year and a concentration of quota holdings is avoided. The advantages of IQ programmes can include a more orderly fishery, since the 'rush for the fish' is reduced, and there may be a smaller number of vessels participating. The disadvantages include an increased

---

Fisheries management in Norway

'The watershed in fisheries management came in 1977, with the establishment of the 200 mile Exclusive Economic Zone (EEZ). With this measure in place, it was possible to establish total allowable catches (TACs) for Norwegian as well as shared resources, and catch control gradually became a central management task … In the 1990s a system of "unit quotas" was introduced within the purse seine and trawling fleets, which is a very close approximation of an ITQ system without actually using the name. Finally, a quota system for the coastal fleet was introduced, functioning partly as an Individual Vessel Quota (IVQ) system and partly as a limited entry regulation in the economically most important cod fishery … The cod fishery is therefore regulated by effort as well as output. In addition, we find an array of technical regulations, pertaining to minimum sizes: legal mesh size, area closures, time closures, bycatch, etc. Consequently, the control system is extensive and costly.'

Hersoug *et al.* (1999)

---

incentive for dumping and high-grading of fish, and possibly a lessened acceptance of adaptive management (Chapter 11). There are also a variety of social, economic and conservation implications of ITQs versus INTQs, which are discussed in more detail in Chapter 14.

5.7.3 *Community quotas*

The basic concept of a community quota is no different from that of an individual quota, or of a sectoral allocation of the TAC (e.g. a part of the TAC allocated to long-liner vessels). In each case, the approach is to subdivide a TAC into pieces. However, the difference is apparent when we recall the growing movement, noted elsewhere, towards community-based management of natural resources in much of the world. This movement recognises the potential benefits of (a) bringing management more to a local level, (b) involving community institutions to improve both the design and the acceptance of management, and (c) drawing on inherent moral suasion within communities to improve compliance. The key aspect of community quotas therefore lies in its integration of the 'individual' quota concept with that of community-based management. This arrangement implies that since control of the quota is at the community level, the objectives being pursued through its use will probably be those of the community as well. (An example of such a community is shown in Fig. 5.5.) See Hatcher 1997 and Loucks *et al.* 1998 for discussion of the community quota approach, among producers' organisations in the UK and groundfish management boards in Canada, respectively.

5.7.4 *Escapement controls*

Suppose that, instead of regulating what is taken out of the ocean for a given stock, we focus on ensuring that enough is left behind to allow for successful spawning. This implies that we focus not on limiting the catch, but on ensuring an adequate *escapement*: the amount of fish 'escaping' the fishery. This is the standard management approach for salmon fisheries.

Fig. 5.5   The community of Sambro in Nova Scotia, Canada, was the first in its region to institute a *community quota* system, within which a local organisation of fishers takes on the responsibility to establish and enforce harvesting regulations under which its members fish for a specific share of the TAC. The local fishers credit the approach with safeguarding the well-being of their fishery-dependent community.

The objective to be achieved is an escapement, although the tools used to achieve that goal are often input controls which limit the number of fishers through limited entry (and possibly area licensing) and limit the days of fishing by the fleet.

5.7.5 Challenges with output controls

Several general principles seem to apply with respect to output controls. First, just as input controls create incentives to increase effort, so too do regulatory measures restricting catch (or the catch mix, in terms of species, fish sizes or fish locations) create economic incentives to go beyond those limits through excessive catches or changes to the catch mix. Second, the incentive to underreport catches rises as quotas are placed on a more and more individual basis, i.e. from global TACs, to sector quotas (e.g. a quota for 'small fixed-gear vessels'), to individual fisher quotas. This is because with individual quotas, a fisher who fails to report a catch keeps his or her own quota higher than it should be, which allows that fisher to catch more in the future than would otherwise be the case. Under a TAC, on the other hand, catches unreported by a particular fisher are not counted against the total allowable harvest, so the latter remains available to catch subsequently. However, since all fishers have access to this, the cheating fisher will probably receive only a small fraction of the 'benefits' from the cheating. Third, a quota limitation, and particularly one at the individual level (ITQs or trip limits), creates the incentive not only to exceed the quota, but also to land the most profitable catch possible. In other words, there is an incentive not to land lower-valued species or lower-valued individuals of the given species. This can lead to anti-conservationist practices at sea: *high-grading, dumping and discarding*. High-grading involves keeping fish only of the preferred species, of the greatest size or of the highest unit value (typically dumping lower-value fish overboard), in order to maximise the *value* of what is reported as caught. Dumping and discarding involve throwing overboard undesired or prohibited fish species (including those for which the quota has been reached) or fish of the 'wrong' size (notably small fish) so as to be able to continue fishing. These considerations, and particularly the various incentives (positive and negative) involved in individual quota schemes, are discussed further in Chapters 14 and 15.

5.8   Technical measures

While effort and catch controls are the subjects of most discussion in the literature on fishery management, measures to limit the 'how, when and where' of fishing are historically the most widely implemented management tools. These are referred to as *technical measures*, and are typically oriented towards meeting biological/conservation goals directly related to specific fish stocks (see, e.g. FAO 1983; Caddy 1984).

5.8.1 Gear restrictions

These management measures restrict the allowable types of fishing. Most common are restrictions on gear attributes, such as a minimum mesh size for nets, or a minimum hook size for fishing lines. Such restrictions are usually implemented to help in meeting biological objectives, e.g. reducing catches of small juvenile fish in order to allow more to mature. Another form of gear restrictions concerns the type of gear itself. For example, bottom trawling may be prohibited in some contexts to protect a sensitive ocean bottom, or to avoid disrupting the spawning process. Note that gear restrictions tend to reduce the catching efficiency of the fishers, i.e. these gear restrictions are *cost-increasing* measures in that it costs more to catch a certain quantity of fish when restricted by such regulations. While this may seem undesirable based on conventional economic logic (the idea that we want everyone to be as efficient as possible), in fact reduced catching efficiency implies reduced pressure on the fish stocks and greater manageability of the fleet, two important indirect conservation benefits of gear restrictions. However, any efficiency-reducing measure creates an inherent incentive for fishers to thwart it, since anyone who can do so will be able to fish more profitably than others, and thus enforcement measures must accompany these restrictions.

5.8.2 Size limits

As noted above, it is often considered desirable to minimise the fraction of small, juvenile fish in the catch, in order to allow these individuals to mature and reproduce. There may also be a rationale for avoiding the capture of particularly large animals, on the basis that these produce a disproportionately large number of eggs and are therefore important to the future of the stock. The application of suitable gear restrictions is often the most desirable means to achieve both of these ends, since if such regulations are effective, the fishery does not even come into contact with the animals to be avoided. In some fisheries, however, gear restrictions may be (1) relatively ineffective in protecting components of the stock, and/or (2) considered unacceptable tools of management for some reason.

A popular alternative in such cases is directly to regulate the size of animals that can be landed. It is then illegal for a fisher to keep any animals of an unacceptable size that are caught; these must be returned to the sea immediately. This is a typical approach in many hook-and-line recreational fin-fish fisheries, as well as in many fisheries for shellfish involving traps (e.g. for lobster) or hand-gathering (e.g. various molluscs). The reasonableness of such measures depends on the fish and the fishery under consideration (King 1995), since it is important that the fishery itself does not cause undue mortality to animals returned to the sea, and that the process of assessing the size of each individual does not unduly reduce the fishery's profitability. Thus, for example, minimum size limits are proven regulatory tools in trap-based crab and lobster fisheries and recreational salmon and trout fisheries. In both cases the volume is relatively low (compared, say, with trawl-based fisheries) and the value (however measured) of individual animals is relatively high (Fig. 5.6).

5.8.3 Closed areas

As with the above restrictions, this form of management is usually implemented to achieve biological and/or ecological goals. The concept involves areas of the ocean being closed to fishing on a permanent, temporary or seasonal basis. Permanent closures may be implemented to provide ongoing habitat protection, for example, to ensure that the habitat where eggs are deposited is not disrupted or otherwise harmed, and/or to protect nursery areas where young fish congregate. Shorter term closures, sometimes called 'spot closures', may be im-

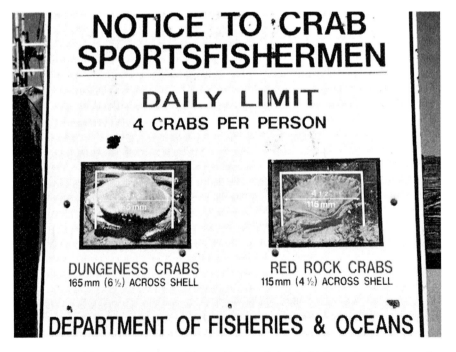

Fig. 5.6 This sportsfishery for crab operates with a combination of a 'bag limit' (four crabs per person per day) and a 'technical measure' in the form of a minimum size limit. The latter ensures that small animals have a chance to mature; this measure is only effective if mortality incurred in returning animals to the sea is very low.

plemented in well-defined and usually small areas of a fishing zone to avoid concentrations of small fish, specific age classes that need protection, or fish of a species being protected. For example, if the goal is to avoid killing small fish, but gear restrictions are not sufficiently effective and it is not appropriate to use a minimum size limit, it may be possible to achieve the desired benefits using a so-called 'small fish protocol'. This involves a process of sampling the size composition of the catch in each sub-area of the fishing grounds, with any given sub-area being closed temporarily if there is an excessive proportion of small fish in the catches within that zone. The sub-area can reopen when testing indicates a low enough level of small fish. (Note that usually the term 'closed area' is used to describe a very specific management measure in a specific fishery; a broader concept is that of a marine protected area, as described below.)

5.8.4 Closed seasons

Closely related to closed areas, a closed season is implemented typically to safeguard the spawning process, and perhaps to restrict fishing at times when juvenile fish are most easily caught. Since fish are often highly concentrated at such times (e.g. during spawning), fishing would probably be highly profitable. Thus, there will typically be an economic incentive to violate the closed season regulation.

5.9   Ecologically based management

The vast majority of fishery management approaches have been, and still are, focused on controlling the exploitation of a single fish stock. Certainly, almost all TACs are set in this way, and indeed regardless of the management method used, the very structure of management (at least in developed nations) is typically on a single-species basis. Yet increasingly, it is being recognised that we must also incorporate predator–prey interactions, environmental conditions and other ecosystem considerations into fishery management. (This is a focus of Chapter 12.)

It may be possible to accomplish this, at least to some extent, by attempting to take such factors into account when setting single-species TACs, effort limitations, etc. This is done already with respect to by-catch impacts on non-target species. In rare cases, there is also explicit attention to predator–prey situations, i.e. systems involving the harvesting of a prey species (e.g. a small pelagic, say herring) and a predator (e.g. a groundfish such as cod). For example, the TAC for herring might be set so as to allow for the importance of herring as a food for cod, and the TAC for cod might be set to reflect the desire to maintain large herring stocks for harvesting.

Furthermore, some management measures have the inherent potential to incorporate ecological considerations. One example is the use of effort controls, which limit the quantities of vessels and gear used in a fishery system over a certain time period. If applied within mixed-stock fisheries, such controls limit the total harvesting pressure imposed on the ecosystem, and thus have an impact on a range of species. This could also be the case for technical measures such as gear restrictions, closed areas and closed seasons, as outlined above. While in the past, most such technical measures were applied only to specific species (for example, an area closure may apply only to fishing targeting a certain stock, or a closed season may be applied only to fishing for a certain stock), if applied uniformly, technical measures will have an impact not on a single species, but also on the entire set of species affected by fishing within the ecosystem.

The above examples reflect a characteristic of *ecologically based management* in that measures are applied not to deal with a single fish stock, but instead are implemented to manage human activity within the ecosystem as a whole (including non-commercial or non-targeted as well as target species). Perhaps the most significant new initiative in this category is the *Marine Protected Area* (MPA, or marine reserve). The MPA is a true example of ecologically based management in that there is no focus on individual species, but instead there is an emphasis on limiting human activity throughout a designated area of the ocean or other water body. The limitations can range from an outright prohibition on the extraction of fish, minerals or any other entity, through to a zoned approach that could allow limited fishing and other activities in certain parts of the MPA. This and other *integrated* approaches to fishery management are discussed in detail in Chapter 12.

5.10 Indirect economic instruments: taxes and subsidies

Economists have long noted that a suitably designed tax can 'induce' the desired behaviour on the part of private individuals and firms by modifying the financial calculations made by those private decision makers. In fishery systems, taxes are usually envisioned as applying to catch (referred to as *royalties* or *landings taxes*, and charged per unit of harvest) or to fishing effort (applied, for example, per day of fishing). This type of indirect regulation can, in theory, be set at the right level to induce the desired behaviour by fishers and bring the fishery to an economic optimum. While such taxes are discussed in most fishery economics texts, they are virtually unknown within most domestic fisheries, largely owing to natural fisher resistance.

While it is not surprising that taxes are unpopular with fishers, it should be noted that some arguments against such taxes are invalid. It is sometimes argued, for example, that fishers should not have to pay a tax on landings or on fishing access since no one outside the fishery must pay such a tax, or that any such tax would penalise the fisher doubly since they also must pay income tax. This is not a valid argument, since a properly set tax only collects *above-normal* profits; precisely what society as the resource owner should be collecting from the fishery as a return on the resource stock. Thus, a royalty is simply a way for resource owners – society – to acquire a reasonable benefit from the natural resource.

While broad-based use of economic instruments is not common in fisheries, taxes on effort are quite common for collecting resource rent from foreign fleets in coastal waters. In addition, royalty collection has been instituted in many individual quota fisheries, where the fisher pays a royalty on the basis of the amount of quota held.

A further form of economic instrument is the subsidy, usually selectively applied to specific inputs, in order to steer the fishery in a desired direction. For example, subsidies on fuel are common in many developing countries, where the policy direction has been towards increasing the modernisation and mechanisation of the artisanal fishing fleet. In some countries, such as Canada, subsidies exist in the form of special support payments made by government to fishers (and fish plant workers) during times of the year when they are unable to take part in the fishery, owing to its seasonal nature. Since these payments are made only to those in the fishery, they represent a form of subsidy; one which can (a) reduce fishing pressure in the short term (within a given year) since fishers do not have to rely solely on catching fish to generate income, and (b) maintain fishery employment, and fishing pressure, over the long term, providing a base income that leads some individuals to remain in the fishery when they would otherwise exit.

5.11 Summary

This chapter has focused on the tactical and operational stages of fishery management that arise subsequent to the strategic stage of policy and planning described in Chapter 4. Under *tactical management*, the discussion focused on issues relating to (a) the selection of desired exploitation levels (fishing effort and catch) compatible with fishery objectives and policy, and (b) the determination of a portfolio of management tools, based on evaluating broad classes of management measures. For *operational management*, discussion revolved around more detailed analysis of specific management measures, and the process for setting suitable levels of each chosen management tool, on time scales ranging from annual (e.g. catch limits) to day-to-day (e.g. fishery openings).

Chapter 6
Fishery Development

6.1 Introduction

Chapters 4 and 5 focused on what might be viewed as the core management activities within the fishery management system: the 'strategic' component of policy development and planning, and the 'tactical'/operational component focused on choosing and implementing a portfolio of specific control measures. In this chapter, we turn to a closely related but distinct component of the system: fishery development. The essential idea of fishery development is to initiate a new flow, or improve an existing flow, of sustainable benefits from the fishery system. This is done not through the implementation of management measures *per se*, but rather through a range of *developmental* measures targeting various stages in the fishery system. It is useful to identify three principal forms of fishery development.

6.1.1 New fisheries

First, there is the rather literal sense of the term: *developing a (new) fishery*. Suppose that a particular nation has a virgin unexploited fish stock located in its waters, and there is an interest on the part of the nation, or a local community therein, to gain some benefits from that resource. In such a case, the process of development could involve, in general terms, two stages:

- assessing the level of exploitation the resource can sustain (e.g. calculating a suitable sustainable catch level and a corresponding sustainable fleet size);
- building up the human and physical capital inputs in order to benefit from the natural resource (e.g. building boats to enable local people, or entrepreneurs, to harvest the fish at the sustainable level).

6.1.2 Increasing benefits

Fishery development is an important process even if we are not developing new fisheries. The second, broader sense of fishery development includes any effort to increase the sustainable benefits from the fishery, not through implementing management measures as such, but through improvements at any point in the system (including the management system as well as the human system of fishers, processors, etc.). This could take many forms, such as:

- assisting fishers to increase the catching power of their vessels, which could range from boat-building subsidies to support for motorising basic artisanal vessels;
- purchase or adaptation of suitable fishing technology;
- training fishers in fishing and fish-handling methods;
- capacity-building for management personnel and management organisations;
- facilitating the establishment of cooperatives and other fisher organisations;
- improvements in post-harvest stages, including market development, processing quality control and product distribution processes;
- construction of the necessary infrastructure, such as building a wharf at which fishers can land their catch, or providing ice facilities to increase the quality and longevity of the catch (and thereby increase fisher incomes);
- environmental protection and enhancement efforts to improve fish stock productivity and ecosystem resiliency.

6.1.3 Integrated development

These possibilities for increasing benefits from the fishery seem to range widely across approaches dealing with the natural system and the human side. However, they have a particular feature in common: all focus *within* the fishery itself, and mainly on the harvesting and post-harvest sectors. In contrast to this traditional and still dominant approach, there has been a trend in recent years among agencies supporting development efforts (primarily bilateral international development agencies) to broaden aid to a '*multi-sectoral*' model. Within this approach, development is targeted not just on the core of the fishery system, but also on the coastal communities and the socioeconomic environment of coastal areas. This has led to a focus on *integrated* coastal development, in which attention is paid to all relevant coastal resources simultaneously, as well as a focus on the people and communities within the coastal area.

This emerging approach is of great importance and will be discussed more extensively in Chapter 12. The remainder of this chapter, however, reflects the reality of most current fishery development efforts, and focuses on the nature and structure of development within the fishery system itself: in other words, the two traditional forms of fishery development described above.

Such development activities, taking place within a given fishery system, can be carried out by local coastal communities, by the relevant sub-national provinces and states, or by national governments. In the case of regional (multinational) fisheries, there may well be regional bodies involved. Finally, in the case of fisheries in developing nations, there is a long history of involvement in development by international donors, whether national (bilateral donors, such as Japan and Denmark, among many others), multinational (see box on p. 108) or non-governmental. This wide range of potential support for a given development initiative implies the possibility of many different configurations (for example, the simultaneous involvement of a local community, a province, a national government and an international agency in a single fishery).

Some terminology on international development in fisheries

Exclusive economic zone (EEZ)
Extended fisheries jurisdiction (EFJ)
Coastal state
Distant water fleet
Official development assistance (ODA)
Economic cooperation between developing countries (ECDC)
Technical cooperation between developing countries (TCDC)
Non-governmental organisation (NGO)
United Nations Conference on the Law of the Sea (UNCLOS)
United Nations Conference on Environment and Development (UNCED)

Some international agencies involved in fishery management and development

United Nations agencies

Food and Agriculture Organisation of the United Nations (FAO)
General Fisheries Council for the Mediterranean (GFCM)
Indian Ocean Fishery Commission (IOFC)
Indo-Pacific Fishery Commission (IPFC)
Western Central Atlantic Fishery Commission (WECAFC)
Fishery Committee for the Eastern Central Atlantic (CECAF)
United Nations Development Programme (UNDP)
United Nations Environment Programme (UNEP)
United Nations Industrial Development Organisation (UNIDO)
Intergovernmental Oceanographic Commission (IOC)

Development banks

The World Bank
Asian Development Bank (ADB)
Inter-American Development Bank (IDB)
African Development Bank (AFDB)

Regional bodies

Northwest Atlantic Fisheries Organisation (NAFO)
Commission for the Conservation of Antarctic Marine Living Res. (CCAMLR)
Forum Fisheries Agency (FFA)
Secretariat of the Pacific Community (SPC)

Southeast Asian Fisheries Development Center (SEAFDEC)
Latin American Fisheries Development Organisation (OLDEPESCA)

Species-specific bodies:

Inter-American Tropical Tuna Commission (IATTC)
International Commission for the Conservation of Atlantic Tuna (ICCAT)
International Whaling Commission (IWC)

Others

International Council for the Exploration of the Sea (ICES)
International Center for Living Aquatic Resources Management (ICLARM)
International Ocean Institute (IOI)

6.2   Objectives of fishery development

As in fishery management, the setting of clear objectives is crucial to successful efforts in fishery development. While the broad idea of fishery development is similar for both developed and developing nations, it is likely that a wider set of goals may be pursued in the latter, with a greater complexity of players. An example of multiple objectives being pursued in a fishery development context is shown in Fig. 6.1.

A key issue to be faced is: whose objectives are we speaking about? This is a particularly relevant issue in fishery development, for example, when (1) a donor agency is providing support to (2) a national government, for development of a fishery that is utilised by (3) the local fishers and their communities along a stretch of coastline. Given that each of these participants has a set of objectives, whose priorities will guide fishery development programmes and projects in such a setting? What is the balance among the funding agency's goals, priorities and areas of concentration, the societal aims of the nation involved, and the needs of local recipients?

In particular, the multilateral international organisations involved in fishery development, specify priorities for aid to the fisheries sector of recipient nations (most often with particular attention to small-scale fisheries). Among the more common of these priorities are:

- improving the understanding of available fishery resources;
- developing sustainable fisheries, in an economically and environmentally sound manner;
- improving management techniques and strengthening fishery data and information services;
- obtaining full benefits from the extension of national jurisdiction (200-mile limit) through planning of fisheries development, management and investment;

Fig. 6.1   Support to the fishing co-operative in Puerto Thiel, Costa Rica, has achieved a variety of goals, including product quality enhancement and income improvement (e.g. through construction of the ice plant shown in the background), and increasing community stability and employment of local women (e.g. through efforts to diversify the local economy).

- increasing the availability of fishery products, in order to improve food security and alleviate poverty;
- increasing fishery contributions to national economies through the generation of income, employment and trade;

A set of strategic objectives for fishery development

Fish for food: helping a hungry world

Providing food to feed the growing global population is arguably the most critical need in the world today. Yet there are only two major means to generate food: through agriculture or through fisheries and aquaculture. For coastal nations around the world, and especially for the people of those nations living near the coasts, fisheries provide a crucial source of protein and nutrition.

Fish for income: economic wealth to meet national goals

Fisheries have the potential to generate wealth both for fishers and for society. The former is especially relevant in domestic fisheries, while the latter arises particularly as royalties collected from foreign fleets. In either case, fisheries development can improve the contribution of the fishery to the general economy.

Fisheries for people: socioeconomic well-being for fishing communities

The socioeconomic well-being of fishers and coastal fishing communities depends on the availability of employment and the distribution of income, as well as on the maintenance and reinforcement of social and physical infrastructures in communities.

Fisheries for the nation: ensuring coastal states benefit from their resources

To ensure that coastal nations realise the potential benefits from fisheries in their coastal waters, there is a need for careful development of domestic fisheries and/or suitable management of foreign fleets. Each of these tasks requires nations to gain full control over fishery resources in their exclusive economic zones (EEZs), with adequate monitoring, control and surveillance (MCS) activity.

Fisheries for the future: developing sustainability of ecosystems and human systems

A sustainable fishery is one which simultaneously maintains the integrity of the marine ecosystem, supports the local human system (i.e. the fishing communities) and maintains the economic viability of the fishing sector. This must be viewed on a long-term basis, so the fishery system as a whole remains healthy into the future.

Fisheries as catalysts: enhancing and diversifying the coastal economy and society

Artisanal fisheries have a long and strong tradition in many coastal communities, both in 'developed' and developing regions, as key parts of the sociocultural fabric. As such, fisheries can form a base for enhancing and diversifying the coastal economy, with development building on existing fisheries through expansion of post-harvest activities and related economic sectors.

- addressing environmental concerns, including controlling environmental degradation;
- reducing post-harvest losses through reduced waste, better marketing and promotion of the role of fish in nutrition;
- improving human resource development, training, transfer of appropriate technology, and South–South technical and economic cooperation (*between* developing countries) to promote the self-reliance of developing countries in management and development;
- encouraging grass-roots participation in development, and the role of women in development.

6.3   Fishery development as a priority

Within the development thrusts of a particular nation or international agency, choices must be made among the many areas that could receive aid. If the fishery sector is seen as risky, with its high level of complexity, both natural and human, and its history of frequent failures in development projects, there may be a temptation for a development agency to attempt to improve its 'success rate' by cutting support to fishery systems. However, there is a long list of strong reasons to provide support for fishery development.

- Since, in the developing world, fishery-dependent nations are often among the poorest, a focus on fishery aid within a development agency will tend to reach those most in need.
- Since fishing communities are often among the poorest of a nation, properly targeted fishery development can reach those most in need of help within a given nation.
- Since fishery systems often play a major role as the 'engine' of the coastal economy, providing local economic benefits, foreign exchange and employment, while helping to maintain coastal communities and cultural traditions, support to the fishery sector has very broad 'spin-off' benefits.
- Aid to fishery systems can improve food security and self-reliance in food production by improving the provision of fish protein through optimised production, by improving management capabilities in the fishery, and by quality control at the post-harvest stage.
- Participation of women in the economy can be enhanced by aid to fisheries programmes, particularly at the post-harvest level where women often play a key but neglected role.
- Fishery support can promote 'sustainable development' goals, by enhancing resource management, especially through traditional management measures. Furthermore, the well-being of fisheries serves as a barometer to gauge both the health of surrounding ecosystems and the health of rural communities lying along the coasts.

The extent to which this rationale is convincing in a given context will determine how much developmental support is targeted on the fishery system.

6.4   Targeting fishery development

Within the fishery system, decisions must be made about the specific groups of beneficiaries towards which fishery development efforts are to be targeted. The appropriate agency must establish priorities among possible fisheries, industry sectors, fishing communities or geographical regions. This determination may be at the level of a national agency, deciding among the various coastal areas or fishery types, or an international body, deciding among candidate nations. For example, the latter choice might be based on an examination of indicators such as population, gross national product (GNP) per capita, fish harvests, fishery employment and fish imports/exports, in order to assess criteria such as national need (e.g. based on GNP per capita) and reliance on the fisheries (e.g. based on the significance of fishery employment in the economy).

   In general, decisions on targeting development may be based on a variety of factors.

- *Needs assessment.* Typically, an assessment of developmental needs will be carried out in candidate fishery systems. In this process, there is likely to be an ongoing tension between a responsive approach, with decision making focused on the stated goals and needs of the recipients (whether whole nations or local fishers, fishing communities and non-governmental organisations) and an approach focused on the fishery or development agency's policies and objectives. Ideally, priorities in fishery development would involve a balance between these. It is also important to monitor carefully whether the local expressions of needs are properly representing the identified target group. Finally, to make the needs assessment as complete and comprehensive as possible, it can be useful to utilise integrated, interdisciplinary study teams.

- *Positive signs.* In determining who are to be chosen as the beneficiaries of fishery development support, it may be useful to take into account the existence of:
  - a strong existing record in carrying out projects within the fisheries sector, including mechanisms to deal with possible technical, institutional or sociopolitical constraints;
  - a high priority placed on fisheries, as indicated in development plans;
  - a high potential for involving fishers, communities and non-governmental organisations in development activities;
  - a likelihood of being capable of maintaining a resilient ecosystem and human system, as well as institutional self-sufficiency for the fishery system.

- *Other considerations.* It is also relevant to examine the following factors:
  - overall development policy, e.g. a decision to focus support on the poor;
  - the desired balance among economic and social benefits from fishery systems;
  - the desire for local control of local resources (where 'local' may refer to the community, provincial or national level);
  - the logical geographical connections between neighbouring jurisdictions (which can be especially important when dealing with migratory fish stocks).

A typology of developing fisheries

The United Nations Industrial Development Organisation (UNIDO 1987) provided an approach to classifying the fisheries of developing countries based on certain key elements. The study grouped countries according to ten fishery development patterns:

- Least favoured countries
- Largely state-controlled fisheries
- Low-priority fisheries
- Labour-intensive fisheries
- Small states with growth potential
- States with large but fluctuating resources, and limited local demand for fish
- *Laissez-faire* fisheries
- Non-industrialised fisheries
- Probable exporters
- Long-distance, state-controlled fisheries

6.5   Small-scale versus industrial versus foreign fisheries

Looking at the national level, fishery development efforts could be focused on three major options:

- domestic small-scale fisheries (including subsistence and commercial);
- domestic capital-intensive industrial/offshore fisheries;
- monitoring, control and surveillance of foreign fleets in the exclusive economic zone (EEZ).

The choice of priorities among these potential areas of attention is a crucial one. Such choices have been faced by many nations over the years. A common situation is that in which foreign vessels exploit resources off the shores of a coastal state, and the choice is between developing a domestic fleet to displace the foreign vessels (which involves a capital outlay), or allowing the foreign fleet to continue but develop a strong system of monitoring, control and surveillance (MCS) to ensure that royalties are fully collected by the coastal state. A choice between the above development paths might take the following points into account.

- If the key goals of development are to increase national utilisation of the EEZ, service export markets and generate foreign exchange, then industrial fleet development may be a priority. If the involvement of industrial fisheries is contemplated, it may be desirable to ensure that (1) benefits accrue to those most in need within the target country, (2) domestic food security is enhanced, and (3) the development of capital-intensive industrial fleets is not excessive, leading to rapid over-expansion and over-exploitation. Since such development is essentially for purposes of wealth generation, a major role may be played by the private sector if there are profits to be made.
- If the primary development goals are (a) to focus support on the poor, and (b) to improve domestic food security and nutrition, whether catches are consumed locally (subsistence) or sold in domestic fish markets (commercial), then small-scale fisheries deserve priority. It is notable that in contrast to the *de facto* priority placed for much of the past few decades on fisheries industrialisation, international aid for fisheries development tends to be focused more on supporting artisanal fisheries. This does not necessarily mean sustaining traditional fisheries, however, since much support has involved efforts to *commercialise* artisanal fisheries. Caution is needed in this; for example, if such efforts are applied to subsistence fishing, fishers who are financially capable of *modernising* their vessels may well increase their share of the harvest at the expense of poorer fishers, thereby possibly destabilising fishing communities.
- If (a) the economics of domestic fishing are unattractive, or (b) domestic investment capital is limited, or (c) domestic fishery infrastructure is poor, or (d) resource royalties collected from foreign fleets are an important source of financial capital (and foreign exchange) for the development process, then a focus on managing and benefiting from foreign fleets may be the preferred option. In such cases, development might focus on MCS of such fleets and the formulation of suitable legislation to accomplish this, in order to gain full benefits from the national EEZ. There also remains an option to develop domestic offshore fleets and/or engage in joint ventures, thereby creating national control, additional employment and improved technical capabilities (Table 6.1).

Table 6.1   Linking fishery objectives and development priorities.

Focus on artisanal fishing if fishery objectives are:	Focus on industrial fishing if fishery objectives are:	Focus on foreign fleets if domestic constraints are:
Poverty reduction Improving food security Improving nutrition	Increasing EEZ utilisation Servicing export markets Generating foreign exchange	Unattractive economics Limited investment capital Poor infrastructure Resource royalties needed

6.6   A typology of fishery development measures

Approaches to fishery development can be categorised within seven functional areas (Charles 1991b), each of which is discussed in more detail below.

(1)   Direct support to harvesting activities.
(2)   Institutional enhancement.
(3)   Training and human resource development.
(4)   Economics and planning.
(5)   Scientific, assessment, statistical and information support.
(6)   Fisheries management and monitoring/control/surveillance.
(7)   Post-harvest support.

6.6.1 Direct support to harvesting activities

The type of support included in this category includes (a) the development of productive harvesting capabilities, notably the provision of appropriate gear and vessels to harvest the resource efficiently, (b) the development and improvement of the infrastructure, such as wharves and transportation services, and (c) measures to improve the physical welfare of fishers, particularly increasing safety at sea. This has generally been seen as the major area of support in the past, i.e. boats, gear, wharves, ice plants and other physical capital, but it is notable that in lists of developmental needs identified by the fishery sector in developing nations, much more attention is paid to the other six fishery development areas (see box).

6.6.2 Institutional enhancement

A key direction for current fishery development lies in addressing the need for development and enhancement of suitable organisations to manage fisheries. These may operate at a national or international level, but particular support is needed to encourage those functioning at a local level: self-reliant fisher and community organisations engaged in planning, management, conflict resolution, marketing, community services and related efforts. Specific needs include improved administrative support and conflict resolution mechanisms. *Integrated* approaches, within a broad context of coastal development, can involve attention to social infrastructure, fisheries infrastructure and social services, as well as involvement with coastal zone management institutions.

A focus on sustainability

Many traditional fishery development activities imply greater 'efficiency' in the fishing process. In some cases, this may be desirable, but it must be recognised that such measures have the potential to produce a completely unsustainable fishery system, leading to stock depletion or collapse (as has happened not infrequently in the past). Thus, with *sustainable development* increasingly being accepted as an essential ingredient in fisheries development, it is important to differentiate between fishery development that fits within the framework of sustainable fishery systems, and that which does not. In many cases, emphasis is being placed on maximising the benefits obtained from a *given* fishery production, rather than trying to increase the quantity harvested, in recognition of resource limitations and the potential for over-expansion in the harvesting sector of most fisheries. This focus on *value added* is accompanied by a focus on the 'soft' side of development (training, technical assistance, human resource development, capacity building and institutional development, etc.) rather than the more capital-intensive fisheries projects, usually focused on the harvesting process, such as boat-building and technological aid, which have had a relatively poor track record.

6.6.3 Training and human resource development

Education, training and extension programmes are important components of fisheries development. These can involve training in such topics as fisheries technology, operation of fishing enterprises, fishery management and extension, organisational and financial management, post-harvest activities and community facilitation. Much of this can be carried out *within* fisher organisations and fishing communities. Training and human resource development are necessarily multifaceted, requiring a balance among recipients (e.g. fishers, processors, women's groups, managers, technicians, extension workers, researchers, community/non-governmental organisation participants), and across programme delivery levels (e.g. industry, community, technical/vocational and professional/academic). Among the areas in which training needs have been identified in recent years by developing nations and donors are middle-level managers, professionals and technicians, personnel to provide advisory services to fishers, fisheries observers, and expertise in maritime law, post-harvest aspects and data management.

6.6.4 Economics and planning

Development in this area involves increased efforts to place fishery programmes within an integrated planning framework by looking at the fishery as a whole, from fish stocks to product marketing, and seeking to understand and incorporate social, economic and socioeconomic aspects of fishery systems. One particular topic of importance economically is the provision of credit and local capital for small-scale fishers. While capital-intensive development has had a poor record, such approaches can play a key role within integrated fishery projects in which the capital component is carefully limited.

6.6.5 *Scientific, assessment, statistical and information support*

This area includes the vast array of traditional science and information considerations, including: the need for marine science research and improved resource assessment; collection, analysis and dissemination of fisheries data; monitoring of catches, fishing effort and market price; and the establishment of a central registry of fishing vessels and improved fisheries statistics in general.

6.6.6 *Fisheries management and monitoring/control/surveillance*

This area focuses broadly on the enhancement of fisheries management capabilities, and developing self-sufficiency in fisheries management. Specific topics include: development of national fisheries policies, management plans and monitoring, control and surveillance (MCS) systems; community-based resource management; development of industrial fleet licensing policies; measures to control and enforce negotiated fishery access agreements and the activities of foreign vessels within coastal EEZs; fish habitat management, and particularly measures to protect fish habitat and fishing areas from environmental damage; suitable fishery legislation for all of the above.

6.6.7 *Post-harvest support*

The key here lies in increasing the value of what is caught, through means such as (a) improved processing, handling and preservation of fish, (b) improved quality control, (c) improvements in physical marketing (e.g. construction of fish markets) and in coordinated marketing (through suitable bodies and facilities), more efficient distribution of the fish and the development of export strategies.

Fishery development needs: the South Pacific

Fisheries in the South Pacific region present a dichotomy between two very different sets of management and development needs:

- measures to support small-scale inshore fishery development;
- surveillance and enforcement (MCS) for offshore foreign fleets.

South Pacific inshore fish stocks, while relatively small, support strong village-based artisanal fisheries and play an important role in meeting goals of poverty alleviation and food security. Offshore, the South Pacific has more substantial resources and provides a large fraction of the world's tuna catches. Most of these resources are exploited by foreign fleets, but access agreements with these fleets, combined with comprehensive MCS, allow the region to obtain significant revenue from these fisheries. (Further discussion on these points may be found, for example, in Herr 1990.)

6.7   Participatory fishery development

An emphasis on a participatory approach to fisheries development, carried out through direct contact with fishers, fishing community groups, cooperatives and non-governmental organisations, is seen as particularly effective. As the FAO (1984) noted, 'An integrated approach through, and with the participation of, fishing communities is often the best way of channelling technical, financial, and other forms of assistance ... The cooperation and participation of fishermen is necessary to ensure the success of small-scale fisheries management schemes.'

A key aspect of the participatory approach lies in ensuring the involvement of women in fishery development. In the past, this was a clear failing, largely as a result of an emphasis on the harvesting sector, where men tend to predominate (with exceptions, such as some South Pacific artisanal fisheries), rather than the post-harvest sector, where women very often play a dominant role (Fig. 6.2). Only recently have development agencies come to realise that (a) women should be more involved in fishery development, (b) an increasing share of develop-

Fig. 6.2   Women play a major role in many fisheries. This Inuit women works in a processing facility in Canada's Arctic region.

ment resources should be targeted on sectors of the fishery in which women work, notably post-harvest activities, and (c) the potential impact of development efforts on women in particular must be evaluated carefully. Given this reality, an FAO (1984) report concluded that 'Fisheries development programmes should recognise that women often play an important role in fishing communities, both in trading and processing, and provision should be made ﹍or enhancing that role.'

6.8 Summary

This chapter has explored a range of themes relating to fishery development, in particular:

- the distinction between the two major forms of fishery development (initiating new fisheries versus improving benefits from existing fisheries), as well as between these and the emerging approach of integrated development (to be elaborated in Chapter 12);
- the range of objectives being pursued in the process of development;
- the merits of fishery systems as recipients of national or international developmental aid, and the issues involved in targeting that development support (for example, between small-scale, domestic industrial and foreign industrial fisheries);
- the seven categories within a typology of fishery development measures;
- the need for participatory approaches to development.

Some of the major aspects of fishery development discussed above are summarised in the *policy flow chart* shown in Fig. 6.3, which focuses on three aspects: considerations in choosing targets in fishery development, important priorities among the various types of fishery development, and several key issues in the development of fishery systems.

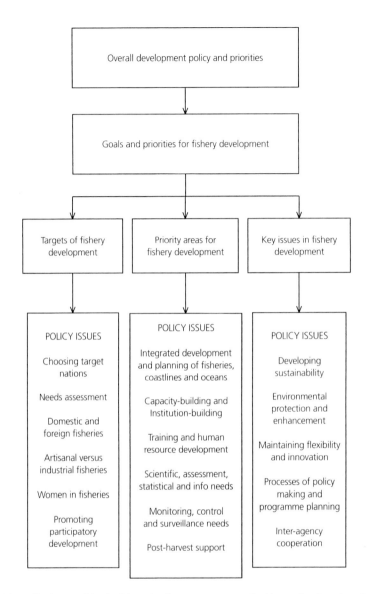

Fig. 6.3   Major policy issues arising in fishery development are summarised in a *policy flow chart* that focuses on three aspects: the choice of targets in fishery development, priorities among the various types of fishery development, and key issues in the practice of fishery development.

Chapter 7
Fishery Research

Fisheries research is that arm of the fishery management system that focuses on accumulating, analysing, synthesising, interpreting and disseminating the knowledge needed for good decision making. Fishery research includes any and all studies with a connection to fish and fisheries, from biological research on plankton through to anthropological studies of human activity in ancient fishing villages. This chapter examines the need for fishery research, the nature of this research, its institutional and disciplinary structure, and the range of practitioners globally.

It is useful here to note that two related terms, *fishery science* and *fishery analysis*, are also used to describe fishery research, or subsets thereof. The former term incorporates fishery studies that follow a 'scientific' approach (Charles 1995a). This should include a broad range of natural science, social science and management science disciplines, although it is sometimes used more narrowly to refer only to sciences such as fisheries oceanography and biology. The latter term, fishery analysis, applies analytical (as opposed to purely descriptive) methods to study fishery systems using whatever tools are appropriate in the circumstances. This would typically include most of fishery science, as well as areas such as policy analysis (the assessment of fishery policy measures).

7.1   The need for fishery research

It is a truism that 'good decisions require good information', and a rather obvious reality that fishery systems pose one of the greatest challenges to obtaining the 'good information' we need. Indeed, the International Ocean Institute (IOI), an international non-governmental organisation promoting the sustainable use of the oceans, noted in its *Halifax Declaration on the Ocean* (International Ocean Institute 1998) that humans face a 'crisis of knowledge' concerning the oceans. Not only are humans remarkably ignorant about the oceans, and in great need of an improved knowledge base, but in fact the very approaches and methods used in studying marine matters need great improvement. These points led the IOI to formulate a set of recommendations, under four key headings.

- 'There is an urgent need to improve our approaches to the development of adequate knowledge bases concerning the ocean.

- The use, synthesis, integration, and dissemination of existing knowledge derived from scientific and traditional, experience-based sources must be improved.
- There is an urgent need to adjust to the reality that, in many cases, uncertainty is irreducible and that we must "live with uncertainty" through the application of the precautionary approach, risk management, and risk-reducing management measures.
- It is imperative that strong institutional measures be put in place at the international level to address issues emanating from the Crisis of Knowledge.'

Clearly, there is a need to understand what goes on in the aquatic environment, so that we can better assess the impact of humans on that environment. Indeed, there is ample evidence demonstrating that fishery management and development initiatives are often counter-productive if undertaken without a reasonable understanding of the fishery system, including biological, ecological, social, economic and technological aspects. Thus, it is only proper that fishery research is a standard component of fishery management systems, whether this is a single one-time resource assessment carried out in preparation for fishery development efforts, or a large and ongoing governmental research infrastructure, as in many developed nations.

While research is by no means a 'frill' in fishery systems, it is also true that in a world of pressing problems, irreducible uncertainties and limited budgets, complete knowledge is simply not possible. It is important to ask the key question: how much research is necessary? In any given situation, a fishery agency must determine carefully the knowledge required, the degree of precision needed, and whether additional research and information would significantly improve the probability of success in management and development.

7.2   The nature of fishery research

The fishery research endeavour may be seen as involving three key components:

- *monitoring*: the collection, on a regular basis, of useful data to assess fishery performance;
- *research (per se)*: the study of fundamental questions, adding to our base of knowledge;
- *decision support*: the provision of science-based assessments to aid in decision making.

Governmental research establishments may engage in all three of these components, while university-based researchers may focus on the middle component (while typically making use of data collected in the monitoring activity). In many cases, the third component (decision support, or what is sometimes called the 'advisory function') is of greatest immediate interest to managers and fishers. Indeed, while the need for data collection and adding to fishery knowledge is widely accepted, the advice and/or assessments provided by scientists about allowable catch levels and the like tends to be most eagerly awaited and most vehemently debated.

Note that the difference between 'advice' and 'assessment' can be critical. Assessments of resource status can be provided in almost any circumstances, assuming that scientists are aware of which indicators are of interest to decision makers and the fishery sector. On the other hand, the provision of advice presupposes that scientists are aware of society's predetermined fishery objectives and are formulating the advice to meet corresponding reference points. It is important to understand, and revise as necessary, the institutional mechanism by which assessments, and possibly advice, are provided.

In government-supported fishery research, an ongoing debate seems to arise (as it does in other research fields) over the desirable balance between *pure* (*basic*) and *applied* (*targeted*) research. The former emphasises studies on the fundamental nature of the fishery and its components, such as fish behaviour and larval drift, while the latter focuses on topical issues identified in the fishery, stock assessment being a major example. An excessive preoccupation with basic research carries the risk that (a) the results may not lead to improved benefits from the fishery at any point in the near future, and (b) problems of immediate concern to fishers and managers are not addressed, and thus support for research diminishes. On the other hand, a research body with too much targeted research may focus excessively on pieces of the fishery puzzle without looking at the big picture, and may lead to low motivation among researchers, since the work may be relatively mundane with little potential to allow for unexpected scientific discovery.

In many cases, particularly in northern countries, the *pure vs. applied* balancing act for fishery research has oscillated over time between encouragement for basic research and pushes for relevant applied research. For example, in recent years, with cuts to research budgets common in many countries, there seems to have been a move towards the applied side, with an emphasis on targeted research projects. On the other hand, following the collapse of the groundfish fisheries in Atlantic Canada, there have been many calls, particularly from fishers, for more attention to basic 'biological' studies of the fish and the ecosystem (in contrast to a previous focus on stock assessment models, which have received some of the blame for the collapse).

There is no universally right or wrong orientation for fishery research. The chosen path will depend very much on the objectives set for the fishery, the identified research needs, and the size and capability of the research infrastructure. However, whatever the choice, it is important to take into account the implications of the choice made. For example, if research project funding depends on the *relevancy* of the work involved, but career advancement and promotion of individual scientists is based on producing '*cutting-edge*' published research, care will be needed to reconcile these potentially conflicting directions. Conversely, if the research budget is shifted towards more basic study, and this leads to an increased level of uncertainty in stock assessment forecasts, care will be needed to avoid a decline in support for research from the fishing sector.

7.3 The structure of fishery research

This section presents an overview of several options for structuring a fisheries research programme. The principal choices seem to be (a) by discipline, (b) by species, (c) by functional area, and (d) by geographical region or ecosystem.

7.3.1 Disciplinary structure

Fishery research, as with so many other areas of study, is often viewed from a discipline-specific perspective. One approach to structuring a research programme is to do so by discipline. A wide range of disciplines are engaged in studies of relevance to fishery systems, including:

- physical oceanography
- biological oceanography
- biology and ecology
- chemistry
- engineering
- food science
- mathematical modelling
- economics
- sociology and anthropology
- political science
- legal studies
- history
- business administration

… and perhaps any other discipline we might consider! This structuring by discipline is the norm in a university setting, but it is unlikely that this would occur in the context of a fishery

---

Multidisciplinary fishery research

Traditionally, fishery research establishments have focused on disciplinary studies, notably in oceanography and biology. The need to move toward more multidisciplinary research has been noted for many decades by many social scientists and natural scientists (e.g. Andersen 1978; Pollnac & Littlefield 1983; Fricke 1985; Pringle 1985; Charles 1991a; Durand *et al.* 1991), as well as by a variety of fishery advisory bodies, and within documents such as the United Nation's Agenda 21.

The most common form of multidisciplinary research, not surprisingly, is that which involves precisely two disciplines. Examples include: (1) combining biology and oceanography to study the interaction of ocean currents and larval movements, (2) linking biology (notably population dynamics) with economics (prices and costs) in bioeconomic studies (e.g. Clark 1985, 1990), and (3) combining social factors (such as income distribution and community welfare) with economic ones (such as labour processes and social institutions) in fishery socioeconomics (Charles 1988). Such approaches have been important in providing a means for the 'cross-fertilisation' of ideas across disciplines (Holling 1978) (see Chapter 12, in particular, for a further discussion of the bioeconomic approach to integrated fishery analysis).

More fully multidisciplinary research, involving a combination of oceanography, biology and other natural sciences with economics, sociology and other social sciences, is less common. However, there are many topics which can benefit from a truly multidisciplinary research approach (Fig. 7.1). For example, studies of the impact of climate change on fisheries can involve oceanographers, biologists, economists, social scientists and others. Studies of fisher behaviour and responses to new management initiatives can utilise the efforts of economists, sociologists and anthropologists, and legal and business analysts, as well as natural scientists. Research into the effects on fishing communities of changing fish stock abundance and changing economic conditions over time may involve specialists in regional analysis and community development, as well as the range of expertise above.

Some moves are evident toward realising this potential for a multidisciplinary approach. There are two major approaches. On the one hand, the fishery-related research of many international development agencies (and indeed the structure of the agencies themselves) is organised on a thematic or ecosystem basis, rather than by disciplines. On the other hand, national fishery research bodies tend to carry out multidisciplinary work by establishing special teams and panels. This is usually done on a temporary basis rather than being incorporated into the structure of the agencies themselves. There continue to be barriers to multidisciplinary research, which can create challenges within organisations (in terms of structure and operation) as well as within research teams (e.g. communication difficulties, due to the prevalence of disciplinary jargon).

Fig. 7.1   While the natural sciences have predominated in fishery research, there is a great need as well for social science research and multidisciplinary studies that help us understand the fishery's human system. In the absence of such efforts, policy measures may be mis-designed or mis-targeted, and fishery management may be inefficient or ineffective – as a result, for example, of unexpected responses to regulations.

research organisation. Instead, groupings of disciplines may be formed, so that, for example, the structure may involve a division into four groups: physical sciences (notably oceanography), biological sciences (including stock assessment), technology (engineering and food science) and social sciences (with economics often being dominant).

The two boxes on pp. 124–125 and pp. 126–127 highlight the need for, and the state of adoption of, multidisciplinary research and social science research.

7.3.2 Species structure

Many governmental research centres, particularly in developed countries, carry out research largely on a species-specific basis. Certainly, stock assessment activities are almost exclusively organised this way – which is not surprising, of course, since the basic idea of stock assessment is to determine the status of a particular fish stock (of a given specie). Thus, one scientist may work on herring assessments, another on shellfish assessments, and so on. This

---

A focus on social science research

Fishery management agencies have traditionally focused almost exclusively on research in the natural sciences, and the acquisition and use of data from the natural world. Indeed, in a fishery context, the word 'research' itself is sometimes seen as synonymous with natural science; economists, social scientists and managers within these agencies are rarely involved in research. Yet socioeconomic factors clearly play a major role in fishery systems, and specifically in fisheries management, as highlighted by Jentoft (1998: p. 178):

- 'Fisheries is an industry and fishing is a human activity, and it is through regulatory measures of fishing behaviour that we attempt to secure the viability of fish stocks. Therefore, the social scientist would argue the obvious: to manage well, you need to know not only fish, but also fishers and fishing.'
- 'In most countries [fisheries management] is a political battlefield of conflicting interests, and management goals have the character of delicate compromises. Consequently, one also needs to understand the political process of fisheries management.'
- 'Third, a fisheries management system is an institutional set-up, and effective management is a question of finding the appropriate organisational mechanisms, i.e. the rules, procedures and incentives, that will help fulfil management goals.'

Thus, social science, i.e. the study of themes such as those described above, clearly has a role to play in understanding fisheries and in implementing appropriate fishery policies (Fig. 7.1). This is by no means a new point, having been noted by a variety of commentators over the years. Consider the following comments of an economist, Ian Smith, from more than two decades ago:

'a necessary precondition [for fishery research] is an understanding, on the one hand, of the resource/fishermen/distribution continuum and, on the other hand, of the linkages among fisheries, fishing communities, and other rural sectors, and institutions, including government.' (Smith 1979: pp. 35–36)

While recognition of the need for social and economic analysis has expanded over the past two decades, such analysis is by no means widespread in fishery research centres. There remains a major research challenge in studying the key social, economic and socioeconomic aspects of the multidimensional fishery system, and in using that understanding to aid in the development and implementation of feasible fishery management plans in a world of complex behaviour and multiple fishery objectives.

What social science research is most useful to fishery decision making? This can range widely, from individual measures (age, education, ethnicity and family structure of fishers) to community measures (community organisation, local employment options and unemployment rates, distribution of income in fishing communities) to *institutional* measures (e.g. transaction costs in rule making and enforcement). The needs will vary with the type of fishery under consideration. For example, it has been suggested (FAO 1985) that, at least in the context of small-scale fisheries, socioeconomic information concerning employment, efficiency, food supply and demand, and the distribution of income and wealth may be paramount in managing fully exploited fisheries, while information relating to improving the standard of living, increasing fish production and expansion of markets for fish may be crucial for developing new fisheries.

is less clear-cut in the case of economic research, where the focus is more on studying the fishers than the fish – in multi-species fisheries, a species structuring makes little sense, but this could occur if fishers themselves tend to specialise on a single-species group.

Note, however, that even in the case of stock assessments allocated among researchers by species, the overall structuring of the research system is not clear-cut. If a nation or region has more than one research centre, there arises a structural choice between (1) a multi-centre team of researchers focused on a particular species or group of species, or (2) the more common arrangement in which the stock assessment researchers, although focused on specific species, work within each centre as a multi-species team. Both arrangements have advantages: the first encourages synergy and coordination among those studying a particular species, while the second encourages cross-fertilisation of methodological ideas among those working on different species. The second approach seems easier administrationally, however, with each researcher being responsible to local leadership at the individual's centre.

7.3.3 Functional Structure

This approach attempts to link research activity with the principal areas ('functions') of a fishery management system. Thus, the research organisation may be structured so that individuals from relevant disciplines form groups focused on such functional areas as:

Stock assessment

If there were no need to limit the catch or the exploitation rate on a fish stock, there would be much less, if any, need for fishery management. Of course, as is clear from historical experience, there is a very great need to manage fisheries, and in particular to limit harvesting. This implies that we must know something about the fish stocks, in order to determine how much fishing pressure they can withstand and remain healthy. The task of stock assessment is to provide that information, using available data and appropriate analysis. As Saila & Gallucci (1996: p. 6) state:

> 'In very general terms, the procedure of fish stock assessment consists of the following elements: *inputs* (fisheries data and various assumptions concerning the data and methodology); a *process* (analysis or analyses of the data); *outputs* (estimates of population or system parameters). These outputs then provide *inputs* to another *process* which consists of predictions under various alternatives, and there is a final *output* consisting of some form of management strategy, which includes optimization of yield or some other objective function(s).'

The inputs to an assessment may be obtained (a) from biological surveys, using such means as trawls, acoustics, tagging studies and counting fences, (b) from biological sampling, e.g. to determine the age and size composition of the catch, (c) from temporally and spatially disaggregated catch and fishing effort data, (d) from fisher experience, provided through log books, meetings/hearings and individual interviews, and (e) from any other means possible, such as information from outside the fishery sector. It is to obtain these data that much of the *monitoring* function of a fishery research agency is focused.

The methods used in stock assessment range widely. Some of the most important methods are described briefly below. For more detailed treatments, see, for example, Gulland (1983), Hilborn & Walters (1992), King (1995) and Gallucci *et al.* (1996).

(1)   Perhaps the most fundamental assessment method is the biomass survey, carried out through a statistically based sampling technique such as a set of bottom trawl tows used in a groundfish assessment. This method can provide an indicator of abundance with which an initial TAC is set for a new (developing) fishery, and eventually a substantial time series on which annual TACs are set.

(2)   Statistical analyses of catch and effort time series, and in particular the generation of a catch per unit effort (CPUE) index, is among the most widespread of assessment methods. This is especially useful in providing an indicator of stock abundance in the absence of biomass data (and for stocks where ageing is not carried out, such as those of some shellfish), and is based on the assumption that abundance is positively related to the rate at which fish is caught in the fishery.

(3)   Analysis of acoustic data provides a measure of fish abundance in the specific location examined. Adopting a suitable set of assumptions about the distribution

of the fish allows this to be expanded to provide a measure of a full stock. Acoustic data are used, for example, in fisheries for herring and other schooling fish.

(4) Virtual population analysis (VPA) and cohort analysis are methods that mathematically combine the above data sources (i.e. survey/acoustic and catch/effort data) to 'reconstruct' an age-structured stock. The idea is to add together actual catches with assumed natural mortality, in order to deduce in retrospect what the stock size used to be. Once a VPA analysis has been validated, it can be used to project stock abundance into the future. The VPA approach is widely used for species such as groundfish in the North Atlantic and elsewhere.

The outputs of a stock assessment can range from a single number (the estimate of current stock biomass, which traditionally has been the starting point in assessing the potential of a 'new' fishery) through to a complex risk analysis of the implications of various harvesting levels. Increasingly, the trend is towards the latter approach, through such means as management procedures (Butterworth *et al.* 1997; Cochrane *et al.* 1998), and incorporating into the assessment the uncertainty inherent in fishery systems (Chapter 11).

The provision and explanation of stock assessments represents one of the major *decision support (advisory)* functions of fishery research. It is important for the research agency to devote sufficient attention not only to generating good assessments, but to communicating these effectively, and in particular to being available to decision makers and fishers to explain the results. The success or failure of the research infrastructure in accomplishing these latter tasks (and in developing an institutional mechanism for doing so) may well have an impact on the sustainability of the fishery itself.

This brief summary is far from doing justice to a subject area that may well have received more research attention than any other in fisheries. A relatively complete treatment of stock assessment methods may be found in the classic works of Ricker (1958, 1975) and Gulland (1983), and in more recent explorations such as those of Hilborn & Walters (1992), King (1995) and Gallucci *et al.* (1996). Treatments of tropical fishery assessment are provided in the last-named work and the FAO manual of Sparre *et al.* (1989). For an impressive study of the history of stock assessment, see Smith (1994).

- habitat protection (including biology, oceanography, engineering and economics);
- stock assessment (including biology, oceanography and possibly social science);
- resource management science (including ecology, management science and economics);
- ocean science (including biological, oceanographic and geological aspects);
- fishery development (including engineering, food science, economics and business);
- international (including legal studies, economics and political science).

Note that this is not meant to be a comprehensive listing of functional areas, since each list will be tailored to the specific circumstances of each fishery. A notable point about structuring by functional area is that it implies inherently interdisciplinary groupings, such as those noted for each entry above. This also has the advantage of being 'targeted' on declared needs in the fishery management system; functional areas can be readjusted and reformulated as needs arise.

The analysis of European fishery research activity by functional area shown in Table 7.1 has been compiled from data in a directory of research centres within the fisheries sector prepared by the Commission of the European Communities (1994). The functional areas considered are: science, management and assessment, economic and social research, upgrading of fishery products, aquaculture and others. For each nation, the data indicate how many research centres within that nation are engaged in each function; as a given centre often engages in more than one function, the sum across a row typically exceeds the total number of centres for that nation. This analysis indicates that aquaculture is the major subject of research attention, but that within the fishery sector, research on the first four functional areas listed above, i.e. those of a scientific nature, and particularly on management and assessment themes, is clearly dominant. Research relating to the last two areas, fishery development and international aspects, is contained largely in the economic/social and fishery products categories, and therefore takes place at a small minority of research locations.

7.3.4 Geographical/ecosystem structure

In a large or relatively wealthy jurisdiction, fishery management is likely to be based on geographical divisions, perhaps with an administrative headquarters in each state or province of the country, or in each major fishery area. In such cases, the overall research framework is rather naturally structured in a similar manner. In such cases, however, the research itself, i.e. the work actually carried out, is more likely to be organised by discipline, species or functional area, as described above. Nevertheless, with the emergence of worldwide interest in ecosystem-based approaches to fishery management, it is worth noting the possibility

Table 7.1 European fishery research centres, by functional area.

	Total number	Science	Management/ assessment	Economic/ social	Fishery products	Aquaculture	Other
Austria	14	10	3				1
Belgium	19	3	10	2		14	1
Denmark	28	2	15	5	15	13	1
Finland	40	25	17	5	1	10	12
France	88	24	37	6		48	8
Germany	50	14	21	5	3	27	8
Greece	33	5	16	6	7	27	5
Iceland	10	4	2	1	1	6	3
Ireland	37	8	18	4	11	26	11
Italy	90	17	49	13	14	60	14
Netherlands	14		11	1	3	10	2
Norway	34	22	4	4	2	10	7
Portugal	23		11	1	1	8	3
Spain	63	11	32	6	9	40	8
Sweden	14	14	1			1	
UK	109	19	58	17	16	62	17
Totals	666	178	305	76	83	362	101
Percentage		27	46	11	12	54	15

of structuring research on the basis of geographical units or ecosystems, with researchers forming a highly interdisciplinary team to address topics of concern in that particular area or ecosystem. The size of geographical areas involved, and/or the choice of spatial scales at which ecosystems are defined, will depend in part on the size of the research infrastructure; without sufficient staff and resources, there is no point in defining units too small to be operational. (For example, governmental fishery research on Canada's Atlantic coast is structured into at least five units (Bay of Fundy, Scotian Shelf, Gulf of St. Lawrence, Quebec and Newfoundland), but this number of divisions may not be feasible for an equivalent geographical area in nations with less to spend on research.)

7.4   Participants in fishery research

There are five principal fora in which fishery research takes place: (a) governmental fishery agencies and/or research laboratories, (b) international agencies, (c) universities and other educational institutions, (d) the fishing sector, and (e) the private sector (e.g. privatised research agencies) and non-governmental organisations (NGOs). Each of these is discussed in turn here.

7.4.1 Research by governments

Many nations have well-established government fishery research centres. Since, in many if not most countries, jurisdiction over fisheries is held by the national government, the vast majority of governmental fisheries research takes place at this level. In some countries, however, where local states or provinces maintain jurisdiction over some aspects of fisheries, research does take place at other governmental levels, albeit to a more limited degree.

As noted previously, fishery research is often heavily oriented towards biological research. This reflects the traditional focus on conservation objectives. Government research centres tend to follow this pattern, emphasising scientific (biological and oceanographic) research, with a mix of basic research (e.g. on fish biology) and activities such as stock assessments which enter directly into the fishery management process. In contrast, few research centres are devoted to (or even include) economic or social science aspects of fisheries research. (For example, of 666 European fishery research centres listed by the Commission of the European Communities (1994), only 11% state that social and/or economic research is included in their coverage.) However, not infrequently there is a presence within the research infrastructure of work on fishery development with the aim of improving the efficiency of the harvesting process and the quality of the end-product. This reflects the perceived requirement that research should not only help in protecting the fish, but should also be of direct economic relevance to the fishing industry and society in general.

From a structural perspective, there is the important matter of how the research arm of the fisheries bureaucracy relates to the management arm. This may depend on historical and practical realities, as well as on (a) the balance desired among the monitoring, research (*per se*) and decision support components, and (b) the extent to which the research arm provides 'advice' on management, in addition to assessments of resource status. The latter will depend in turn on the actual and desired linkages between research and management, and the extent to which fishery objectives are clearly specified for use in the research process. For

---

Research in South Africa: the Sea Fisheries Research Institute

Research at the SFRI falls under four sub-directorates: Resource Biology, Resource Assessment and Modelling, Fisheries Environment, and Whole Systems (as outlined below). (In addition, there are two other sub-directorates, Engineering Services, and Technical Management and Development, which 'are responsible for developing in-house instrumentation and software'.)

A. Resource biology

- *Inshore resources.* 'Research into the biology of rock lobster, abalone, seaweed and a variety of linefish forms the basis for making recommendations in the management of both commercial and recreational fisheries.'
- *Offshore resources.* 'Management of the commercially important pelagic fish (e.g. anchovy and pilchard) and demersal fish (e.g. hake, sole, horse mackerel and king-klip), as well as squid and a variety of line-fish species, relies on an understanding of the biology of each species.'

B. Resource assessment and modelling

- Surveys and fish behaviour
- Stock assessment

C. Fisheries environment

- *Physical and chemical oceanography.* 'Ship-based measurements, satellite imagery and automatic monitoring instruments … provide information on factors such as sea temperature, salinity, currents, wind speed and direction.'
- *Biological oceanography.* 'Research into primary and secondary production levels … In addition, monitoring and research of red tide is conducted in order to warn the public against consuming toxic shellfish.'
- *Pollution.* 'The impact of various sources of marine pollutants is studied in order to develop sound management approaches and clean-up strategies.'

D. Whole systems

'… interactions among populations and between these populations and the environment are studied.'

Sea Fisheries Research Directorate (1999)

---

example, a research body that focuses mostly on monitoring and fundamental studies, with little decision-making support, may need few connections to other arms of the fisheries department.

Related to this is the fundamental matter of where fishery research should take place: should it lie within the government fishery department (as is the case in most nations) or be external, either in a quasi-independent agency of government or in the private sector? This issue can be seen largely as one of objectivity versus responsiveness. The 'internalised' approach has the advantage of being more *integrated*, encouraging better mutual understanding of the various aspects of a fishery system among scientists, managers, economists and others within a single government fisheries department. As a result, the research establishment may be more likely to engage in relevant work, respond to current needs and use the crucial knowledge base of management more efficiently. On the other hand, an 'external' arrangement for research provides a clear separation between those providing scientific information and those using that information to manage fisheries (Hutchings *et al.* 1997). The idea with such an arrangement is to avoid the potential loss of scientific integrity that could arise if a government fishery research structure is influenced by political or bureaucratic pressures, leading to manipulation (or even withholding) of research results to fit within government policy. It seems that there is no single 'best' way in which to structure fishery research institutions. This is an important issue that must be taken into account in the design of fishery management systems.

7.4.2 Research by international agencies

A number of multilateral international bodies and regional organisations play a role in fishery research. Within the former group, targeted research and data collection/compilation are carried out (or sponsored) by such major agencies as the Food and Agriculture Organisation of the United Nations (FAO) and the World Bank, as well as by the International Centre for Living Aquatic Resources Management (ICLARM), a member organisation of the Consultative Group for International Agricultural Research (CGIAR). ICLARM has a particular focus in Southeast Asia, but also has research operations in the South Pacific, the Caribbean and Africa, and plays a role not only in research itself, but also in information dissemination through a journal (Naga: The ICLARM Quarterly) and an impressive library. Regional multinational organisations involved in carrying out and/or supporting fishery research include the European Union, the Southeast Asian Fisheries Development Center (SEAFDEC), the Forum Fisheries Agency (South Pacific), the Caribbean Community and Common Market (CARICOM, Caribbean), the International Council for the Exploration of the Sea (ICES, in Europe and elsewhere) and a variety of others.

Within a given nation, these various organisations typically play a supporting role (with emphasis on research planning, coordination and communication) as either a funding body or a partner with a local research body. On an international level, however, many of these agencies (such as FAO and ICLARM) play a leadership role in defining directions for research (as well as fishery management) and in facilitating that work through conferences, workshops and the flow of information. Thus, the profile of such agencies, most of which have relatively modest numbers of staff, tends to be out of proportion to any simple measure of research productivity (such as the number of research papers published).

7.4.3 *Research by universities*

Universities are the principal sources of fishery research outside of government. While the quantity of biologically oriented fishery research carried out in universities probably exceeds the quantity of social science research, universities nevertheless produce the vast majority of economic and social science work, given the relative lack of attention to this area by government agencies. Not surprisingly, given a lesser need for strongly applied work, universities are home to many of the theoretical developments arising in fisheries research.

A wide range of universities around the world have at least some level of research activity on fishery topics. It would be impossible to determine a full list of such institutions, but it is clear that there is a particularly large number in North America and Europe, with the Commission of the European Communities (1994) listing well over 300 European universities carrying out such research. There are also many in Australia, New Zealand and Japan, as well as a growing number of significant university centres in developing nations, such as the University of the Philippines, the University of the South Pacific (Fiji), the National University (Costa Rica) and the University of Cape Town (South Africa).

Support for fishery research in universities varies from targeted funding, as is the case for Sea Grant universities in the United States, which the federal government funds as centres of excellence in the field (see, e.g. King 1986), to non-targeted funding received by fisheries researchers who compete against other academics for general research grants (as in Canada, where university-based fishery researchers compete with those in other fields for funding from the Natural Sciences and Engineering Research Council and the Social Sciences and Humanities Research Council). These approaches each have their advantages: the former more easily encourages interdisciplinary marine research and education (since funds can be so targeted), while the latter promotes strength within a discipline (since it is usually a disciplinary peer group that judges the research proposal).

7.4.4 *Research by the fishing sector*

Although resource users in fishery systems have accumulated a large body of knowledge (notably *traditional ecological knowledge*; TEK), in most cases there has been little effort to involve these users in determining research priorities or in the research activity itself. Indeed, in most nations, the vast majority of fishery research takes place within government (and universities), and although fishers may often express the desire to participate in such research, such cooperative ventures are still uncommon.

This is changing, however, with the increasing recognition that (a) fishers have a base of useful knowledge which is continually updated through direct experience at sea, and (b) support for management is enhanced if fishers are involved in dealing with the information available. This recognition has led to recommendations such as the following (FRCC 1994: p. 118):

> 'It is important that a genuine thrust be made to give a more effective role in fishery science to those with practical experience and knowledge in the fishery, and the role must be rigorous and transparent.'

There are a variety of significant moves toward participatory research involving fishers and industry. Some partnerships are now being institutionalised. For example, community-based fishery management in developing countries (for example, in some Caribbean and Southeast Asian small-scale fisheries) often has a built-in participatory research component (e.g. IIRR 1998). A participatory approach is also widespread in some industrially oriented fisheries. For example, in Australia and New Zealand, fishers often not only participate in research, but also design and fund much of the work as well. In Europe, individuals from the fishing industry serve on committees, nationally and within the European structures, to advise on research.

On Canada's Atlantic coast, some of the annual groundfish surveys are carried out by government scientists in cooperation with various associations of trawlers and fixed-gear fishers. There are various formal organisations which are interested in research, including the Fishermen and Scientists Research Society (a voluntary organisation of fishers and scientists: see King *et al.* 1994), the Eastport Peninsula Lobster Protection Committee in Newfoundland (which works with university and government scientists to carry out research on lobster stock conservation and management) and the Bay of Fundy Fisheries Council (which undertakes both biological and socioeconomic studies within an ecosystem context).

It should be noted that not all involvement by fishers and industry is in the form of *participatory* research. Independent efforts are also undertaken by industry members (notably processing companies and fisher unions), as well as by native fishery groups and others, to analyse and provide a critique of fish stock assessments produced by government, in order to formulate preferred fishery management plans and to investigate problems of habitat degradation.

7.4.5 *Research by the private sector and non-governmental organisations (NGOs)*

It has been noted that traditionally, the vast majority of fishery research has been carried out by government research centres and by academic institutions. Two areas where this situation is changing most noticeably relate to the research roles of (a) private companies and consultants, and (b) non-governmental organisations.

With respect to the former, the growing presence of the private sector in research can be traced to a trend towards government downsizing and privatisation, which has had two related impacts. First, some national and international agencies now lack the staff resources to carry out research internally, and thus turn to consultancies with private sector companies. Second, and more fundamentally, a philosophical orientation toward privatisation has led some governments to dismantle their research infrastructure and let out contracts to conduct stock assessments and the like, or to maintain the infrastructure but force it to operate on a business basis. This trend is apparent in New Zealand and Australia, for example, as well as in Great Britain to some extent.

With respect to NGOs, there has been a noticeable trend in recent years toward greater public participation in matters pertaining to the world's oceans. As a result, many environmental organisations and related legal and community-based agencies, at the local, national and international levels, have become involved not only in advocacy efforts, but also in underlying research. Internationally, this includes such well-known organisations as the WWF

(Worldwide Fund for Nature or World Wildlife Fund), Greenpeace, the Third World Network and the World Conservation Union (IUCN). At a regional level, one finds a range of NGOs with fishery interests throughout the world. In Europe, for example, a significant number of organisations in the compilation of NGOs by the European Centre for Nature Conservation (www.ecnc.nl/doc/europe/organiza/europorg.html) have a fishery interest. At the local level, the range of NGOs concerned with marine and fishery matters is far too great to list here. While many of these are small groups, there may be many that engage in significant local-level research activity; two examples are the Ecology Action Centre in Canada (e.g. Fuller & Cameron 1998) and the Tambuyog Development Center in the Philippines.

Finally, one particular NGO deserves special mention. The International Ocean Institute has been conducting research on fisheries and ocean matters, as well as running training programmes and annual conferences and publishing the Ocean Yearbook, for many years. Founded by Elisabeth Mann Borgese, the IOI now has centres in many parts of the world, including south Asia, Central America, North America, southern Africa, western Africa and the South Pacific. A particular focus of the IOI lies in research on law and policy relating to peaceful and sustainable ocean use.

7.5   Summary

This chapter has reviewed various aspects of fishery research, which is a part of the management system that plays a key supporting role. In particular, the discussion here has explored:

- the need for research;
- the orientation of research;
- the structure of research;
- the participants in research.

Particular attention has been paid to the latter two elements. Figure 7.2 summarises the various participants in research, the key activities involved and the options for structuring the research.

It has not been possible in this brief overview to examine specific research projects or the details of individual countries. Instead, an attempt has been to provide a flavour of the diverse research activity worldwide, a sense of the 'generic' structure of that research and a discussion of some widespread trends in research such as the gradual move toward multi-disciplinary approaches.

Participants in research

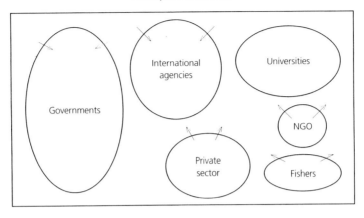

Research activities

Monitoring   Research   Decision support

Research structure

by discipline   by species   by ecosystem   by function

Fig. 7.2   The various participants in research, the major activities involved and four key options for structuring the research. Arrows for participant groups attempt to indicate the growth or decline in importance and/or involvement of those groups in research activities. NGO, non-governmental organisation.

Chapter 8
Dynamics of the Fishery System

How do the various components of fishery systems change over time? How do these components interact with one another over time? These questions are dealt with through *dynamic*
analyses of the fishery system. Such analyses are to be found in studies of both the natural
and human sub-systems. They have a particularly long history in the natural sciences, such
as biology and oceanography. For example, fish population dynamics, i.e. the processes of
change inherent in fish stocks, has been a key focus of biological studies, both in seeking to
understand the causes of fish population changes and in providing support for setting biologically sustainable harvests. This research has involved the study of basic biological aspects of
the fish (such as size-at-age, intrinsic growth rates and carrying capacities), as well as more
mathematical approaches such as virtual population analysis, simulation modelling, time
series analysis and a variety of statistical techniques gaining widespread use (see, e.g. Hilborn
& Walters 1992; Gallucci *et al.* 1996).

On the human side, social scientists have also paid attention to fishery dynamics, albeit
generally in a less quantitative or predictive manner than natural scientists. The focus of attention has been particularly on processes of change in fishery labour forces, fishing communities and management institutions (e.g. Andersen 1979; Bailey 1988; Apostle *et al.* 1998; and
the articles in Charles *et al.* 1994). Until the 1970s, economists were not known for their attention to dynamics in fisheries, probably because the economics tool-kit lacked the means
to solve messy problems of economic optimisation over time. This changed with the application of mathematical approaches such as optimal control theory to problems in fisheries economics, and in particular the development of bioeconomics (Clark 1985, 1990). The use of
dynamic models has now become widespread in fishery economics (e.g. Hannesson 1993).

This chapter examines fishery dynamics and explores how the fishery system, and its
components, change over time. The chapter is organised into three major segments.

- First, a recurring theme of the chapter is discussed: the range of time scales at which fishery
  dynamics operate, from short-term to long-term, and the importance of each.
- This is followed by a component-by-component examination of intrinsic fishery system
  dynamics in each of the components identified in earlier chapters. Specifically, this involves the natural system (notably single- and multi-species population dynamics), the
  human system (such as effort, labour, capital, fleets, technology, communities and markets), the fishery management system and the information base.

- Finally, an examination of integrated *system dynamics* is illustrated through a detailed description of *open-access dynamics*, a process by which a fish stock and fishing fleet may change over time in the absence of controls over fisher participation.

In order to focus on the underlying forces driving the dynamics of the fishery system, throughout most of the chapter (with the exception of a few brief discussions) a bold assumption is made of complete *certainty*. This implies in particular that (a) when fish stocks, fish prices and other variables change over time, such dynamics are *deterministic* in nature, with no random fluctuations, and (b) all participants in the fishery have full knowledge of all aspects, so there is no uncertainty in the form of imprecise parameters or ignorance of the underlying structure of the system. This assumption is bold in that uncertainty is in fact ubiquitous in fishery systems. The nature and implications of such uncertainty are explored in Chapter 11. A related assumption is also made here that all dynamics are 'well behaved', in that given a suitable model, changes in the variables over time can be fully predicted. A brief discussion is provided of what can happen when this latter assumption fails (if, for example, the dynamics are *chaotic*).

8.1   Time scales

It is important, in examining fishery systems and each of their components, to be conscious of the many time scales over which change occurs. Five major time scales can be noted:

- sub-daily to weekly;
- monthly to seasonal;
- annual;
- inter-annual;
- decadal or longer.

The nature of each of these, and relevant examples of each, are discussed below.

8.1.1 Sub-daily to weekly

Within the course of a fishing day, industrial fishers can move from one fishing location to another, searching for fish concentrations or maximal catch rates, while small-scale fishers may head out to sea, fish for a few hours and then return to port as part of a daily routine. In some fisheries, management also operates on a daily or sub-daily time scale regarding decisions about opening or closing the fishery. For example, salmon fisheries in British Columbia, Canada, operate through openings determined on a day-to-day basis. The time scale is shorter for the herring roe fishery of the same location, in which openings occur in more localised geographical areas when the roe content of the herring reaches desirable levels. In this case, the fishery remains open until the allowable catch is taken, which may take as little as a few minutes or as much as a few hours. (Clearly, there are immense information requirements in operating on such a short time scale!) Fisheries that function in keeping with tidal cycles also operate on a short time scale. The post-harvest sector can also do so, for example, in terms of the rapid processing of fish as it is landed, and the equally rapid search for markets when fish is to be sold fresh.

8.1.2 Monthly to seasonal

Within the course of a fishing season, temporal constraints may exist. For example, migrating fish stocks move across the range of the stocks, managers open and close the fishery in appropriate areas, and fishers shift between open fisheries, either physically moving from a fishery that closes to one that is opening, or perhaps changing gear to fish for a different species. On a seasonal basis, fishers and those in post-harvest activities will shift between occupations, temporarily leaving and re-entering the fishery as opportunities change. Natural cycles within the biophysical environment (temperatures, salinity and currents varying by season, for example) are crucial at this time scale, and may well underlie fishery dynamics through their impact on such factors as fish migrations and life-cycle development.

8.1.3 Annual

At an annual time scale, fishers make decisions about whether to take part in the fishery (which may involve an annual licence), how many crew to hire, what fisher organisation to join that year, etc. In many fisheries, harvesting plans are developed, whether by government or by fishers themselves, on an annual basis. In Europe, an elaborate annual process takes place involving total allowable catch (TAC) determination, allocation among nations, and the development of national fishing arrangements. In the Atlantic Canadian groundfish fishery, 'conservation harvesting plans' are developed by sector-based groupings of fishers and then submitted to the government fishery agency for approval. It should be noted that there may be a tendency to see management plans developed on an annual basis as unchangeable within the fishing season; something which is dangerous from a conservation perspective (Chapter 11).

8.1.4 Inter-annual

In the longer term, over time periods ranging from a few years to a decade or more, fishers make investment and career decisions – increasing their financial stake in the fishery, or perhaps exiting completely. The institutional structure of the fishery system changes over this time period, with fisher organisations forming and dissolving, new forms of management appearing, new legislation being passed by governments, etc. On this time scale, strategic and tactical plans are reviewed and re-worked. At the strategic level, this could involve re-assessing the objectives being pursued and the overall policy framework. ('Has a consultative approach to management solved the problems we discovered with command-and-control methods? If not, then change is needed.') At the tactical level, the portfolio of management methods selected and used within the operational aspects of management are reviewed. This could lead to a shift in the mix of methods, e.g. a shift away from effort controls to catch controls, or vice versa. At this longer time scale, fishery management must also respond to major trends impacting on the fishery, such as changes in market conditions or environmental conditions, and new international conventions.

8.1.5 Decadal and longer

The longest time scale considered here is one that may cover one to several decades. Dynamics at this scale can have a major impact through (a) wide-ranging phenomena such as global warming and economic globalisation, and (b) high levels of uncertainty in the dynamics, and the corresponding severity of potential management failures. This time scale crosses human generations and thus is 'inter-generational' from a human perspective. Indeed, fishers often express concern about future generations, particularly in the sense that their children should have a chance to go fishing in a healthy fishery system. Dynamics at this time scale therefore underlie the concern for conservation in fisheries, reflecting what sustainable development is all about: using resources to meet the needs of the present generation while leaving enough to meet the needs of future generations. In practice, however, fishery management rarely operates explicitly at this time scale.

8.2 Fishery system dynamics: component by component

The above time scales arise in each component of the fishery system. We turn now to an examination of the dynamics of these various system components. First, it is useful to note some parallels in dynamics among the various key components of the fishery system.

For example, it is standard practice to discuss the building of new vessels or the addition of new fishing gear as *investment* in the fishery's physical capital. However, the same concept of investment has been applied in recent years to the fish stocks. In this case, conservation measures are investments in boosting the *natural capital*, i.e. the abundance of the fish stocks themselves as well as the health of the ecosystem in which the fishery is located (Chapter 2). The clearest example of *investing in natural capital* is a moratorium on fishing. This focus on stock rebuilding is analogous to a company's decision not to issue dividends to shareholders one year, in order to put all available funds into a particular investment. Similarly, it is also possible to speak of investing in *human capital* through capacity-building, training and education programmes, and investing in *social capital* through the encouragement of suitable management institutions and fisher organisations.

There are also parallels in examining the constraints inherent in the dynamics of fish, people and fishery institutions. For example, consider the capability to control the various elements of the fishery system. This controllability is limited by tradition, culture and social attachments in the case of the people involved, by non-malleability in the case of physical capital (since disinvestment is difficult for the forms of specialised capital which are typical of fisheries), and by high levels of uncertainty and interconnectedness in the case of the fish resources. Thus, amid the many differences among the components of a fishery system, there are at least some commonalities in aspects of the dynamics.

8.3 Dynamics of the natural system

The best studied dynamics within fishery systems are undoubtedly the population dynamics of the fish, arising from the processes of survival and reproduction. This section reviews some key features of single- and multi-species dynamics, as well as a brief discussion of the dynamics of ecosystems and the biophysical environment. At this point, it is worth reiterating what was noted earlier: while uncertainty is a key aspect of the dynamics of all fisheries, to simplify the presentation here, its treatment is relatively brief, with a detailed discussion being left until Chapter 11.

8.3.1 Single-species dynamics

Whether to satisfy our curiosity about the workings of nature or to aid in the process of stock assessment, it is useful to have some idea (and predictive capability) of how individual fish species vary over time, affected by a combination of reproduction, survival, growth and natural mortality. Suppose that before the start of this year's fishing season, a management plan must be developed which requires a prediction of the available biomass. In simple terms, the population dynamics may determine this year's biomass as a combination of (a) the net survival of fish from last year, modified by the individual growth rate, plus (b) the new '*recruitment*' of adult fish this year, resulting from the eggs laid by the adult spawning (reproducing) stock in past years, with intervening mortality arising from environmental/ecological effects.

> Biomass this year =
> (Average growth rate) × (Survival of last year's biomass) + (Recruitment of new fish)

These population dynamics become simplified in a number of special cases.

(1) If the fish stock has non-overlapping generations, there is no survival from year to year, and only the final term in the above equation remains. This does not apply to most fish stocks (where *age structure* is important), but is often the case, for example, with shrimp stocks and with pink salmon on the Pacific coast of North America, which follows a strict 2-year life cycle. In the latter case, for example, the above expression for this year's biomass becomes

> Biomass this year = Recruitment produced by adult spawning stock 2 years ago

To use this equation, it is necessary to know the underlying connection between the size of the spawning stock and the subsequent recruitment – a connection referred to as the *stock–recruitment relationship*. If the relationship can be determined, it becomes possible to predict average future stock sizes. This typically involves a positive connection between the number of spawning fish in the past and the number of new fish expected now. The mathematical form of the stock–recruitment relationship is discussed in the box on p. 143, for those interested.

(2) Simplified population dynamics are also produced if we assume a single pooled fish stock, with (a) a constant survival rate, independent of the age of the fish, and (b) so-called *knife-edge* recruitment, in which all fish recruit to (i.e. join) the adult (or *fishable*) stock at one specific age. (Note that the latter assumption is usually unrealistic since, in reality, fish mature at varying rates, and recruitment to a fishery often depends more on

The stock–recruitment relationship

Population dynamics can be depicted, using our best knowledge of the dynamic process, in the form of mathematical 'models'. Suppose that we use the symbol X_t to depict the number of adult fish at the start of fishing season 't' (where t is a number denoting the year), and H_t to denote the harvest of fish during that fishing season. For simplicity, let us assume that the fishing season is short relative to the calendar year (so that natural mortality during the season can be ignored), and that spawning takes place immediately after the fishery. Then the adult fish stock remaining after fishing has taken place in year t is $X_t - H_t$, which is the spawning stock. Thus, the spawning stock 2 years ago can be written $X_{t-2} - H_{t-2}$ which, in the pink salmon example, is the stock that must generate the new adult fish (the recruitment) this year.

What will the new recruitment be? This depends on the specifics of the stock–recruitment relationship. We usually write this as a function 'F', so that in symbols, the deterministic connection between spawners 2 years ago and the resulting recruitment this year is

$$X_t = F(X_{t-2} - H_{t-2})$$

Note that this relationship, although overly simplified, captures the key conservation issue in the sustainable use of renewable resources: if the harvest 2 years ago (H_{t-2}) was excessive, the spawning stock ($X_{t-2} - H_{t-2}$) would have been less than would otherwise have been the case, impacting negatively (on average) on the resource available this year (X_t).

fish size than age, with the relationship between size and age varying over time.) Given these assumptions, the dynamics can be written as

Biomass this year
= (Average individual growth rate) × (Survival of last year's biomass)
+ (Recruitment of new fish due to reproduction 'a' years previously)

Figure 8.1 shows an example of fish population dynamics that reflect such a model. To generate this figure, the dynamics were assumed to be such that this year's stock combines survival from the past year with recruitment from 2 years previously, while the harvesting process was based on a fixed annual harvest rate (i.e. catching a constant percentage of the stock each year). A brief discussion of the model details for this approach is given in the box on p. 145.

(3)   A third option for simplification is to assume the recruitment of new fish into the fishable biomass to be constant, and perhaps equal to an historical average level, say over a previous 5-year period. This approach has been used in many fish stock assessment applications, notably when attempts to statistically 'fit' a stock–recruitment relation-

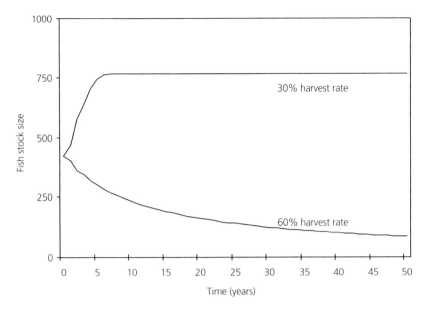

Fig. 8.1   An example of the dynamics of a fish stock driven by natural population dynamics together with fishing activity. The population dynamics incorporate survival from the previous year and recruitment resulting from reproduction 2 years ago. The harvesting process is based on a fixed annual harvest rate, as indicated in the two scenarios. Note that at a low harvest rate the stock rises, while at a high rate the stock falls; in both cases, an equilibrium is reached.

ship have failed. In such cases, an implicit assumption that no such relationship exists leads to the assumption of constant recruitment. This has been common, for example, in shrimp fisheries (where the life cycle of the animal is a single year and environmental factors often seem to dominate recruitment) and in North Atlantic cod fisheries (where virtual population analysis methods have dominated assessments). In such a case, the dynamics might be written as

> Biomass this year
> = (Average individual growth rate) × (Survival of last year's biomass)
> + (Constant recruitment)

However, when a constant recruitment assumption is made, attention is usually focused on the age structure of the fish stock. The box below provides some technical details of a possible approach.

It should be noted that the drawbacks to assuming constant recruitment have proved to be considerable (as will be discussed later in this book). An implicit assumption of no stock–recruitment relationship can lead to management that neglects the importance of maintaining a strong spawning stock. Some success has been had in developing models that combine stock–recruitment and key environmental factors, so it may be that further research in this direction will lead to less use of a constant recruitment assumption.

Age-structured population dynamics models

The two approaches discussed above are elaborated here. The second option involves assuming (a) that the number of adult fish surviving to the present year is a constant fraction 's' of those fish left after fishing took place last year, $X_{t-1} - H_{t-1}$, and (b) that young fish 'recruit' to the adult stock precisely at age 'a' (measured from the year of reproduction). Then the recruitment of new fish this year is given by applying a stock–recruitment function F, as discussed in the preceding box, to the surviving fish 'a' years earlier, in year $t - a$, i.e. $X_{t-a} - H_{t-a}$. In symbols, this dynamic connection between past and present can be written

$$X_t = s \cdot (X_{t-1} - H_{t-1}) + F(X_{t-a} - H_{t-a})$$

The third option above involves an *age-structured* model in which each 'age class' of fish is treated separately. In this case, it is helpful to examine two measures of the fish stock: the adult biomass and the 'fishable biomass', i.e. the biomass that is actually vulnerable to fishing. For simplicity, suppose that at an age a^*, fish become mature *and* are first vulnerable to the fishery (although not all fish of that age will be vulnerable). Then the adult stock is comprised of all age classes from a^* on, while the fishable biomass is made up of certain fractions of each age class from age a^* on. Define $X_{a,t}$ as the biomass of adult fish of age 'a' at the start of year t, $F_{a,t}$ as the fishable biomass of age-a fish at the start of year t, and $H_{a,t}$ as the number of that age class caught during that season. Then $X_{a,t} - H_{a,t}$ is the number of fish of age 'a' at the end of year t.

- Applying the constant-recruitment assumption, assume that $X_{a^*,t} = C$, where C is the constant recruitment of fish entering the stock at age a^*.
- Define s_a as the age-dependent survival rate of fish from age $a-1$ to age a, so that a fraction s_a of the $X_{a-1,t-1} - H_{a-1,t-1}$ fish of age $a-1$ at the end of year $t-1$ will survive to remain in the stock in year t.
- Determine the fishable biomass by taking a fraction of the adult stock at each age, the fraction being given by a term in a so-called *partial recruitment vector*, giving for each age 'a', the proportion (p_a) of fish of that age that are actually vulnerable to harvesting (i.e. in the fishable stock). The proportion typically is around 0% for very young fish, rises to 100% at higher ages, and may decrease at very high ages if old fish become less available to the fishery.

Given all this, the population dynamics for this age-structured model states that the adult biomass $X_{a,t}$ of fish of age 'a' in year t, and the corresponding fishable stock $F_{a,t}$, are given by

$$X_{a,t} = (s_a)(X_{a-1,t-1} - H_{a-1,t-1}); \quad F_{a,t} = (p_a)(s_a)(X_{a-1,t-1} - H_{a-1,t-1}) \quad \text{for } a > a^*$$
$$X_{a^*,t} = C; \quad F_{a^*,t} = (p_{a^*})C$$

Uncertainty and random fluctuations

This chapter focuses on analysing and understanding the underlying dynamics of the fishery system. To accomplish this, the discussion of dynamics throughout the chapter is carried out without considering the presence of random fluctuations and uncertainty. In reality, however, every fishery system is subject to fluctuations (in fish survival rates, fish prices and many other quantities), and their presence implies that we will never see in the real world the sort of 'smooth' curves, regulated by simple, well-determined functions, shown in Fig. 8.1. Compounding this is the fact that, even if we were to account for fishery fluctuations, the presence of other forms of uncertainties, arising from a lack of full knowledge of the system, means that we cannot expect to know the correct parameters to use to mimic the dynamics of fish populations and other system components. Thus, uncertainties complicate immensely the analysis of fishery dynamics.

The challenges faced by uncertainty in fisheries are the subject of detailed attention in Chapter 11. At this point, we simply illustrate how fluctuations due to uncertainty may appear in a graphical setting. Figure 8.2 shows the same scenarios as those in Fig. 8.1, but with random fluctuations incorporated. Note that there is a trend towards an equilibrium, but the effect of the fluctuations is to prevent such an equilibrium from being reached. This indicates an important inherent characteristic of random effects: their presence makes it impossible to manage a fishery to maintain an equilibrium. Recognising the presence of uncertainty also requires us to accept a more changeable, less steady world.

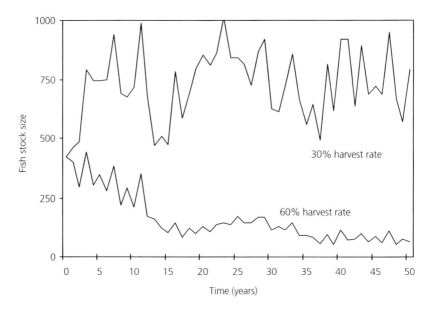

Fig. 8.2   A single-species scenario in which the fish stock is harvested as in Fig. 8.1, but the recruitment process in the population dynamics involves random fluctuations. As a result, the equilibria of Fig. 8.1, while approached, are never reached. Note that fluctuations are assumed to be proportional to the biomass, so the lower the stock size (e.g. under high harvesting pressure), the less noticeable are the effects of randomness.

Chaos

A topic of considerable debate among fishery researchers in recent years has been the possibility that fishery dynamics may be *chaotic*. In this context, the term *chaos* implies that, rather than smoothly approaching an equilibrium, as in Fig. 8.1, or even fluctuating around such an equilibrium, as in Fig. 8.2, the dynamics seem haphazard, with no underlying equilibrium and no regular pattern. This phenomenon of chaos has become well established in mathematical analysis and in the dynamics of fluids, where it has been studied by physicists, astronomers and many others. (See Gleick 1988 for a popular account of the subject.)

It is not clear whether such a situation arises in any real-world fishery system. Nevertheless, it is a simple matter to produce a 'chaotic fishery' in the artificial world of computer simulations. Figure 8.3 shows a chaotic scenario obtained using similar single-species population dynamics to those in Fig. 8.1, but with a higher intrinsic population growth rate.

In this case, there is no regularity to the dynamics: patterns may sometimes appear, but they do not persist, and certainly no equilibrium is reached. On the other hand, it is important to note that there can be order underlying a chaotic system. Just as a stochastic system may be 'attracted' to (but not reach) an underlying equilibrium, a chaotic system may focus on what is called a 'strange attractor'. This entity, visible in

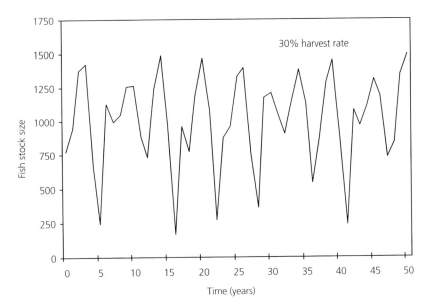

Fig. 8.3   An example of *chaotic* dynamics for a single-species situation. This figure is comparable to the case of a 30% harvest rate shown in Figs 8.1 and 8.2, but the fish stock's intrinsic growth rate here is a higher 4.2, versus 2.0 in those cases. (In a simple model such as this, the phenomenon of chaos often appears as the growth rate is increased.) Note that the dynamics are highly irregular, and that any patterns that may seem to appear from time to time do not persist. It is not clear whether chaos occurs in real-world fisheries.

what is called 'phase space' rather than in diagrams such as Fig. 8.3, depicts the fuzzy constraints that may guide the macroscopic behaviour of the system dynamics.

Future studies will probably clarify whether or not chaos really does arise in fishery systems, but in the meantime, even the idea of such chaos has led to much discussion of its possible implications on the manageability of the fishery (see, e.g. Wilson *et al.* 1994; Fogarty *et al.* 1997, and various papers referenced therein).

8.3.2 *Multi-species dynamics*

It is relatively rare to find a fishery that exploits only a single species. Although one species may be considered of most importance, others are usually caught at the same time. For example, fishers in a shrimp fishery may be entirely focused on the catch of shrimp, yet the majority of the catch by weight is often comprised of other species. In such a case, these other species are considered 'trash fish' at worst, or lower-valued *by-catch* at best. In many other fisheries, the variety of species caught are of somewhat comparable value. In the North Sea, for example, various flatfish, cod and other groundfish are all important to the fishers. In this case, it is useful to be able to predict changes in species abundances over time: in other words, to predict the multi-species dynamics.

To assess and predict change over time in multiple species, suitable models are needed. An example of such a model for two-species predator–prey dynamics is shown in Fig. 8.4. In this particular case, the so-called Lotka–Volterra equations are used to mimic the up-and-down

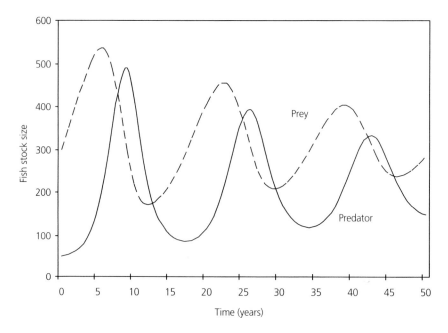

Fig. 8.4   An example of two-species Lotka–Volterra predator–prey dynamics (with no fishing activity). Cycles of the predator and the prey occur, gradually declining in magnitude as an eventual equilibrium is approached.

cycles of the predator and prey, which gradually decline in magnitude as an equilibrium is approached. (Note that there is no human harvesting in this example.)

The key issue in considering this challenge is to determine whether interactions amongst species are:

- *ecological interactions* based, for example, on predator–prey effects, competition among species for habitat or food, or other food web interactions (such as the species being in the diet of a common predator, e.g. seals), or
- *technical interactions* arising when the species are caught together in the fishing gear.

Ecological, particularly trophic, interactions have frequently been studied by biologists. There are many famous laboratory experiments and observational studies exploring the interactions of a predator and a prey (see box below). However, within an aquatic environment, such interactions remain poorly understood.

Technical interactions have received considerable attention by fishery scientists, since problems of unwanted by-catches, and in particular the impacts of dumping fish of the 'wrong' species or size, can have major conservation implications. This arises particularly because of the intrinsic 'mixing' that occurs in an aquatic environment. This can also arise in a terrestrial environment – analogies might be of a trapper catching an unwanted animal or a forestry firm discarding trees of a sub-optimal size – but overall, since controls over harvesting tend to be easier in such situations, the problem receives less attention than in the fishery context.

Predator–prey interactions

An interesting, and frequently referenced, example of predator–prey interaction is that of the Canadian lynx and the snowshoe hare. From records of lynx pelts obtained in northern Canada and dating back to the early nineteenth century, it was apparent that the lynx followed a 9–10-year cycle of abundance. Ecological studies also showed a similar 10-year cycle for the snowshoe hare. This initially suggested the existence of a classic predator–prey interaction in which an increase in the predator drives down prey abundance, which in turn reduces the predator population, allowing the prey to increase again, leading to a cycle in populations. However, while the hare is indeed the major prey of the lynx, it turned out that the hare's cyclic abundance was driven not by lynx predation, but rather by the impact of the hare on its own food supply. At high levels of abundance, the hares depleted their food supply, causing population declines, regeneration of the vegetation and a repeat of the cycle. This therefore illustrates a 'carnivore–herbivore–vegetation' system, and also shows that caution is needed in interpreting ecological data, since we would be in error to conclude here that the lynx–hare dynamics reflect a classic predator–prey cycle.

Sources: Odum (1983); Jackson & Jackson (1996)

8.3.3 Dynamics of ecosystems and the biophysical environment

Chapter 2 presented a discussion of fishery ecosystems, their inherent characteristics and the various forms in which they appear. It was noted that defining precisely what constitutes the relevant ecosystem in a given situation can be difficult, since there is a variety of spatial scales that can be used, and an issue of setting appropriate boundaries. This challenge is particularly great in an ocean environment.

Let us assume, however, that a suitable ecosystem has been defined. How does that ecosystem change over time? Certainly, changes in the presence of various species, as addressed by multi-species dynamics, represents one specific form of ecosystem dynamics. Intensive fishing activity and the phenomenon of *fishing down the food chain* (Pauly *et al.* 1998) can have major ecological impacts. For example, in the northwestern Atlantic, off the coast of North America, fishing appears to have led to major changes over time in absolute and relative species abundances, changes in species composition at the various trophic levels, and even changes in the relative biomasses of the various trophic levels, although interestingly, *total* biomass in the system seems to have been fairly constant over time.

Other aspects of ecosystem dynamics could include those induced by natural or human impacts on the physical nature of the aquatic habitat (e.g. by trawling on the ocean bottom), and those caused by external forces in the biophysical environment. Let us turn now to the latter impacts, beginning with an illustration.

At least until the collapse of Canada's Atlantic groundfish in the 1990s, the best known case study of fishery collapse was the anchovy off the coast of Peru in the late 1960s and early 1970s. The cause of the collapse, in what had been the world's biggest fishery, seems to have been an excessive level of exploitation that left the stock unable to absorb the 'shock' of a natural perturbation in ocean conditions known as *El Niño*. This phenomenon generally occurs every few years, and involves the temporary altering of ocean currents and consequent warming of the ocean off that coast. This, in turn, dramatically affects fish stocks such as the anchovy, and has impacts over much of the Pacific basin and beyond.

In the case of the anchovy, Hilborn & Walters (1992: p. 19) note that El Niño 'apparently had two effects: it initially concentrated the fish close inshore where they were highly vulnerable to the fishing boats, and then it caused poor juvenile survival for the offspring of the remaining spawners. The net result was a general recruitment failure.' The devastation of the Peruvian anchovy collapse clearly demonstrates that one cannot understand the behaviour of fish stocks off the coast of Peru, or manage the fisheries, without considering the impact of El Niño.

The El Niño phenomenon, and its broader coupling with the southern oscillation (ENSO), are clear examples of dynamic change in the biophysical environment. These dynamics relate to what are already inherently dynamic features of the aquatic system, namely *changes* in *flows* (the movement of water) such as ocean currents, upwellings and tides. However, the matter of time scales is useful in differentiating among these phenomena: specifically, El Niño operates on a relatively long time scale (and has a correspondingly large-scale impact), whereas upwellings and tides vary on a much shorter time scale.

Another important large-scale example of biophysical dynamics is that of global climate change (e.g. Mann & Lazier 1996), which may have diverse impacts on aquatic ecosystems worldwide. At the more local level, an example of biophysical dynamics would be variations in river outflows. Such a phenomenon in the St Lawrence River, on Canada's Atlantic coast, leads to varying salinity levels in the Gulf of St. Lawrence ecosystem, with resulting impacts on fish stocks. Finally, biophysical dynamics also relate to aspects of aquatic systems that are stationary by nature, such as the shoreline, where discussion of *dynamics* may focus on changes in physical features, such as the occurrence of beach erosion.

8.4   Dynamics of the human system

Relative to the study of fish population dynamics, considerably less attention has been paid to addressing the dynamics of the fishers and other components of the human system. Yet an understanding of such changes over time is essential both to the study of fishery systems, and to their management and conservation. In this discussion, we look at the dynamics of (a) fishing effort, (b) fishery labour, (c) fishery capital, (d) technology, (e) fishing fleets, in terms of movement within the season, (f) fishing communities and (g) the broader socioeconomic environment.

8.4.1 Effort dynamics

The number of fishers participating in the fishery, and the aggregate fishing effort, can vary over time according to such factors as:

- the perceived profitability of fishing versus other economic activity, and in particular the effects of changes in stock sizes;
- the traditional practices of the fishers, perhaps reflecting religious or cultural norms;
- policy measures and management restrictions, as well as government actions to reduce the fleet by 'buying back' fishing vessels;
- external 'forcing' factors, i.e. changes elsewhere in the economy or society, notably in the availability of non-fishing opportunities (affecting labour mobility).

An early theoretical study on this theme, by Smith (1968), focused on the first of these factors, using a model of the joint dynamics of a fish stock and the corresponding fishing effort to examine how fishers might adjust their aggregate fishing inputs (effort) in response to changes in the level of available profits in the fishery. Empirical work has also been carried out on this theme (e.g. Opaluch & Bockstael 1984), examining the range of objectives being pursued by fishers, and the process by which decisions are made concerning the desired levels of harvesting effort to meet these goals.

8.4.2 Labour dynamics

Closely related to effort dynamics is the matter of labour dynamics, i.e. the processes by which workers, whether inside or outside the fishery, shift over time:

- between occupations (reflecting *occupational* mobility);
- between locations (involving *geographical* mobility).

Various authors have emphasised the importance of examining fishery labour dynamics in order to understand the behaviour of fishery systems and to help in determining suitable management policies (see references in Charles 1988). For example, Terkla *et al.* (1985) argue that 'understanding labour adjustment processes is likely to be crucial for implementing efficient and equitable management policy' throughout the fishing industry, and Smith (1981) noted the important relationship between labour mobility and the economic state of fisheries in developing countries.

Fishery labour processes are intimately related to the overall socioeconomic environment (Panayotou 1982). Specifically, in many fisheries, fishers have few alternative employment options outside the fishery, and thus low occupational mobility. However, several studies (Ferris & Plourde 1982; Panayotou & Panayotou 1986) have found that fisher labour dynamics are sensitive to market signals and economic incentives, which can substantially increase occupational mobility. This reinforces the point made by Munro (1990) that fishery management must anticipate the impacts of changes in the employment options available to fishers (whether positive economic diversification or unanticipated factory closures and other downturns) by looking *beyond* the fishery at the dynamics of the larger economic system.

Deliberate policy actions can also influence labour dynamics. One such policy direction lies in efforts to increase employment options in coastal areas. This is an important option, given the results from Terkla *et al.* (1985) that in many fishing ports, 'labour outmigration is low because of strong attachment to community and family', and those of Panayotou & Panayotou (1986) that labour is 'quite mobile between occupations but less so between locations'. The latter authors suggest that based on this result, efforts at economic development may be best focused on promoting non-fishing employment alternatives in those areas where geographical mobility is most limited.

Another example of a policy measure influencing labour dynamics occurred in Costa Rica, where government policy to encourage the development of agricultural exports provided incentives for local farmers in inland areas to cut down rain forest, so as to 'improve' the land, which then could be sold to ranchers, coffee growers or others. The small-scale farmers, having sold the land, found it increasingly difficult to find more land – in other words, it became more difficult to obtain a livelihood inland – so many migrated to urban areas and to the coast (Fig. 8.5). Those in the latter group began fishing, notably in the Gulf of Nicoya on the Pacific coast. This increase in the coastal labour force, together with an institutional environment of inadequate fishery controls, led directly to increased exploitation of coastal shrimp and fish resources, and declines in abundance (Charles & Herrera 1994).

8.4.3 Capital dynamics

Human inputs to the fishery naturally include the *physical capital* in vessels and gear, as well as the labour involved. Thus, we need to examine capital dynamics in conjunction with labour dynamics.

Changes in fleet size and catching power reflect both investment decisions and technological change. Continuing with the example of Costa Rica described above, Fig. 8.6 shows the changes in fleet sizes over time for two types of artisanal fishing vessels in the Gulf of Nicoya. The less technologically advanced type of vessel (*botes*) grew in numbers fairly steadily over many years, while numbers of the more advanced type (*pangas*) rose rapidly in the mid- to

Fig. 8.5   The dynamics of people in the fishery (labour) and the associated fleets (capital) can be affected by external factors - such as labour migration. These cattle stand on deforested land in Costa Rica that may have previously supported some of the coastal fishers who left inland areas to migrate to the coast.

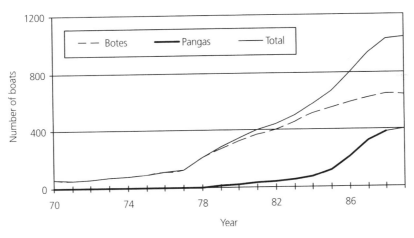

Fig. 8.6   Capital dynamics for two types of artisanal fishing vessel in the Gulf of Nicoya, Costa Rica. The fleet size of the less advanced type of vessel (*botes*) expands in numbers fairly steadily over many years, while the fleet of more advanced *pangas* rises rapidly in the late 1980s. Both fleets level off in size toward the end of the 1980s.

late-1980s. This latter growth was connected, at least in part, to the economic incentives and the consequent labour dynamics noted above. Increasing concern about the impacts of un-controlled growth in the fleets led to the imposition of limited entry licensing, which slowed the expansion in boat numbers by the end of the 1980s (see Fig. 8.6).

A key issue in capital dynamics relates to the phenomena that lead to capacity expansion and consequent problems of over-capacity. Consider, for example, a fishery system in which the fish stock has been depleted, leading to a logical response from management to implement conservation measures in order to rebuild the stock. If successful, such measures may produce a perverse sequence of outcomes:

(1)   an initial increase in the stock;
(2)   an induced increase in fishing effort, producing temporarily above-normal profits;
(3)   increased investment, due to the incentive to increase one's share of revenues;
(4)   a corresponding increase in overall costs, and consequent dissipation of rents;
(5)   political pressure to allow even greater harvest levels, in order to maintain incomes;
(6)   a reversal of the stock rebuilding, combined with a chronically unprofitable fishery.

Note that this unfortunate (and undesired) sequence of events occurs not because of a failure of the rebuilding effort, but because of the *success* of that effort. Indeed, once the fish stock has been 'rebuilt' to higher levels, it is actually possible to sustain greater effort for some time, without excessive depletion. However, this success can lull the industry and the government into a false sense of security, until the expansion reaches a point where the stock declines and a 'crisis' sets in. This phenomenon has occurred in many fisheries, including, for example, Canada's Pacific salmon fishery and Atlantic groundfish fishery (see Parsons 1993). However, in most cases it seems that the outcome of the process was not foreseen, perhaps due to a lack of an integrated analysis. This illustrates well the need not just to look at the population dynamics of fish stocks, but also to integrate the dynamics of human factors (labour and capital) as well as the dynamics of management decision making (including the setting of objectives and constraints).

An important consideration in examining capital dynamics is the *malleability* of capital. If capital is malleable, this implies that investments are reversible, and that fishers can receive a reasonable resale value for capital, when it exits the fishery. On the other hand, if, as is common in many fishery situations, the fleet of specialised vessels has few alternative uses, the capital is non-malleable (investment is relatively irreversible), and thus reducing the existing capacity is made much more difficult. The dynamics of this problem have been analysed in detail by Clark *et al.* (1979), Charles (1983a,b) and others.

There have been many studies of capital and investment dynamics in fisheries. This work comes from two rather different perspectives. On the economics side, such studies date back at least to Gordon's (1954) classic study of the process by which open-access fisheries become over-capitalised (discussed below). A particular focus has been on examining the development of excess fishing capacity in fisheries. On the social science side, researchers looking at capital dynamics tend to focus on the evolution of capitalism as a mode of operation in the fishery, and on the interaction between social structures, government regulations and the investment behaviour of fishers (e.g. Marchak *et al.* 1987; McCay & Acheson 1987).

Modelling studies include *optimisation* analyses (oriented toward determining investment levels that maximise specified societal objectives, e.g. Clark *et al.* 1979; Charles 1983a,b) and *behavioural* studies of fisher investment focusing on understanding fisher objectives and predicting investment dynamics. Examples of the latter include the empirical examples of Lane (1988) on trollers in the British Columbia salmon fishery, and Tettey & Griffin (1984) on investment patterns for American shrimp fisheries in the Gulf of Mexico.

Growth of the Chinese fishing fleet

'The quantity of marine fishing vessels seems to be approaching a maximum. The fleet will not be able to continue to increase its capacity as rapidly as in the last decade. From 1990 to 1995, 29 745 new motorized fishing vessels were added to the fleet, an annual growth rate of only 2.3%, compared with a growth rate of 17.2% between 1980 and 1990. The new fishing vessels were more powerful; mean power per vessel increased from 27.8 kW in 1990 to 35.8 kW in 1995. This can be considered the fishers' response to obvious depletion of major commercial fish stocks in offshore waters. They started building larger, more powerful vessels to exploit fishery resources in offshore and foreign waters ... Hundreds of fishing vessels are operating off West Africa, North and South America, the Middle East, and in the South Pacific, far from their traditional fishing grounds. The number of vessels fishing in distant waters has increased steadily in recent years ... It can be anticipated that high-power vessels will become more popular and fishing pressure in foreign waters will become more intensive in the next century.'

Zhong & Power (1997)

8.4.4 Technological dynamics

Advancements in fishing technology have been the source of long-standing processes of change in the fishery. These include both gradual change, over the course of centuries and decades, and more rapid changes. Technological change in fisheries has been dramatic at times, but it has an impact on the sustainability of the fishery primarily when management fails to recognise that change in its assessment of the fish stocks and in designing its management plans. For example, adoption of the *turbo trawl* by otter trawlers in Canada's Gulf of St. Lawrence produced a major increase in catching power. Failure to monitor and account for such changes on the part of government managers is considered a factor in overestimates of biomass levels and consequent excessive catches.

It is instructive to look at technological change in the context of fishery development over a long time scale. For example, harvests of Northern cod, the now-collapsed cod stock in NAFO (Northwest Atlantic Fisheries Organisation) area 2J3KL on Canada's Atlantic coast, grew gradually from the mid-1600s to the late-1800s, reaching a level of approximately 250 000 tonnes (Hutchings & Mycrs 1995). From that time to the mid-1900s, the catch tended to remain around that level, apart from natural fluctuations and short-term ups and downs. Then in the mid-1900s, the catch built up dramatically, making Northern cod, for a time, among the biggest fisheries in the world. This period corresponded to a particularly rapid process of technological change – the introduction of factory freezer trawlers into distant-water fishing fleets around the world – which had great conservation impacts in many fisheries. Following these years of very high catches, the stock collapsed in the 1970s, rebuilt somewhat in the early 1980s, then collapsed again. Technological change, and a lack of monitoring or control thereof, was certainly among the causes of both collapses.

Technological change is widely viewed as irreversible. The trend towards increasing technological sophistication is global in nature, although there is some heterogeneity. For example, while in commercial fisheries, fishing techniques, vessel design and notably the use of electronic gear have all evolved dramatically over the course of the past few decades, this process has tended to have less impact on subsistence fisheries, which often use traditional technologies for local food production.

Finally, it is worth noting that technological change also has an impact on the dynamics of other components within the fishery system. For example, McCay (1979) found that the introduction of new long-liner technology in the fishery of Fogo Island, off the northeast coast of Newfoundland, had an impact on labour dynamics. The new fishery affected the level of earnings, the sharing of revenues between owner and crew, returns to capital and labour mobility. These impacts, in turn, produced changes in fishery participation levels, to the extent that at times difficulties arose in recruiting vessel crews, despite high unemployment levels in the region.

8.4.5 Fleet dynamics

It is important to understand *where* and *when* fishers choose to operate within the fishing season. As Hilborn & Walters (1992) note, 'it is foolish to study only the prey in the predator–prey system … it is equally important to monitor and understand basic processes that determine the dynamics of the predator – the fishermen'. The matter of within-season change is often referred to as *fleet dynamics*, i.e. the movement of boats into and out of the fishery, and between fishing grounds, over the course of a fishing season. Such dynamics are by no means well understood, but there have been some efforts in this area, e.g. Hilborn (1985), Allen & McGlade (1986) and Sampson (1994).

Changes in fleet activity within the fishing season can occur in response to many factors, such as changes in perceived fish abundance and distribution, in fish prices or in fishery management measures. Furthermore, fleet dynamics of fishers can take place on various time scales, such as the two described below.

- Within the course of a given day: for example, if customary practice involves fishing in the morning hours, and onshore work in the afternoon. This type of *satisficing* behaviour was observed among the fishers of Puerto Thiel in Costa Rica (Charles & Herrera 1994).
- Within the course of a fishing season: for example, if the fish stock, and therefore the profitability of fishing, decline over the season, or if management invokes a 'closed season' to allow uninterrupted spawning activity. Hilborn (1985) looked at fisher dynamics on a within-season time scale and explored the movement of fishers between fishing grounds over the course of a single season.

8.4.6 Dynamics of communities and the socioeconomic environment

Social scientists have produced a wealth of studies about the structure and dynamics of fishing communities. The communities vary widely, from those of indigenous peoples in the South Pacific (e.g. Ruddle & Johannes 1989) and in North America (e.g. Pinkerton 1989), to communities based on commercial fisheries ranging from the Mediterranean (e.g. Frangoudes

1996) to Canada and Norway (Apostle *et al.* 1998) to the Philippines (Ferrer *et al.* 1996; Graham 1998). Some of the points examined include demographic features, particularly processes of growth and decay in communities, the evolution of community-based fishery management institutions, and responses of fishing communities to externally driven forces, whether environmental, economic or social.

Moving beyond the core of the fishery system, the socioeconomic environment can be viewed as the outer layer of the human sub-system, lying on the edge of whatever boundaries are drawn around the fishery. The socioeconomic environment is to the human part of the fishery what the biophysical environment is to the fishery ecosystem. As described in Chapter 3, this environment includes economic, social and institutional aspects at larger spatial scales than the fishery or the local community, i.e. those at the regional and national levels, ranging from legislative realities to fish and labour markets to sociocultural considerations.

An example of dynamics in the socioeconomic environment that impact on the fishery could be a major change in the coastal economy, such as the closure of a coastal factory or the decline of coastal forestry. Such a change could have an impact on fishery systems through changes in the labour force, with potentially a cascade of further implications: e.g. greater availability of labour, lower wages, higher profits, more investment, and eventually greater pressure on fishery resources.

These processes of change affect economic, demographic, social and political aspects of the fishery, with the change being driven by 'external forces' rather than internal mechanisms. It is crucial to note that external driving forces operate on a variety of time scales. Consider, as an illustration, the market for fish. We can point to three different time scales of change in the market-place.

- *Fish prices (short-term fluctuations: < 1 year).* Prices for fish can vary widely over short time frames, due to variations in the availability of the fish itself (e.g. whether stocks are abundant or not, whether the fishing season is open or closed in the relevant area) and the availability of substitutes (including other species of fish, and other protein alternatives, such as meats). In a sense, price fluctuations have a similar role with respect to the economic operations of fisheries as short-term environmental fluctuations have on fish population dynamics. In other words, while price fluctuations can be global in terms of spatial scale, they operate on a short temporal scale, representing a rapidly changing forcing term in the fishery; changes can be on the scale of days to weeks, or can be seasonal, following the course of fishing seasons. Typically, one has some idea of the cause of a given price fluctuation, but global markets are sufficiently complex that often this is more of a supposition than a proven effect. In general, price fluctuations are reversible and simple to detect. It should be noted that in certain cases, such as a local fishery serving a specific local market, price fluctuations are integral to the fishery system, rather than representing a fundamental externally driven change.
- *Demand shifts (medium-term: annual to decadal).* While price fluctuations are relatively short-term, the consumer side of the fishery also changes on longer time scales. The *demand schedule*, which depicts how much (of any given good) would be purchased by consumers at each possible price, may change over time, on a time scale longer than that of simple fluctuations, in response to changing consumer preferences or changing income levels. For example, if consumers gradually become wealthier, an increasing total amount

of income may be available for spending on fish, and the demand curve may move outwards. Demand may also undergo rapid *regime shifts*. For example, in recent decades, shifts over a short period of time have resulted from publicity about health issues. An awareness of the relative healthiness of fish (compared with red meat) led to a substantial 'outward shift' in the demand curve, while a shift inward followed publicity over ocean pollution and possible contamination of fish in the sea.

- *Globalisation (long-term: decadal)*. While many fishery systems have relied on large-scale trade for a very long time, globalisation of the economy as a whole has accelerated remarkably in recent decades, forcing substantial dynamic change in many fisheries. This structural change, together with trade liberalisation in the global economy, has had major impacts at various levels. First, globalisation is affecting fish price levels and competition between seafood and other food categories. Friis (1996) has commented that 'international deregulation will lead to a further fall in food prices, including those for fish and fish products' (p. 177) and furthermore, that even with such a price decrease, the reality of competition with non-fish foods means that 'whether consumption levels [for fish and fish products] can be sustained … remains very doubtful' (p. 181).

  Second, globalisation affects the nature of seafood markets. For example, consider the emergence of a world market for *white fish*, a generic term for a set of species, such as Alaskan Pollock, Icelandic cod and North Sea haddock, that are to a considerable extent interchangeable (substitutable) in the market-place. This market has created a situation in which the ups and downs of any one *white fish* stock, even collapses as with the Northern cod off the coast of Newfoundland, may have little overall impact on world markets. Furthermore, not only is there now little geographical attachment between producer and consumer, but globalisation has also brought a 'weakening of ties between catching and processing' (Jónsson 1996: p. 190). Processing companies, formerly reliant on 'local' *white fish* stocks (notably groundfish), now process fish brought to their plants from anywhere in the world. This shields these plants (and the employment they produce) from local variability in stock sizes, but at the same time, there may be less incentive on the part of such processors to conserve the local resource, on which they are now less reliant.

8.5   Dynamics of the management system

A systems approach to fisheries encourages us to examine the dynamics of all components of the fishery system. Certainly, the management sub-system (including scientific research, fishery managers and the legislative framework) is a key component of that system. While discussion of change in fisheries has traditionally focused on the population dynamics of fish stocks, and to some extent on human dynamics, it is also apparent that the fishery management component of the system changes over time. This regulatory component fits within a dynamic system alongside the fish, the fishing fleet, the fishers and the fishing communities. The dynamics of regulation in fishery systems can occur at various scales:

- changes in the objectives being pursued, and the priorities among these (e.g. shifting from a food production orientation to a preference for rent maximisation and export orientation in the face of a structural adjustment or austerity programme);

- changes in policy directions, such as the preference of one fishery type over another (e.g. recreational or native/aboriginal versus commercial, small-scale versus large-scale, etc.);
- changes in the strategic level of management (e.g. implementation of new property rights and/or co-management arrangements – see Chapters 13 and 14);
- changes in the tactical level of management (e.g. shifting from effort controls to catch controls, or vice versa);
- operational changes (e.g. a change in the allowable mesh size of nets, introduction of a closed season, etc.);
- changes in the institutional structure of fishery management (e.g. a change in the governmental body responsible for management).

In recent years, it has been realised that good fishery management requires not only the setting and enforcing of regulations, but also the ability to predict fisher response to these regulations. Anderson (1987: p. 126) notes that 'if a fishery agency hopes to regulate with any degree of accuracy it has to know what effects different policies will actually have on industry behavior'. Hilborn & Walters (1992: p. 104) point out that a lack of understanding of such responses 'led to management strategies and regulatory schemes that ignore the dynamic responses of fishermen to changes in stock size and to management itself. These responses can dampen or even reverse the intended effects of regulation …'. This has been the case in many fisheries. In Canada, for example,

- efforts to restrict groundfish vessel length, in order to limit capacity expansion, were thwarted by fishers through the construction of *wider* vessels, so capacity expanded nonetheless;
- limitations on the number of traps allowed per fisher in a lobster fishery induced more frequent hauling of the traps (increased use of labour) and may also have led to a change in the location of fishing, with increased effort further offshore;
- single-species catch limits, imposed without an understanding of built-in incentives to thwart such controls, led to misreporting, under-reporting, dumping, discarding and highgrading.

Improving our understanding of interactions between management agencies and the fishing industry requires not only an understanding of the dynamic behaviour of those involved in the fishery system, but also studies of *management dynamics*, examining the regulatory framework as an endogenous part of the overall fishery system, and integrating interactions between fishers and regulators.

8.6   Information dynamics

The reality of fishery systems is that what we do not know about them far exceeds what we do know. Given this reality of extensive uncertainties, it is important to make the best use of any and all information, to learn over time and to adapt to changing circumstances. The processes by which new information becomes available occur at various time scales. For example, in most major fisheries, there is an informal *annual* (or more frequent) updating of the information base. This revised information is used to adjust fishery management measures, most notably the TAC, from year to year. In some fisheries, however, information flows at a shorter time scale. In the salmon fishery on the west coast of North America, for example, management is by means of *weekly* openings, so information must be compiled and analysed *daily*. This is necessary because indications of an unexpectedly weak or strong run of fish can lead to shorter or longer openings in a given week.

The process of information collection and knowledge updating can be referred to as *information dynamics*. Note that although we can develop processes to collect certain data daily, weekly, monthly, annually or on an irregular basis, we obviously cannot know in advance what the new numbers will be. Furthermore, the data we do obtain will be uncertain, owing, for example, to unavoidable sampling errors. Despite these uncertainties, we can usefully *model* information dynamics, simulating the process of information acquisition (Charles 1992a). This can aid us in assessing the value of information collection processes (and indeed whether the benefits of such information outweigh the costs of collecting it) and improve the manner by which information is used in fishery management decisions.

8.7   Fishery system dynamics

How do the various components of the fishery system interact with each other? More specifically, how does the state of one component, the fish stock, say, induce dynamic change in another component, such as the fleet or the management system? These questions have been explored in various multifaceted approaches (e.g. BEAM 4; see FAO 1999b). The reader is referred to the Appendix of this book for a presentation of integrated 'biosocioeconomic' analysis, using models that mimic the dynamics of both fish and fishers in the fishery system. The Appendix is organised so that the reader has the choice between a non-mathematical exploration of how models are developed and corresponding results produced, and a more technical focus on the model details.

Here we focus on a graphical approach to examining dynamic interactions between the fish stock, the fishery capacity (comprised of fishers and fleets) and the fishing effort. This is based on a key building block in a *bioeconomic* understanding of the fishery system, the Gordon–Schaefer diagram, named after an economist, H.S. Gordon (1954), who modified the analysis of a biologist, M.B. Schaefer (1954) by blending together biological (population dynamics) and economic factors.

8.7.1 Fishing-induced dynamics and sustainable yields

We begin with Schaefer, who recognised that there was not just one *sustainable yield*, i.e. a harvest that can be taken year after year indefinitely into the future, but an infinite number, depending on the level of annual fishing effort. Consider the case of a virgin unexploited fish stock, which for simplicity we assume to be a single-species fish stock, with no random fluctuations in the stock from year to year, and for which we have a perfect knowledge of population dynamics.

Suppose that the catching power of the corresponding fishing fleet is zero. In other words, there are no fishers and no boats, and thus no fishing effort is possible. Is there a sustainable yield in this odd 'fishery'? Yes, it is precisely zero. Maintaining that level of effort each year, the biomass will remain in equilibrium, with nothing being caught this year, next year, or any year in the future. That defines exactly what is meant by a sustainable yield. Figure 8.7 shows the time series of the biomass and of this very simple catch pattern over time.

Now suppose that instead the fleet consists of a very small number of fishing vessels, with a correspondingly low catching power, and that this catching power is exerted regularly each year, implying a certain constant (but modest) level of fishing effort. This harvesting activity will lead to a small reduction in the fish biomass over time (Fig. 8.8), with the catch time series adjusting accordingly, based on the assumption that harvests are jointly proportional to effort and biomass. Note that the fishery system reaches an equilibrium; the catch in equilibrium is the sustainable yield corresponding to that effort level.

Now consider a very powerful fishing fleet. Given a high enough fishing effort, the percentage of the fish stock taken each year will be beyond the reproductive capability of the stock

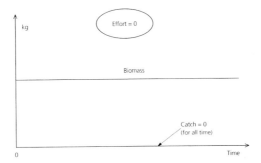

Fig. 8.7   Time series of the biomass and catch for a fishery in which the catching power of the fishing fleet is zero, i.e. no fishers, no boats and thus no fishing effort. The biomass remains in equilibrium and the sustainable yield is zero in this 'non-fishery'.

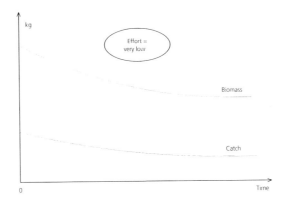

Fig. 8.8   If a fishery involves a small fleet with low catching power, and this generates a certain constant (but modest) level of fishing effort, this harvesting will lead to a small reduction in the fish biomass over time, and a corresponding decline in the catch time series, until an equilibrium is reached (with the catch at this point being the sustainable yield for the particular effort level).

to absorb. Thus, the stock will decline steadily until it disappears (potentially in extinction). In such cases, initially catches may be very large, but the *sustainable* yield is zero, i.e. from a *sustainability* perspective, it is irrelevant that high catches are obtained in an initial period of 'mining' the resource, since the harvest that can be taken *year after year indefinitely into the future* (the definition of sustainable yield) is zero. The relevant time series are shown in Fig. 8.9.

Suppose that E_{max} is the minimum of all those high effort levels that, exerted year after year, deplete the stock to zero. This means that at that effort level, and any higher effort, the sustainable yield is zero. Now consider the above three scenarios as points on a graph of sustainable yield versus fishing effort (Fig. 8.10), indicating the harvest that can be taken each year indefinitely into the future, at each effort level. The first point is (0,0), the second point is at positive but low effort and yield levels, and the third is at $(E_{max},0)$.

What about all other fishing effort levels? Joining the points with a curve, we see that the graph starts at (0,0) and rises initially, but must eventually decline to $(E_{max},0)$. Thus, the

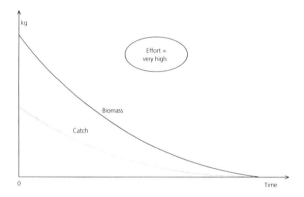

Fig. 8.9   If a fishery system has a very powerful fishing fleet exerting a high fishing effort, catches may be very large initially, but the fish stock may decline steadily until it reaches zero (extinction). In such cases, the *sustainable* yield is zero.

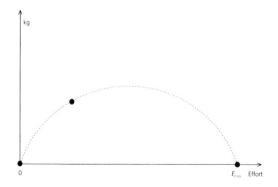

Fig. 8.10   The three sustainable yield scenarios of Figs 8.7–8.9 are shown as points on a graph of sustainable yield versus fishing effort. They appear in order, from left to right, with E_{max} representing the minimum of all those high effort levels that, exerted year after year, deplete the stock to zero. A curve joining the points starts at (0,0), rises initially, but eventually declines to $(E_{max},0)$, forming the full Schaefer yield–effort curve.

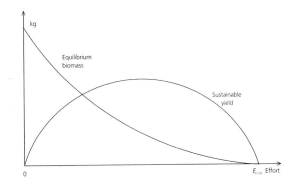

Fig. 8.11   The equilibrium *biomass* for each effort level together with the sustainable yield. This graph summarises the population dynamics of a fish stock and how it responds to fishing activity. The peak of the sustainable yield curve corresponds to the *maximum sustainable yield* (MSY) for the fishery.

sustainable yield must be positive for all effort levels between 0 and E_{max}. This insight lets us create the full yield–effort curve that Schaefer derived years ago (Fig. 8.10).

Note that every point on this graph shows, for a given level of fishing effort repeated annually forever, the resulting sustainable yield. Each point also corresponds to a time-series graph of biomass and catch (as in Figs 8.7–8.9).

Finally, the equilibrium *biomass* for each effort level can be displayed, together with the sustainable yield, in a full Gordon–Schaefer diagram (Fig. 8.11).

This graph now summarises the population dynamics of a fish stock and how it responds to fishing activity. In particular, it shows how, if a fleet capacity *E* operates each year, the eventual biomass will be *B* and the sustainable yield will be *Y*, the catch that can be taken every year into the future. The peak of the curve corresponds to the *maximum sustainable yield* (MSY) for the fishery. In theory, this important value (and the effort needed to attain the MSY) can be derived from a statistically estimated yield–effort curve, but it must be kept in mind that the curve itself can only be derived using *equilibrium* yield levels, not from merely plotting annual catch and effort data (since these are not necessarily reflective of an equilibrium of biomass and catches).

8.7.2 *Open-access dynamics*

The yield–effort curve tells us (in theory at least, and without dealing with uncertainties) what sustainable yield can be obtained if we repeat indefinitely a *given* level of annual fishing effort, but what level of fishing effort will be adopted in a particular fishery? To examine this question, we need to add some economics, and that is where Gordon comes into the picture, converting Fig. 8.11 into an economic analysis.

To do this, we first note that *yield* is measured in biomass or fish numbers, which, when multiplied by the price of fish (per kg or per fish), becomes the *total revenue* (or *gross income* or *landed value*) received by the fishers. Specifically, assuming a constant price (*p*), the total sustainable revenue produced annually from a sustainable yield *Y* is $TR = pY$.

Now, it costs something for fishers to catch that fish. The costs faced by fishing vessel owners include not only the actual costs incurred (such as fuel), but also the *opportunity costs* of fishing – what was foregone by putting capital into fishing rather than into some other economic activity. Specifically, opportunity costs reflect what the vessel owners *could have* received as a return on investment if the capital invested in fishing vessels had instead been invested outside the fishery. By including opportunity costs, this analysis contrasts with typical financial analysis since 'normal' (reasonable or acceptable) profits are already included in the costs. We return to this point in the discussion below. (Note that the idea of opportunity costs also applies to labour in fisheries: the opportunity cost of labour is a measure of what fishers *could have* earned in their best alternative employment.)

With this approach, let us assume that each unit of fishing effort costs a certain amount c. This might be, for example, the *unit cost* for an average vessel to fish for a day. Then at an effort level E, the total cost of fishing is $TC = cE$.

Note that both TR and TC are functions of effort E. For TC, as effort increases, total cost increases linearly. For TR, as effort increases, the TR curve matches the yield curve, first rising and then falling. In fact, we can write $TR = pY(E)$ explicitly. If we now combine the TR and TC curves on a single graph (Fig. 8.12), this new graph reflects both economic and biological aspects of the fishery; in other words, it is a *bioeconomic* graph.

A key element in Fig. 8.12 is the difference between TR and TC. Since TC already includes *normal profits*, the fishers would be content to fish even if $TR = TC$. In such a situation, their earnings (TR) would provide them with normal profits (included in TC), in keeping with those profits and wages available elsewhere in the economy. If, on the other hand, $TR > TC$, then there is revenue available over and above *normal* profits.

The fishery economist refers to this quantity, $TR - TC$, as the *resource rent*, a measure of the economic returns (benefits) accruing to the resource owners, just as wages are paid to labour and dividends are paid to capital. The argument is usually that resource rent could be collected by government for the benefit of society as a whole. However, here we are assuming an open-access, unregulated fishery, in which there is no rent collection by society. This means that these rents remain with the fishers: not only do they receive *normal* profits (and wages), they keep these *above-normal* amounts as well.

Now let us return to the matter of how an effort level is chosen, or evolves, in a fishery system. We focus on an open-access fishery, which implies that there are no controls over

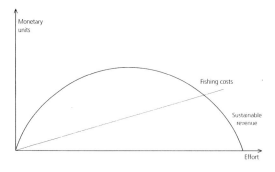

Fig. 8.12   A Gordon–Schaefer diagram is produced from a yield–effort diagram by transforming sustainable yield into sustainable revenue, and adding a cost line. This is now a *bioeconomic* graph that reflects both economic and biological aspects of the fishery. Note that the difference between revenue and cost is the *resource rent* (see text for details).

fishing effort; specifically, there are no institutional arrangements for exclusion (to keep out outsiders desiring to enter the fishery) or self-regulation (to limit their collective fishing activities to avoid excessive levels).

Suppose initially that a biological equilibrium has been reached in which there are few fishers and boats in the fishery. This could correspond to point A in Fig. 8.13. Note that at this point, there is plenty of revenue to be obtained at not much cost; the resource rents are positive ($TR - TC > 0$). Since these rents are not collected by society, they remain as above-normal profits for the fishers, whose profitability now looks very appealing relative to other economic activities. This attracts new entrants (and more effort by each existing fisher), just as a gold rush attracts people from other activities.

Thus the fishing effort expands upward from the level at point A. In this process of *open-access dynamics*, at what point will the expansion end? Consider point B. Here the above-normal profits are even greater, so that still more entrants are attracted to the fishing industry, and the effort level rises again. Thus, while B is again a biological equilibrium (in theory, being a point on the *TR* curve), it cannot be a *bioeconomic* equilibrium. What about point C? Here the total costs, *TC*, are higher, but the total revenue, *TR*, is the same as at point B. Thus rents are now smaller, but nonetheless positive. There are still above-normal profits, so the fishery is more attractive than other investments, and effort will continue to expand to above C.

At point E, however, the effort level is so great that costs are high and revenue low ($TR < TC$). Not only have above-normal profits dissipated, the fishers are now receiving *less than normal* profits. New entrants are no longer attracted to the fishing industry, and some existing *marginal* fishers will exit, since other economic activities now appear preferable.

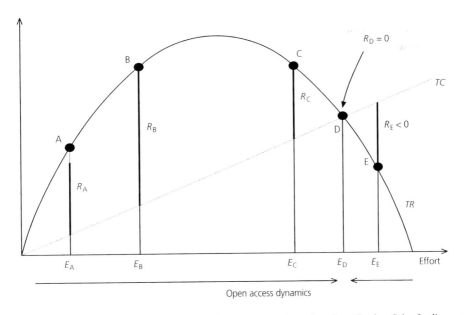

Fig. 8.13 The behaviour of a fishery system driven by *open-access dynamics* using a Gordon–Schaefer diagram. Fishers respond to the existence of above-normal profits (rents R), shown as bold vertical lines between the revenue (*TR*) and cost (*TC*) lines. The fishery therefore shifts over time from an initial equilibrium at point A, to point B, and then to point C, and eventually to the *open-access* or *bionomic* equilibrium D (above which rents are negative).

While all points on the *TR* curve represent biological equilibria (since each represents a sustainable yield), it is now clear that for the open-access fishery, only one point, D, is a natural *economic* equilibrium as well. Only at this point is there no incentive to expand or contract fishing effort (or for fishers to enter or exit the fishery), since profits are *normal* and equivalent to those elsewhere in the economy. (Recall that *TR* = *TC* at this point, so it appears that fishers are just breaking even, but in fact *TC* already incorporates a 'normal profit'.) This equilibrium is referred to as the *open-access* or *bionomic* equilibrium: the point to which *open-access dynamics* drives the fishery over time. At this point, fishers continue to profit-maximise (by assumption), but potential rents have been dissipated owing to participation by a relatively large number of firms; the long-run average unit revenue produced from a unit of fishing effort merely equals the average cost of that unit.

The bioeconomic equilibrium is the only point on the Gordon–Schaefer graph that represents both a biological and an economic equilibrium. At any effort level other than D, we can expect that regulation will be needed to maintain that effort level. For example, a combination of limited-entry licensing with effort or catch controls might be used to keep the fishery at point C, where in theory there will be rent generated. This could be used by society, or left to the fishers as above-normal profits. (The phenomenon of *rent seeking* can occur when certain groups of fishers lobby to exclude others, thereby capturing the above-normal profits for themselves.) In contrast, if higher employment is desired, subsidies (on fuel costs for example) could be used to maintain the fishery at point E. It must be noted, however, that actually maintaining a fishery at or near a fixed point is not something that has met with success in many real fishery systems.

This description of open-access dynamics has formed the basis of much theoretical discussion and a variety of policy measures in fisheries worldwide. However, it is important to reiterate the many assumptions on which these dynamics are based. First, there are assumptions of total certainty (no randomness) and complete knowledge (everyone knows all about biological and economic aspects). Second, it is assumed that fisher entry and fishing intensity are totally uncontrolled – this is the underlying idea of open access. Third, it is assumed that at any given level of fishing effort, a biological equilibrium is first reached, and the resulting dynamics of fishing effort are entirely determined by the flow of resource rents. The latter assumption can be questioned on various biological and economic grounds. For example, as noted above, biological equilibria will not exist in the face of uncertainty, and in reality, fishers respond to objectives other than just resource rent. These caveats do not remove the usefulness of discussing the idea of open-access dynamics, but they do call for caution in taking the results too literally.

8.8   Summary

In earlier chapters, emphasis was placed on examining the *structure* of fishery systems, and each of the constituent components. This chapter has been devoted to the *dynamics* of the components and the fishery system overall. The discussion began by reviewing the various time scales at which change occurs in fisheries. This was followed by a component-by-component examination of:

- dynamics in the natural system, ranging from single- and multi-species population dynamics, to the dynamics of ecosystems and the biophysical environment, along with brief discussions of interactions between uncertainty and dynamics, and the possibility of chaotic dynamics;
- dynamics in the human system, affecting fishery inputs (effort, labour, capital, fishing fleets and technology) as well as communities and the socioeconomic environment;
- dynamics in the fishery management system, at the strategic, tactical and operational levels, and the idea of information dynamics, i.e. incorporating the learning process in fishery analysis.

Following these component-specific discussions, approaches to addressing fishery system dynamics were considered. The discussion in the text illustrated such integrated ideas with a Gordon–Schaefer approach, examining the nature of *open-access dynamics*. On the other hand, the book's Appendix provides a related exploration of dynamic 'biosocioeconomic' fishery models, using simulation methods to examine both behavioural and optimisation forms of such models.

Chapter 9
Case Studies of Fishery Systems

This chapter presents two brief real-world examples to illustrate a component-based approach to the analysis of fishery systems. Of the cases discussed here, one represents a developed-nation situation (the groundfish fishery system on the Atlantic coast of Canada), while the other is a developing-nation case (the fishery system in Costa Rica's Gulf of Nicoya). In both cases, the principal aim of the case study is not to provide an exhaustive description of the specific fishery, but rather to indicate how an approach structured on the basis of the fishery components described in earlier chapters can provide an integrated look at a specific fishery system.

9.1   Case study 1: Canada's Atlantic groundfish fishery system

9.1.1 Introduction

The cod fishery off the Atlantic coast of Canada, and in particular that for Northern cod off the coast of Newfoundland, collapsed dramatically in the early 1990s. This has now emerged as perhaps the principal example of fishery collapse in present-day classrooms, taking over from the famous Peruvian anchovy fishery collapse of the early 1970s. So many questions arise. To what extent was the collapse a natural event, and to what extent was it due to human actions? If the former, was it the ocean environment, predators such as seals, or a lack of food for the cod? If the problem lay on the human side, was it one of poor science, ineffective management, political interference or bad luck? Specifically, if allowable catch levels were set too high, was this owing to the scientific process, to politics or to the management system? Whatever the catch levels set, was government enforcement able to keep the fishery within these limits? What role did fishing fleets play in the groundfish collapse? Which fleets were most responsible: foreign or domestic, small-scale or industrial, trawlers or hook-and-line fishers?

The collapse has been the subject of much research attention, with studies covering the spectrum from biological (Hutchings & Myers 1994, 1995) and economic (Hannesson 1996) to social (Rogers 1995) and attitudinal (Charles 1995b). It is remarkable to note, in the light of the magnitude of the collapse, that many groundfisheries have continued to operate, and indeed some fisheries that were closed have now reopened.

While some of the insights from recent studies of the Canadian cod fishery are discussed later in this book (Chapters 11–15), at this point we focus on examining the structure of this topical fishery system, including aspects that have changed since the early 1990s, and others that have not. The presentation here follows the order in which fishery system components have been presented to this point in the book, running from the fish, the ecosystem and the biophysical environment, to the fishers, the post-harvest sector and the socioeconomic environment, to the fishery management system, including policy and planning, management, development and research (for further details, see Charles 1997a).

9.1.2 The fish

Groundfish off Canada's Atlantic coast form 'a community of species typical of the latitude' (FRCC 1997a). These species include cod and other gadoids (e.g. haddock, pollock), halibut and various flatfishes (e.g. plaice, flounder) and redfish. Historically, the groundfish resources in Atlantic Canada have been among the most plentiful in the world, providing about two-thirds of the total Atlantic Canadian harvest by weight, and 40% or more of the landed value. However, this contribution has decreased greatly since the collapse or serious decline in many groundfish stocks in the early 1990s. In 1995, for example, groundfish harvests produced only 8% of the total Atlantic Canadian landed value (Department of Fisheries and Oceans 1996).

Cod was the species most attractive to fishers arriving in the region centuries ago, and remained, until recently, the mainstay of the groundfishery. However, with many cod fisheries having closed over the past 3–4 years, the species composition in the groundfish catch has changed dramatically. For example in 1995, more redfish was caught than cod, and the value of the Greenland halibut catch exceeded that of cod (Department of Fisheries and Oceans 1996). Many other species play more localised roles. For example, haddock and pollock are important along the Scotian Shelf of Nova Scotia, and redfish and white hake are important in the Gulf of St. Lawrence. Greenland halibut is a major species off Newfoundland and Labrador, where a lumpfish fishery also operates. Some species also have particular relevance to specific gear types; for example, redfish to otter trawlers, and halibut to hook-and-line fishers.

9.1.3 The ecosystem and the biophysical environment

From a physical and oceanographic perspective, three major ecosystems have been identified off Canada's Atlantic coast (FRCC 1997a).

(1)   The northern shelf areas off Labrador and the southern and eastern sides of the island of Newfoundland (notably the Grand Banks) are affected principally by the Labrador current bringing cold water form the north.
(2)   The Gulf of St. Lawrence, a semi-enclosed body of water, is affected strongly by the seasonal flow from the St. Lawrence River, but also receives waters (usually as a deep layer) flowing inward into the Gulf.
(3)   The Scotian Shelf, the component of the continental shelf off Nova Scotia, is affected near to shore by waters moving south from the Gulf of St. Lawrence, and further from shore by a mixing of water from the (warm) Gulf Stream and the (cold) Labrador current.

At a less aggregated scale, and from a more ecological perspective, many smaller ecosystems can be identified throughout the region. One notable example is the Bay of Fundy, located at the southern end of the Scotian Shelf but often viewed as a separate ecosystem (certainly by those living within its boundaries). The distinctiveness of the bay is partly due to its unique tidal driving force (with the highest tides in the world) and partly due to the particular mix of species in the bay (for example, it is a congregating point for many species of whales). The uniqueness of the bay has been reinforced by recent genetic analyses indicating that some species (such as cod) that have been managed in aggregate as 'Scotian Shelf' stocks in fact have discrete within-bay sub-stock components. This range of information is clearly crucial in deciding upon the boundaries of a given fishery system, and has led to recent efforts to manage fisheries in the bay through 'ecologically based management' and a new fisher-run organisation, the Bay of Fundy Fisheries Council.

9.1.4 The fishers

Fishers in the commercial groundfishery are highly heterogeneous, and use a large variety of vessel sizes (ranging from under 10 m to well over 30 m in length) and gear types. The latter fall into two principal groupings (see Chapter 3 for descriptions of these gear types):

- *fixed gear*, including gill nets, long-lines, hand-lines, traps and weirs;
- *mobile gear*, particularly otter trawls, as well as Danish and Scottish seines.

In recent years, *gear conflicts* (Chapter 13) between fixed and mobile gear sectors have emerged strongly in the groundfishery, but the more traditional dichotomy is between 'inshore' and 'offshore' fisheries. As their names suggest, these fisheries were originally defined on the basis of how far from shore fishing took place, with inshore fishers operating relatively close to their home port, and offshore vessels ranging throughout the Atlantic region. There are many other differences as well. The inshore fishery is tied to coastal communities, while the offshore has been the domain of large companies (especially two corporations, National Sea Products and Fishery Products International). The inshore is relatively labour-intensive (with thousands of fishing vessels), while the offshore is much more capital-intensive.

Bureaucratically, size of the vessel is the determining factor in the inshore–offshore split, with a length of 30 m (100 feet) as the nominal dividing point, but in reality vessels of 19–30 m (65–100 feet) in length are not inshore boats but are called *near-shore* vessels, and those of 14–19 m (45–65 feet) in length, in a fleet that has developed largely since 1977, are often referred to as *midshore* vessels. Both of these have the capability to fish in offshore areas, and seem qualitatively different from the 'true' inshore fleet, which many consider to be confined to vessels under 14 m (45 feet) in length.

Fishers have organised themselves into a variety of gear-based and community-based organisations, as well as unions and cooperatives. In Newfoundland, the vast majority of fishery workers (inshore and offshore, harvester and plant worker) belong to a single union, the Newfoundland Fish, Food and Allied Workers Union. However, in the other Atlantic provinces, there is a proliferation of small organisations, sometimes representing only the fishers of a particular gear sector in a particular community, and many fishers remain completely outside any organisation.

It is also important to note that native (indigenous) fishers (the MiK'maq and Maliseet first nations) have established traditional rights to fish in the region, and will probably play an increasing role in the groundfishery. Indeed, a Supreme Court decision in 1999 affirmed the treaty rights of these first nations to fish not only for food, but also commercially.

In addition, recreational fishers are engaged in catching small quantities of groundfish (although to a much lesser extent than salmon). Finally, in Newfoundland in particular, there is a tradition of a so-called food fishery in which coastal residents and others catch fish for their own domestic consumption. This might be considered a hybrid of subsistence fishing (producing domestic food) and recreational fishing (since it is as much for enjoyment as for food itself).

9.1.5 *The post-harvest sector and consumers*

Historically, many inshore fishers rely on processors as an outlet for their catches. In addition, the entire offshore component of the groundfish fishery is dominated by processors. Processing varies with the species and the location where caught. For example, much of the lobster caught in the Gulf of St. Lawrence is canned, while most of that caught on the Scotian Shelf is exported live to markets in the United States and Europe (Pringle & Burke 1993). In either case, the process involves fishers selling to fish buyers (middlemen) or directly to processors.

Of groundfish, much of that caught by (or bought by) the large offshore companies is either frozen in blocks for export or processed into boxed products (e.g. fish sticks) for retail markets. Other groundfish may undergo minimal processing into a headed and gutted form, for sale to a fresh fish market. In Nova Scotia, much of the latter is processed by fishers themselves and then shipped to the major market in Boston, USA, for sale by auction. (See Doeringer & Terkla 1995 for an analysis of industry structure and trade matters in the Atlantic Canadian, as well as the New England, fisheries.)

9.1.6 *Fishing communities and the socioeconomic environment*

While in Canada as a whole, fish production contributes only in the vicinity of 1% of national employment and gross national product (GNP), the economic impact in Atlantic Canada is more substantial. Indeed, the fishery takes on a major role in exports from much of the region. However, the regional economic role is just one of many. The fishery lies at the heart of social, cultural and historical aspects of the region, so that the overall political importance of fishing in Atlantic Canada tends to far exceed any economic indicators. There tends to be a remarkably great intrinsic interest in news stories about the Atlantic fisheries, perhaps reflecting a sense of the history and tradition inherent within them. There is also a relatively large number of federal politicians based in fishery-dependent constituencies.

However, all this may not imply a full understanding at the national level. Newell & Ommer (1999: p. 4) remark on 'how little public awareness exists about the long-standing importance of fisheries in Canada's social and economic development ... The social, cultural and economic significance of Canadian small-scale, sustainable fisheries consequently has been grossly underestimated.' The fact is that the groundfishery in Atlantic Canada has a history of over 500 years, and in the case of Newfoundland in particular, which was settled originally owing to the proximity of fishery resources, there is a very strong cultural attachment to the fishery.

Looking at the reality of Atlantic Canada, fishers live in over 1000 fishing communities along an often isolated coastline (Fig. 9.1). These communities are spread across the government's four fisheries administrative regions, as shown in Table 9.1 from Moore *et al.* (1993).

Not all of these communities are based on groundfish fisheries, but certainly groundfish, along with lobster, are the resources that have traditionally formed the backbone of the fishery. It is notable that, as in many other coastal locations, many other economic activities are dependent, at least in part, on the fishery. This ranges from closely connected activities such as boat-building and aquaculture (which can benefit from scientific expertise in the fishery) to a sector such as tourism (for which the attraction of travelling through small-boat coastal fishing communities is of major importance). Thus, the groundfish and lobster fisheries clearly represent the 'engines' of many communities, and of parts of the regional economy more broadly.

Fig. 9.1   Fishing communities are scattered along the coastline across Atlantic Canada, and are important not only to the fishery but also to tourism. Shown here is one of the most famous of these communities, Peggy's Cove in Nova Scotia. (Photo by Kay Lannon.)

Table 9.1   Fishing communities in Atlantic Canada.

Fishing region	Number of communities	Percentage of communities
Newfoundland	405	38
Scotia Fundy	279	26
Quebec	110	10
Gulf	280	26
Total	1074	100

9.1.7 *Fishery policy and planning*

Policy objectives

There has been very little formal debate and discussion concerning the objectives to be pursued in Canada's Atlantic fisheries. In one of the many major studies carried out on the fishery, Kirby (1983) suggested a set of three objectives, in order of priority:

(1)   ongoing economic viability, implying 'an ability to survive downturns with only a normal business failure rate and without government assistance';
(2)   maximum employment, 'subject to the constraint that those employed receive a reasonable income';
(3)   Canadian harvesting and processing of Canadian fish, 'wherever this is consistent with objectives 1 and 2'.

On the one hand, these probably still reflect *de facto* goals of the government, but on the other hand, it is clear that they have not been achieved. In particular, the groundfishery collapse led to massive government assistance and greatly reduced fishery employment, which is contrary to the first two priorities. Since the collapse, government has placed emphasis on the goal of conservation, or, as one minister stated, 'conservation, conservation and conservation'. This is meant to connote the idea that the overriding objective is to ensure sustainability of the resource base. However, the manner by which this goal is meshed with socioeconomic pursuits remains somewhat unclear.

Management institutions

Canada's national government has responsibility for the management of ocean fisheries, and in particular for the conservation of ocean resources, whereas the provinces have responsibility for any land-based fishery activity, including fish processing, and for aquaculture. At the federal level, fishery systems are governed by the nation's Fisheries Act. This comprehensive and powerful legislation has been in place since 1868. The Act gives government strong powers to 'make all and every such regulation or regulations as may be found necessary and expedient for the better management and regulation of the Fisheries' (United Canada's Act 1859; see Gough 1993), as well as to control pollution of the marine environment.

Since 1996, the federal government has made efforts to streamline and 'modernise' the Act. This reflects a trend, as in many fishery jurisdictions, in the government's approach to management away from a 'top-down' model of central governmental control. Initially, the shift was to a consultative model, in which government discussed management measures with the industry prior to implementation, but did not delegate decision-making power, and as a result, did not see fishers accepting government-imposed regulations. More recently, co-management approaches have emerged (Chapter 13), driven in part by a downsizing of the Department of Fisheries and Oceans (DFO), and in part on a recognition of the need for the involvement of those being regulated, the fishing industry, working together with government jointly to develop and enforce regulations. New legislative efforts by the Canadian government are meant to provide a vehicle for co-management with commercial and native fishers as well as others, and to reinforce a process of fishery enforcement by administrative sanction (using a more rapid process of administrative hearings for non-criminal offences, with penalties lower than in court, but conviction more likely).

Also of great potential impact on the fishery system is a new Oceans Act, providing government with the capability to manage multiple and conflicting ocean uses. This allows for the declaration of marine protected areas (MPAs), whether as a fishery conservation tool or to protect endangered species or habitats, and for integrated coastal zone management, in order to improve manage interactions between commercial, sport and native fisheries, as well as aquaculture and non-fishery activity along the coast.

Administratively, federal management takes place through the Department of Fisheries and Oceans, which has headquarters staff in Ottawa, and three administrative regions for the Atlantic coast:

- *Newfoundland*, including most fisheries off Newfoundland and Labrador;
- *Laurentian*, i.e. fisheries off the coast of Quebec, and other related fisheries;
- *Maritimes*, i.e. fisheries off Nova Scotia, New Brunswick and Prince Edward Island; this administrative region is subdivided into two areas, Scotia–Fundy (the Scotian Shelf and the Bay of Fundy) and Gulf (the southern part of the Gulf of St. Lawrence).

Note that these administrative regions do not fully match provincial boundaries, and nor do they match the definition of groundfish stock areas, which were set out on the basis of zones established by the international body for fishery management in the region, the Northwest Atlantic Fisheries Organisation (NAFO), or more precisely, its predecessor, ICNAF. Furthermore, and more problematic, neither the administrative regions nor the NAFO zones are compatible with the broad ecological zones discussed above.

For example, it would be natural for fisheries in the Gulf of St. Lawrence to be managed together as an ecological system. There are certainly similarities in oceanographic and fish stock conditions between the southern and northern areas of the Gulf, NAFO zones 4T and 4RS3Pn, respectively. However, responsibility for fish stocks in that ecosystem is divided among DFO regions. Despite efforts at coordination, this has led to complaints from the industry that fishers coming from different regions to harvest fish in the middle of the Gulf may in fact be subject to different rules, depending on the administrative region in which they happen to live.

9.1.8 Fishery management

Management of Atlantic Canada's groundfishery involves a combination of measures, including the use of total allowable catches (TACs) to limit harvests, limited entry licensing (to limit participation in the fishery), gear restrictions, such as mesh and hook sizes (to enhance the selectivity of fishing), and closed areas and closed seasons (to protect spawning and/o. nursery grounds). Of these, quota management, i.e. the setting and subdividing of the TAC, has been the cornerstone of groundfish regulation since the 1970s, when Canada was one of the first nations to introduce the approach.

It would be difficult to claim that Canada's groundfish fishery management has been successful, given the two major collapses of the fishery in the second half of the twentieth century.

The first occurred in the early and mid-1970s, driven largely by heavy fishing pressure by foreign vessels, which continued up to 1977. With more conservative management and some luck (in the form of strong year classes recruiting to the fishery), groundfish stocks were rebuilt in the late 1970s and early 1980s. However, Canadian stock assessment and management measures proved incapable of monitoring and controlling fishing pressure. Indeed, an analysis of the Scotia–Fundy groundfish management region (Halliday *et al.* 1992) found that of all the TACs set over the 1977–1989 period, two-thirds exceeded the levels set in the government's own policy.

This led up to the second groundfish collapse, in the late 1980s and early 1990s, which arose from very high levels of domestic (and in some cases foreign) fishing mortality; levels too high to allow the fishery to withstand the onset of less favourable environmental conditions. The most dramatic problem arose with the Northern cod stock (in NAFO area 2J3KL), which had been supporting a fishery that was among the world's largest, and of great importance to the Canadian province of Newfoundland and Labrador. In the 1980s, many inshore fishers expressed concern about the decline they perceived in this stock, but it was not until 1992, with the failure of the corporate offshore trawler fishery to find fish, that the federal government closed the fishery. Strong shock waves resulted from the closure, as those in Newfoundland came to realise that the mainstay of their fishery was gone. A 2-year moratorium was declared on harvesting, which was extended until 1999 as the stock failed to recover. From 1993 on, fisheries were closed for most other cod stocks and a considerable number of other stocks throughout most of Atlantic Canada. By the end of the 1990s, many of these fisheries had been reopened, albeit at relatively low levels of harvesting. In particular, a small commercial fishery for Northern cod reopened in inshore areas in 1999, although the negative impacts of this fishery on stock recovery are not yet clear.

In the wake of the groundfish collapses and closures, beginning in 1993, the Minister of Fisheries and Oceans revised the management process for the groundfishery. One major change was in the process of conservation decision making. Prior to 1993, the approach to determining conservation measures involved three stages: (1) government scientists, through a structure known as the Canadian Atlantic Fisheries Scientific Advisory Committee (CAF-SAC), published their advice on groundfish conservation measures, notably a set of recommended TACs; (2) a forum of selected members of the fishing industry, the Atlantic Groundfish Advisory Committee (AGAC), debated this advice, as well as allocation issues, providing a second round of advice to the government; (3) the Minister of Fisheries and Oceans made a final decision.

In 1993, both CAFSAC and AGAC were dissolved, and the Fisheries Resource Conservation Council (FRCC) was established (composed of individuals from the fishing industry and universities appointed by the Minister of Fisheries and Oceans, together with provincial delegates and federal fisheries staff) with the explicit and sole mandate to advise the Minister on conservation measures. Now the process is as follows: (1) scientists provide information and analysis (rather than advice) about stock status; (2) this scientific information is considered by the FRCC, together with input from public consultative hearings, at which any stakeholder (fisher, fishery organisation or the general public) can take part; (3) the results of the FRCC's deliberations are provided in public recommendations to the Minister, who then makes decisions on TACs and other conservation measures.

Once the Minister of Fisheries and Oceans announces catch limits (TACs) and other conservation measures for the groundfishery, the TAC is divided into allocations for each sector of the fishery, defined in terms of location, gear type and vessel size category, e.g. vessels using mobile gear, of 14–19 m (45–65 feet) in length, fishing in NAFO area 4X. A schematic diagram of the quota management system is shown in Fig. 9.2.

Previously, the DFO developed fishing plans for each of these sectors, after nominal consultations with the industry. Now, each sector must develop its own Conservation Harvesting Plan (CHP): a detailed set of agreed management measures. Recommendations of the FRCC outline broad guidelines for the CHP, and the DFO may provide more precise requirements, but the onus is placed on each sector of the fishery to determine if and how it can fish within allowable limits. With this responsibility comes the flexibility to develop management measures that best suit the fishers of that sector. On the other hand, as a safeguard, DFO fishery managers must agree that the CHP will meet official conservation requirements before a CHP is approved and fishing is allowed. In addition to noting the sector's total available catch quota and allowable fishing gear (e.g. mesh or hook sizes), the CHP also incorporates new arrangements for at-sea and *dockside catch monitoring: small fish protocols* to close the fishery temporarily if undersized fish are being caught, and so on.

Groundfish fishers in parts of Atlantic Canada have developed their own version of local level management, through so-called *community quotas*. Once an overall TAC has been subdivided into sector allocations (by gear and vessel size), some sectors have chosen to allocate their global quota on a community basis, dividing up the coastline into self-identified sections (often on a county-by-county basis) so that the available harvest can be managed locally. This approach, pioneered in the small community of Sambro (near Halifax, Nova Scotia), has since spread throughout the small-boat fixed-gear fishery in the Scotian Shelf and Bay of Fundy areas (see, e.g. Loucks *et al.* 1998).

9.1.9 Fishery development

Priorities for 'development' in the Atlantic Canadian groundfish fishery have varied considerably over time, although one constant has been the goal of improving the well-being of small-boat inshore fishers. This goal has been pursued through efforts ranging from (a) helping fishers to organise into associations, unions and the like, to (b) improving fisher safety at sea, to (c) increasing the catching power of vessels and gear. For example, Fisheries Loan Boards have provided financing for investment by fishers, and technologists within federal and provincial governments have worked on and introduced new fishing and electronic gear. In recent times, more attention has been paid to improvements in gear to make it more conservationist, for example, through efforts to reduce the impact of bottom trawlers.

Jurisdiction for development efforts is split between the federal and provincial levels. In particular, the latter deals with labour matters, notably the organising of fishers, as well as the post-harvest components. Thus, many provincial governments encouraged extensive growth of processing plants in recent decades, although with the downturn in the groundfishery, much of that expansion has been reversed.

Today, there remains a legacy of past development initiatives that has both positive and negative elements. For example, on the one hand, considerable excess capacity is now present

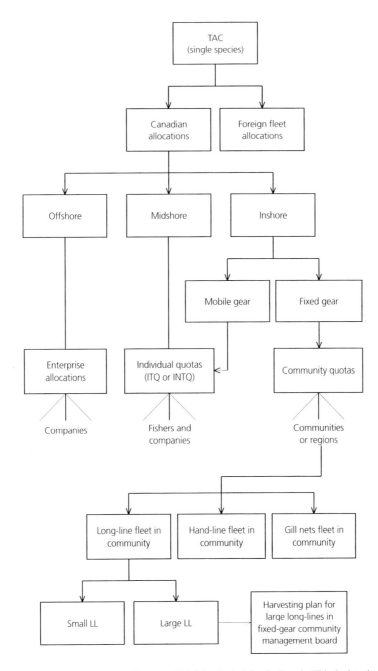

Fig. 9.2   The quota management system for groundfish fisheries in Atlantic Canada. This depicts the subdivision of a single-species total allowable catch (TAC), first into allocations to each participating nation, then subdividing the domestic component into large-scale sector-based divisions, followed by further subdivision according to vessel size and gear type, and in some cases, into community quotas or individual quotas. ITQ, individual transferable quota; INTQ, individual non-transferable quota.

in both the harvesting and processing sectors, while on the other hand, there have been improvements in fishing gear that may be important to conservation in the future.

9.1.10 *Fishery research*

A large part of the research on Atlantic Canadian groundfish fisheries is carried out within the Science Branch of the Department of Fisheries and Oceans. Scientists are considered to be part of the regional administrative structure within which they are located (e.g. the Newfoundland or Maritimes regions), but also part of a national DFO scientific bureaucracy with headquarters in Ottawa. Research within the Science Branch takes place primarily in five research centres:

- The Maurice Lamontagne Institute in Mont Joli, Quebec;
- The Northwest Atlantic Fisheries Centre in St. John's, Newfoundland;
- The Bedford Institute of Oceanography in Dartmouth, Nova Scotia;
- The Gulf Fisheries Centre in Moncton, New Brunswick;
- The St. Andrew's Biological Station in St. Andrew's, New Brunswick.

Research activities range from oceanographic and environmental quality studies, to fish biology, to stock assessment. These activities vary from highly *applied* research (as in the case of stock assessment) to relatively *pure* research (as with some oceanographic studies). The structure of research might best be described as being first *by geographical region* (ecosystem), second *by function* (e.g. habitat science, stock assessment, ocean science, fishery development), and third *by species* (e.g. with individual stock assessment scientists assigned to specific species).

In addition to governmental activities, extensive research is carried out at universities such as the Memorial University (St. John's), Dalhousie University (Halifax), the University of Quebec (Rimouski) and the University of New Brunswick (Fredericton and Saint John). Private sector consultants are also involved in some applied research projects. The most notable development, however, has been the appearance of *cooperative fisher–scientist research* initiatives. A major example is the initiation by Nova Scotian inshore fishers and government scientists of the Fishermen and Scientists Research Society (King *et al.* 1994) to promote cooperative scientific research efforts, as well as to educate fishers on stock assessment and other scientific activities. Related to this are the so-called *sentinel fisheries*, which have been established throughout much of Atlantic Canada to monitor stock status through low-level fishing activity designed by government scientists and carried out by fishers. These various endeavours have had a remarkable impact in forging a new-found positive interaction between fishers and scientists.

9.2   Case study 2: the fishery system in Costa Rica's Gulf of Nicoya*

9.2.1 *Introduction*

This section presents a brief case study of a fishery system in the Gulf of Nicoya, on the Pacific side of Costa Rica. The region is characterised by a tropical mangrove environment, a common-property artisanal fishery (faced with serious issues of stock decline and increasing user conflicts) and an established although limited form of governmental fishery regula-

---

*Material in this section is drawn from research, notably that in Charles and Herrera (1994), carried out jointly with Professor Angel Herrera of the National University (Heredia, Costa Rica).

tion. The Gulf is of major importance both as a mangrove environment (seen as of sufficient importance to receive protection through national legislation) and as the country's principal artisanal fishing area.

In the Costa Rican context, as in many developing nations, pressure on the fishery as *employer of last resort* is enormous. A systems approach therefore reinforces the importance of combining fishery management, regulation and development with economic diversification through formulation of non-fishery employment alternatives, to help cope with changes in the fishery and the marine environment. This theme is explored in this section, following a brief overview of elements within the natural, human and management sub-systems.

9.2.2 The natural system

The Gulf of Nicoya, located in a dry tropical climate zone, extends 80 km inland from Costa Rica's Pacific coast. This estuarine region has one of the greatest mangrove abundances, combined with high levels of pollution and among the highest levels of exploitation by small-scale artisanal fishers. For physical, biological and management reasons, its area of 1500 km² is divided into an 'outer zone' and an 'inner zone'. The discussion here focuses on the latter (northern) portion, where relatively calm and shallow waters combined with suitable soil conditions provide a favourable environment for mangroves. Indeed, over 42% (112 km) of the coast in this region, and over 15 000 ha of coastal area, is covered with mangroves (Solórzano *et al.* 1991). In the Gulf, the species of greatest commercial interest is the white shrimp (three species of the genus *Penaeus*), but on a smaller scale, there is also a fin-fish fishery for such species as corvina (croaker), red snapper and snook, and some specialised fisheries such as that for the bivalve 'pianguas' (genus *Anadara*), which grows on the roots of certain mangroves (genus *Rhizophora*).

9.2.3 The human system

The Gulf of Nicoya fisheries are small-scale and artisanal in nature. The total number of fishers is approximately 5000–6000, comprised of fishers from many small and large communities, with a typical fisher earning revenue from harvesting shrimp, fin fish and possibly other marine resources. Using data from one community in the region (Puerto Thiel), and making a set of assumptions about fishing activities, Charles & Herrera (1994) estimated monthly net income levels at US$88–176, a figure above the 'extreme poverty' level of US$88, but below the official poverty level of US$261. While clearly low, this is seen by many fishers as sufficient to provide a minimal standard of living for their families. The fishing behaviour of many traditional fishers in the Gulf is typically non-intensive, with operational decisions made on a satisficing basis, in which fishing activity is oriented towards catching 'sufficient'. This behaviour has significant implications for the pursuit of sustainability, tending as it does to level out the fisher income distribution, thereby easing pressure on the resource and reducing conflict within the community.

The shrimp fishery now dominates economically, but this situation dates back only to the early 1980s, when its introduction completely transformed fishing in the zone (Charles & Herrera 1994). Historically, the fishery was a traditional artisanal open-access one, using long-lines and gill nets to focus on fin fishes. It experienced relatively low profitability and required little in the way of government management efforts. Then, driven by the higher returns produced from shrimp relative to fin fish, migration increased from interior zones into traditional coastal fishing regions. The number of licensed fishers grew from approximately 2000 in the early 1980s to reach almost 3000 today, while approximately 3000 more fish illegally without licences. The capital stock of boats fishing in the Gulf has also increased dramatically. The number of small *botes* increased steadily since the mid-1970s, while the number of larger *pangas* grew rapidly from 1984 to 1988, reflecting a process of technological change in the fishery (Fig. 9.3).

Artisanal fishing in the Gulf of Nicoya is highly variable. This appears particularly as a cyclical variability within each year. Shrimp harvesting occurs over most months of the year, except during closed seasons; it is particularly successful between December and February. During the closed seasons for shrimp, fish catches tend to reach their peaks, owing to a change in the gear used to long-lines and larger mesh sizes (although fish by-catches have also been significant during the shrimp season). Shrimp gill nets are the principal fishing technology, except in the closed season for shrimp, when long-lines and large-mesh gill nets transform the fishery into a more selective activity targeted exclusively on fishes.

Looking beyond the fishers, there is relatively little processing of harvests in the Gulf. In particular, shrimp is sold onto the world market directly, while fish is often sold locally to tourism facilities, or transported to the capital city, San José, for sale. Marketing facilities and methods are relatively advanced compared with elsewhere in Costa Rica. For example, the fishing cooperative in Puerto Thiel has developed strong marketing channels for their harvests, notably through export avenues for the lucrative shrimp catches.

More than 20 small communities border the Gulf of Nicoya (see Gonzalez *et al.* (1993) for an extensive analysis of these communities). The residents of many of these depend largely on the fishery for their livelihood, although there is also some work in local agriculture, and seasonal migrations to work in coffee and sugar cane harvesting in interior zones of the coun-

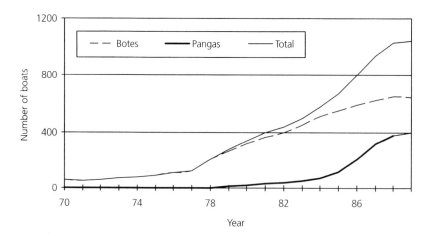

Fig. 9.3   Changes over time for two types of artisanal fishing vessel in the Gulf of Nicoya, Costa Rica. The fleet size of the less advanced type of vessel (*botes*) expanded slowly over many years, while the fleet of the more advanced *pangas* rose rapidly in the late 1980s. Both fleets levelled off in size towards the end of the 1980s, as new management measures were introduced.

try. Faced with increased concern over the state of Gulf of Nicoya fish stocks, there have also been some initiatives to diversify the base of local economic activities.

The socioeconomic environment relevant to the Gulf of Nicoya fishery includes legislative aspects (such as national legislation governing coastal use), direct economic development measures (such as recent export promotion incentives), indirect impacts of non-fishery activities (such as the impacts of agricultural run-offs and pollutants entering the Gulf), and impacts of macroeconomic policies at the national level (such as structural adjustment measures) that affect both the health of the fishing sector itself and the extent to which the traditional 'social guarantees' (the country's social security system) are able to support fishing communities.

9.2.4 The fishery management system

Traditionally, there has been rather little government regulation in the common property regime of the Gulf of Nicoya fishery. In particular, the rapidity of change in the fishery system over the past few decades, described above, was not accompanied by adequate regulation or planning of marine resource use; nor were there substantive social controls over fishing in the Gulf. By 1985, a clear decline in shrimp harvests was noted. This led to a number of management measures, particularly the introduction of an annual 3-month closed season for shrimp (April to June). The closed season is still the major conservation management tool in the inner Gulf region, although recent research on shrimp spawning has been incorporated into management through the introduction of two separate closed seasons each year.

Limited entry licensing measures, gill net mesh size limitations and other regulations (Herrera & Valerín 1992) were also introduced around the same time as the closures. In the late 1980s, capacity expansion began to slow and was largely halted in 1989–1990 by a ban on new boats in the area. This reflects a major change in the fishery, from what appears to have been shrimp 'resource mining' pre-1988 to the present lower level of combined shrimp/fish harvesting. However, while shrimp provides the dominant source of fishery income overall in the Gulf of Nicoya, fin-fish catches have increased to represent nearly 50% of total income. This may in part reflect a general increase in fish prices, but it also could be a signal of problems with the shrimp stock, in terms of increasing difficulty in locating declining stocks.

Development efforts in the Gulf of Nicoya fishery system include initiatives to establish fisher cooperatives and associations, the construction of infrastructure such as wharves and ice-making plants (typically with aid from foreign donors), employment diversification projects, educational and training initiatives, and efforts to improve the environmental quality in the Gulf in order to increase the productivity of the resources. Costa Rica has received a wide range of fishery development aid from agencies in the United States, Canada, Europe, Japan, Taiwan and elsewhere.

In contrast to the Canadian case study above (and the situation in most developed nations), fishery research in Costa Rica is largely carried out by universities. There is very little capability for such work within the government structure, and few funds to hire private sector consultants (except in conjunction with foreign-funded projects). Thus, research takes place principally at the National University and the University of Costa Rica, both of which have coastal research stations to support such work. (See Charles *et al.* (1996) for a review of the considerable variety of research published relating to Costa Rican fisheries.)

Part II

Towards Sustainable Fishery Systems

Chapter 10
Sustainability in Fishery Systems

This chapter explores the idea of sustainability in fishery systems and how we can assess the sustainability of a fishery. The discussion here does not deal with specific management and policy measures that can serve to promote sustainable fisheries (that being the major underlying theme of all subsequent chapters), but instead focuses on the fundamental matters of what we mean by, and how we can 'measure', fishery *sustainability*.

10.1 The evolving nature of sustainability

Perhaps the most fundamental aspect underlying most fisheries theory and practice is that of determining the *sustainable yield*, i.e. a harvest that can be taken today without being detrimental to the resource available in future years. (This concept was discussed in detail in Chapter 8.) In many types of fisheries worldwide, the focus has been on determining a sustainable yield in the form of a total allowable catch (TAC). In this pursuit, one can seek the 'maximum sustainable yield' (MSY), i.e. the most fish that can be caught each year, year after year, or a lower catch level that balances the multiplicity of objectives in the fishery system (Charles 1992b, c). In any case, fishery science has evolved as essentially a *science of sustainability*, with considerable emphasis on the determination of sustainable yields (Schaefer 1954; Beverton & Holt 1957; Ricker 1975; Gulland 1977).

It has become apparent, particularly in recent times, that a focus on sustainable *yield* has a major shortcoming in its intrinsic emphasis on the physical *output* from the fishery. While the balancing of present and future catches is important, there is more to a healthy future than simply a large fish stock. It is also important to pay attention to sustaining the *processes* underlying the fishery. This realisation has led to the introduction into fishery discussions of the need to pursue *sustainable fisheries* (e.g. National Research Council 1999). This typically implies attention to the health of the aquatic ecosystem, and to the integrity of ecological interactions (through the ecosystem approach; see Chapter 12).

What is sometimes missing from these discussions is attention to the state of the human system. This is where the concept of *sustainable development* becomes important. This term was popularised by the Brundtland Report (World Commission on Environment and Development 1987), and subsequently the United Nations Conference on Environment and Development (held in Rio de Janeiro in 1992). Since then, the pursuit of sustainable development has become a *de facto* requirement of public policy; long recognised as fundamental to human societies around the world (Peet & Peet 1990), 'sustainability' has taken on a major role in public debate internationally.

What is meant by *sustainable development*? The international research and policy literature has exploded with a variety of definitions and interpretations. Certainly the best known is that of the World Commission on Environment and Development itself: 'development that meets the needs of the present without compromising the ability of future generations to meet their own needs' (1987: p. 43). However, the range of views is so wide (see, for example, those in Pezzy (1989) and the Inter-American Institute for Cooperation on Agriculture (1991)) that there is no single agreed definition. Despite this, there is wide recognition of the need to view sustainability broadly, in an 'integrated' manner that includes ecological, economic, social and institutional aspects of the full *system* – in this case, the fishery system. (Thus, sustainability could not be seen, for example, as merely 'environmentally sensitive economic growth'.) In this light, sustainable development has been defined as:

'... positive socioeconomic change that does not undermine the ecological and social systems upon which communities and society are dependent.'

Rees (1988)

'... the persistence, over an apparently indefinite future, of certain necessary and desired characteristics of the sociopolitical system and its natural environment. Both environmental/ecological and social/political sustainability are required for a sustainable society.'

Robinson *et al.* (1990)

The present discussion adopts this integrated view of sustainability, focusing not on sustainable output, but rather on the sustainability of the fishery system as a whole. Adopting such a perspective is helpful in providing us with a better view of the problem of sustainability, but by no means does it lead us easily to solutions. Indeed, a vast literature has emerged discussing and debating various aspects of sustainability in local, regional, national and global systems, not just in fisheries, but in all aspects of natural and human systems. The pages of the excellent journal *Ecological Economics* are full of such discussions and debates, and a multitude

---

Old perspective:
Sustainable yield
(Focus on physical output – sustaining fish harvests)

New perspective:
Sustainability
(Focus on the system – healthy ecosystems and human systems)

Multiple objectives
(A balance of resource conservation and human concerns)

of books have been written on these topics (e.g. Jansson *et al.* 1994; van den Bergh 1996; Costanza *et al.* 1997b).

Coverage of sustainability requires attention to both qualitative approaches, notably the many management approaches and policy directions that might serve to promote sustainability, and to quantitative approaches, particularly the formidable challenge of assessing and predicting sustainability. The management and policy aspects will receive detailed coverage in Chapters 11–15, while this chapter focuses on matters of measurement, assessment and prediction. These themes have been examined in various fora (although rarely in terms of fishery-specific applications). Examples of literature on the topic, from a variety of perspectives, include Kuik & Verbruggen (1991), Hammond *et al.* (1995), Munasinghe & Shearer (1995) and Atkinson *et al.* (1997).

Despite this abundance of discussion on how sustainability can be measured, very little has been applied to fishery systems, or even related coastal and watershed systems (Charles 1997b). This chapter focuses on one such effort: the development of a systematic 'sustainability assessment' approach for fishery systems. Within this integrated approach, sustainability involves direct resource conservation, but also recognises that, since fishery exploitation levels can vary over a wide range and still technically achieve 'conservation' (biologically sustainable yields), varying impacts on the broader ecosystem and on the achievement of human goals must be taken into account in deciding upon a harvest strategy. Thus, a multifaceted view is required.

It should be noted that discussions of sustainability are increasingly being linked with the critical concept of *resilience*. As noted in Chapter 1, a *resilient* system (Holling 1973) is one that can absorb and 'bounce back' from perturbations (shocks) caused by natural or human actions. The idea of resilience, while first formulated with ecosystems in mind, is just as relevant elsewhere in the fishery. For example, resilience is important to components of the human system, such as fishing communities, where it implies a capability to persist in a 'healthy' state whatever the state of the natural system and the socioeconomic environment. The management system must be designed with resilience in mind as well: if something unexpected happens (as is bound to be the case from time to time), can the management system still perform adequately? Such questions will be examined in detail in subsequent chapters. At this point, we will keep in mind the extent to which indicators of sustainability also provide information about resilience in the fishery system.

10.2 A framework for sustainability assessment in fishery systems

The idea of sustainability assessment is to evaluate, both qualitatively and quantitatively, the nature and extent of sustainability in a given resource system. This might focus on a present-day system or a proposed future activity:

- evaluating a current situation (for example, the sustainability of an existing fishery system, or a coastal or watershed system) as a form of 'status report', perhaps, for example, involving the assessment of both ecological and human carrying capacity;
- predicting *a priori* the consequences of a proposed activity, such as a new coastal fishery or a proposed fishery management approach, in terms of enhancing or reducing sustain-

ability; this builds on analogous approaches in environmental impact assessment, evaluating the 'impact' of proposed human activities.

The sustainability assessment approach involves four steps (e.g. Charles 1995c, 1997b, c).

(1)   Deciding on a set of relevant sustainability components for the fishery system, which together reflect the overall idea of 'fishery sustainability'.
(2)   Developing a concrete set of criteria that must be evaluated in assessing each component of sustainability (a *sustainability checklist*).
(3)   Determining a corresponding set of quantifiable *sustainability indicators*, reflecting the measurable status of each of the criteria, and allowing comparisons between criteria. For example, in a coastal fishery system, suitable measures of both ecological and human carrying capacities might be determined, based on the natural and socioeconomic environment, respectively.
(4)   Formulating suitable means to aggregate the indicators into *indices of sustainability*, perhaps one for each component of sustainability (if the indicators within a given sustainability component are at least somewhat comparable), or to otherwise facilitate comparison across indicators, recognising that comparisons of fundamentally noncommensurable indicators should be left to policy makers as a 'political' task.

10.3 Components of sustainability

As described initially in Chapter 1, the process of sustainable development can be viewed as being based on the simultaneous achievement of four fundamental components of sustainability: ecological, socioeconomic, community and institutional sustainability (Charles 1994). These components are described in turn below, elaborating on the brief introduction given in Chapter 1.

- *Ecological sustainability* incorporates (a) the long-standing concern for ensuring that harvests are sustainable, in the sense of avoiding depletion of the fish stocks, (b) the broader concern of maintaining the resource base and related species at levels that do not foreclose future options, and (c) the fundamental task of maintaining or enhancing the resilience and overall health of the ecosystem. While the first of these three areas has traditionally received most attention in fisheries, the latter has become the focus of extensive research (and of specific journals such as *Ecosystem Health*), with discussions on both the question of what exactly is meant by ecosystem 'resilience' and 'health', and the management and policy means to achieve ecological sustainability.
- *Socioeconomic sustainability* focuses on the 'macro' level, i.e. on maintaining or enhancing overall long-term socioeconomic welfare. This socioeconomic welfare is based on a blend of relevant economic and social indicators, focusing essentially on the generation of sustainable net benefits (including resource rents), a reasonable distribution of those benefits amongst the fishery participants, and maintenance of the system's overall viability within local and global economies. Each indicator in this grouping is typically measured at the level of individuals, and aggregated across the given fishery system.

- *Community sustainability* emphasises the 'micro' level, i.e. focusing on the desirability of sustaining communities as valuable human systems in their own right, and more than simple collections of individuals. Hence, emphasis is on maintaining or enhancing the 'group' welfare of human communities in the fishery system by maintaining or enhancing, in each community, its economic and sociocultural well-being, its overall cohesiveness, and the long-term health of the relevant human systems (Fig. 10.1). Why worry about human communities within a *sustainable fishery* policy? The idea is that sustainable fisheries go hand in hand with sustainable communities. Not only are the communities important in their own right, but since sustainable communities are by definition self-supporting, it is in the interests not only of the communities themselves, but of citizens/taxpayers everywhere.

- *Institutional sustainability* involves maintaining suitable financial, administrative and organisational capability over the long term, as a prerequisite for these three components of sustainability. Institutional sustainability refers in particular to the sets of management rules by which the fishery is governed, and the organisations that implement those rules: the bodies and agencies that manage the fishery, whether at the governmental, fisher or community level, and whether formally (e.g. the legal system and governmental agencies) or informally (e.g. fisher associations and non-governmental organisations). A key requirement in the pursuit of institutional sustainability is likely to be the manageability and enforceability of resource-use regulations. (The crucial need for institutional sustainability was first pointed out to me in the early 1990s by fishery researchers in Costa Rica, who noted that in a developing nation context, it is all too common for development projects to grind to a halt when aid money runs out and aid workers leave, unless strong –

Fig. 10.1   Community sustainability is a key component of a sustainable fishery system, certainly in small-scale coastal fisheries such as that of Soufrière, in St Lucia.

and sustainable – measures are in place to maintain the new infrastructure and local expertise. Back in the developed world, it soon became clear that with reductions in government and collapses in management institutions, the same need for institutional sustainability exists worldwide.)

The first three of these sustainability components can be viewed as the fundamental 'points' of a sustainability triangle (Fig. 10.2). The fourth, institutional sustainability, interacts amongst these, potentially affected (positively or negatively) by any policy measure focused on ecological, socioeconomic and/or community sustainability.

Overall sustainability of the fishery system can be seen to require simultaneous achievement of all four components. Thus, a proposed fishing activity or fishery management measure will be unacceptable if it produces an overly negative impact on any one component. In other words, overall system sustainability would decline through a policy that increases one element (e.g. socioeconomic sustainability) at the expense of excessive reductions in any other.

For example, marine resources may collapse from the over-expansion of industrial fisheries, while socioeconomic and community sustainability may be threatened by poor fishery planning. In a famous exercise in bioeconomic modelling, Clark (1973) showed that, in a fishery focused solely on maximising economic efficiency, the natural capital may be 'liquidated', with the fish stocks driven to extinction, as the proceeds are invested elsewhere. This action would not maintain ecological or community sustainability, and would therefore be deemed unacceptable.

Given the four sustainability components, the challenge lies in operationalising the practical assessment of sustainability in fisheries systems through the development of suitable *checklists* and *indicators*. If each of the components is viewed as crucial to overall sustainability, it follows that a sustainable development policy must serve to maintain reasonable levels of each.

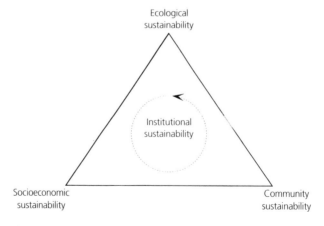

Fig. 10.2 The sustainability triangle forms the basis of a framework for sustainability assessment, based on three fundamental components, ecological, socioeconomic and community sustainability, and a fourth component, institutional sustainability, that interacts with and underlies the pursuit of the other three.

Components of sustainability

Ecological sustainability
(*Avoid foreclosing future options*)
Socioeconomic sustainability
(*Sustainable and equitable economic and social benefits*)
Community sustainability
(*Valuing community as more than a collection of individuals*)
Institutional sustainability
(*Long-term capabilities/resource system manageability*)

An alternative sustainability framework: sustainability attributes

In the *attributes* approach, a set of desirable social objectives (the attributes) is first determined, and sustainability then refers to a situation in which none of the attributes decreases over time, or alternatively, that a 'proxy' indicator for the set of attributes is non-decreasing. See Pearce *et al.* (1990) for a conceptual discussion of this approach.

An example of a specific multifaceted set of attributes has been proposed by Rees (1988) and looks simultaneously at the biophysical environment, the human environment and the relevant political/economic institutions.

Biophysical environment

(1) Recognition of ecological boundaries and adaptive and interactive properties of ecosystems.
(2) Recognition of the need to merge human activities within natural cycles.
(3) Activity based primarily on renewable resources.

Human environment

(1) Satisfaction of basic human needs.
(2) Achievement of equity and social justice.
(3) Provision for self-determination.

Political/economic institutions

(1) Long-term perspective predominates.
(2) Multiple goals (social/environmental/economic).
(3) Adaptive (institutions designed to respond and evolve …).
(4) Responsive to crises at different levels.
(5) Systems-oriented (awareness of interactions, trade-offs …).
(6) Interactive (open/fair/informed/empowering decision making …).

10.4 Sustainability checklist

With the necessity of sustainability established in public policy, resource managers engaged in fishery decision making must address some key questions.

- Is the resource system sustainable?
- If not, in what areas is improvement needed?

These can be addressed by considering the four components of sustainability within a multi-faceted 'checklist' framework. In developing such a checklist, it is necessary to determine precisely what sustainability criteria are required in order to assess a fishery system. These criteria must incorporate aspects of the ecosystem, the macro-level socioeconomic structure, the micro-level well-being of local communities (Fig. 10.2) and the institutional integrity of the system.

Table 10.1 gives a possible checklist of criteria for ecological, socioeconomic, community and institutional sustainability, relating both to the fishery system directly impacted and to related support activities. This framework is meant to be broad in scope; not all items in the checklist will be relevant for a particular fishery under consideration. Naturally, the checklist approach can easily be modified for specific circumstances.

10.5 Sustainability indicators

The sustainability checklist provides a framework by which to highlight 'trouble spots' in fishery systems (or aquatic systems more generally). However, this framework does not in itself allow an assessment of the *extent* of the problems. The next step in sustainability assessment is therefore to build on a set of criteria, such as those listed in the checklist in Table 10.1, in order to develop a *quantitative* set of indicators, the *measuring* tools of sustainability within a fishery system. In such an approach, each relevant sustainability criterion is quantified appropriately, whether through an *objective* variable, which is in some sense observable or measurable (such as a human population or a biomass level), or through a *subjective* measure which is amenable to evaluation (perhaps on a scale from 1 to 10).

The development of quantitative sustainability indicators in fishery systems has largely been restricted to those dealing with environmental or *ecosystem health* aspects. There have been relatively few efforts to develop sets of sustainability indicators in fishery systems that are quantitative *and* reflect an integrated, interdisciplinary, multidimensional view of sustainability, including assessment of both the natural and human sub-systems. However, there is an increasing focus on this approach; examples include Dunn (1996), Charles (1997c, 1998c), Chesson and Clayton (1998), Food and Agriculture Organisation (1999d) and Charles and Lavers (2000). There is a more substantial literature involving the integrated treatment of sustainability concepts, policy issues and the social and economic valuation of aquatic resources and ecosystems, but this generally does not include an indicators approach. (For a review of the literature on indicators in aquatic systems and beyond, see Charles 1997b.)

Table 10.1   A sustainability checklist.

Ecological sustainability

(1)   Are exploitation levels (catches) on directly impacted species such that ecosystem resilience is maintained (or at least not reduced excessively)?

(2)   Are indirect biological impacts reasonably understood to the extent required to ensure sustainability?

(3)   Are impacts on the ecosystem as a whole reasonably understood to the extent required to maintain overall resilience?

(4)   Are alternative systems of management and/or utilisation available so that pressures from any increased demands placed on the system do not increase beyond management capabilities?

(5)   Are imposed stresses and rates of change likely to be within the bounds of ecosystem resilience?

Socioeconomic sustainability

(1)   Will the activity increase the aggregate long-term rate of employment?

(2)   Will the project enhance economic viability in the local and regional systems?

(3)   Are possible impacts on input and output prices understood?

(4)   Is resource depreciation, and changes in natural capital more generally, incorporated into national accounting practices?

(5)   Are the current and projected levels of distributional equity in the system sufficient?

(6)   Will long-term food security and livelihood security be maintained or increased, as measured in both average and minimal terms?

Community sustainability

(1)   Is the project likely to maintain or increase the long-term stability of affected communities?

(2)   Does the local population have access to the resource base?

(3)   Is the local population integrated into resource management and development practices, with traditional management approaches utilised to the extent possible?

(4)   Are traditional value systems of importance to the community maintained?

(5)   Are local sociocultural factors (such as tradition, community decision-making structure, etc.) incorporated?

(6)   Are traditional resource and environmental management methods utilised to the extent possible?

(7)   Are there adverse impacts, at any level or in any component of the system, that unduly affect particular components of the community (e.g. youth, particular religious groups, etc.; gender-related impacts)?

Institutional sustainability

(1)   Will the long-term capabilities of corresponding institutions be increased?

(2)   Is financial viability likely in the long term, or does the intrinsic importance of the system justify ongoing support from society regardless?

It is worth noting that substantial progress has been made in developing sustainability indicators for other sectors. This includes work on agricultural systems and on other resource systems, including forests, rangelands and wildlife, with analyses ranging from the local ('micro') to the regional or industry levels (Charles 1997b). For example, this might involve indicators of sustainability for forestry operations in a specific location, or for wildlife management in a specific park. Also relevant are sets of quantitative sustainability indicators developed at the international or national levels and relating to alternative national accounts (natural resource accounting, or 'green' accounting), and alternative indicators of human progress and well-being, to enlarge on gross domestic product (GDP) measures.

Returning to the fishery system, an important measure that has traditionally been used in discussions of sustainability (and in particular on conserving the fish resource) is the *maximum sustainable yield* (MSY), discussed earlier in this chapter. Specifically, the system will not be sustainable if the harvest exceeds the MSY. The MSY is quantifiable, at least in theory, and provides a simple 'bottom line' for resource managers, summarised in a single number. However, the sustainable yield deals only with single-species conservation and ignores the many other components of fishery systems. Despite the long history of emphasis on sustainable yields, the idea of capturing the extent of sustainability in a single indicator seems unsuitable.

Instead, as discussed above, modern discussions of sustainability adopt a larger view, broadening out from single-species thinking to *ecological sustainability* and to other components of sustainability. An example of this approach is shown in Tables 10.2–10.4. These build on the criteria in Table 10.1 to provide sets of quantitative sustainability indicators, one for each of the three categories ecological, socioeconomic/community and institutional sustainability. Also shown are the range of each indicator and an explanation of the situation leading to a minimum value of the indicator. Note that each indicator is designed as a dimensionless quantity, usually obtained by normalising relative to a historical average value or a maximum level.

It should be noted that the particular criteria selected here, and the corresponding indicators chosen, are somewhat arbitrary. While there are undoubtedly many other means of quantifying series of indicators, these serve to provide an idea of the proposed approach, as well as a base from which to extend and improve the set of indicators.

Table 10.2 Ecological sustainability indicators.

Sustainability criteria	Indicator	Range	Indicator at minimum if
Catch level	(MSY – catch) / MSY	$-\infty$ to 1	Catch exceeds maximum sustainable yield
Biomass	Biomass (relative to historical average)	0 to ∞	Total biomass or spawning stock biomass below a critical low level
Biomass trend	Multi-year average annual percentage rate of change	–1 to 1	Biomass declining rapidly (or predicted to do so due to lack of recruitment)
Fish size	Average fish size (relative to historical average)	0 to ∞	Average size at capture very low relative to optimal capture size
Environmental quality	Quality (relative to historical average) + (% rate of change)	0 to ∞	Environmental quality low and declining
Diversity (harvested species)	(No. of species) /(hist. avg.) + (diversity)/(hist. avg.)	0 to ∞	Number of species in catch and diversity index both low relative to historical levels
Diversity (ecosystem)	(No. of species) / (hist. avg.) + (diversity) / (hist. avg.)	0 to ∞	Number of species and diversity index both low and declining
Rehabilitated areas	Area rehabilitated as % of total area	0 to 1	Increasing area closed due to contamination, etc.
Protected areas	Area protected as % of total area	0 to 1	Decrease in areas protected from exploitation
Ecosystem understanding	Level of understanding relative to full knowledge (subjective)	0 to 1	No clear understanding of resource and its ecosystem

MSY, maximum sustainable yield.

Table 10.3 Socioeconomic/community sustainability indicators.

Sustainability criteria	Indicator	Range	Indicator at minimum if
Community resiliency	Index of diversity in employment	0 to 1	Lack of livelihood alternatives (low diversity in employment)
Community independence	Percentage of economic activity based locally	0 to 1	High dependence on external economic forces
Human carrying capacity (livelihood)	Current (or potential) sustainable employment (relative to population)	0 to 1	Sustainable economic or employment base is substantially below current (or predicted) population
Human carrying capacity (environment)	Natural absorptive capacity/ human waste production	0 to ∞	Generation of human wastes far exceeds the absorptive capacity of the environment
Equity	Ratio of historical to current Gini coefficients of income and/or food distribution	0 to ∞	Dispersion in income and/or food supply is substantially above traditional norms
Sustainable fleet capacity	Ratio of capacity for harvesting at MSY to current capacity	0 to ∞	Current capacity exceeds that required to harvest at MSY
Appropriate investment	Investment/capacity (when stock < optimal)	−1 to 1	Investment above replacement level when stock is fully exploited, or > 0 when stock is declining
Food supply	Food supply per capita (relative to minimum nutritional needs)	0 to ∞	Food available per person is below minimum nutritional requirements
Long-term food security	Probability of sufficient food being available over next 10 years	0 to 1	Stability of food supply is low, or food supply is declining rapidly

MSY, maximum sustainable yield.

Table 10.4 Institutional sustainability indicators.

Sustainability criteria	Indicator	Range	Indicator at minimum if
Management effectiveness	Level of success of stated management and regulatory policies	0 to 1	Existing management structures are insufficient to control exploitation levels and regulate resource users
Use of traditional methods	Extent of utilisation	0 to 1	Traditional resource and environmental management methods not utilised
Incorporating local input	Extent of incorporation	0 to 1	Management/planning activity does not incorporate local sociocultural factors (tradition, community decision making, ecological knowledge, etc.)
Capacity-building	Extent of capacity-building efforts	0 to 1	Lack of capacity-building within relevant organisations
Institutional viability	Level of financial and organisational viability	0 to 1	Management organisations lack long-term financial viability, or there is a lack of political will to support such structures

Carrying capacities as sustainability indicators

As indicated in Tables 10.2 and 10.3, the concept of *carrying capacity* provides possibilities for quantitative sustainability indicators within both the natural and the human system. The idea of carrying capacity is most well established and most prominently applied in ecological studies, where, for any given population of interest, the carrying capacity is that population level that is sustainable indefinitely within the given environment (in the absence of exploitation). However, the carrying capacity concept is applicable not only to renewable resources themselves, but also to the human populations exploiting those resources. Indeed on a global scale, the concept is well utilised in discussions of human population pressure on global resources, where debate relates to the planet's human carrying capacity. However, for the present discussion it is more relevant to note that in a fishery system, just as the natural environment determines the carrying capacity of the resource, so too does the socioeconomic environment influence the carrying capacity of human activity in the system.

   While there has been relatively little research to date on integrating the human and biological aspects of carrying capacity in renewable resource systems, a number of approaches have potential.

(1)   Aggregated approaches focus on determining a single measure of the overall ecological carrying capacity for the human endeavour under discussion. An important variation of this perspective is in fact a converse perspective known as *ecological footprint* analysis (Rees & Wackernagel 1994; Folke *et al.* 1998), which involves assessing the per capita human use of land and other resources in a particular ecosystem. This approach provides an indirect means to deduce how much human impact can be tolerated sustainably within the area in question.
(2)   Disaggregated approaches (using checklists of ecological, socioeconomic, community and institutional sustainability, as described earlier) could provide a set of indicators of socioeconomic, community and institutional carrying capacity analogous to that of ecological carrying capacity.
(3)   Dynamic approaches move beyond the above 'snapshots' of sustainability, assessed at a particular point in time, to focus on dynamic adjustment processes related to the carrying capacities of both the resource and the human population, and to shifts in the fishery system between sustainability states (e.g. from one of non-sustainability, or one that is sustainable but unproductive, to a state in which sustainability has been improved). These are much more complex and data-intensive, but potentially of great practical importance.

The choice amongst these carrying capacity approaches will logically depend on the data availability, the need to deal with temporal changes, and the capability for engaging in modelling efforts to address the development of sustainability indicators.

10.6 Indices of sustainability

When the values of all the sustainability indicators have been determined for a given fishery system, some insight can be obtained into where sustainability seems to be present or absent. For example, as one possible outcome, the system may seem sustainable from the perspective of one indicator, e.g. ecosystem resilience, but not from that of another, e.g. human equity.

To further synthesise the results, it would be helpful to aggregate across the various sustainability indicators, creating aggregate indices of sustainability. This seems most logical when dealing with indicators within a particular component of sustainability (as shown in Tables 10.2, 10.3 and 10.4, respectively). Table 10.5 shows one approach which aims to achieve this. Assuming a quantitative value has been determined for each indicator, *weights* might be selected for each, and the weighted values combined in a suitable manner into three indices:

- an ecological sustainability index;
- a socioeconomic/community sustainability index;
- an institutional sustainability index.

Table 10.5 Sustainability assessment framework.

Sustainability component/indicator	Value	Weight	Weighted value
Ecological sustainability			
Catch level			
Biomass			
Biomass trend			
Fish size			
Environmental quality			
Diversity: harvested species			
Diversity: ecosystem			
Rehabilitated areas			
Protected areas			
Ecosystem understanding			
Index:			
Socioeconomic/community sustainability			
Community resiliency			
Community independence			
Human carrying capacity (livelihood)			
Human carrying capacity (environment)			
Equity			
Sustainable fleet capacity			
Appropriate investment			
Food supply			
Long-term food security			
Index:			
Institutional sustainability			
Management effectiveness			
Use of traditional methods			
Incorporating local input			
Capacity-building			
Institutional viability			
Index:			

To calculate these indexes, there are two principal approaches to aggregation through averaging, depending on how inter-comparable the indicators are. One option is to use a weighted arithmetic average (with suitable weights assigned to each indicator). This has the property that a low value for any one indicator can be 'compensated' for by an equivalently high value for another equally weighted indicator. The second option is a geometric average, possibly weighted. This has the feature that an extreme (low or high) value for a specific indicator will have a greater influence (relative to the arithmetic average) on the overall level of sustainability.

Another possibility is that one or more indicators within a certain sustainability component are considered of critical importance in determining sustainability. For example, it may

---

Challenges in applying sustainability assessment: an example

While the focus of this chapter is on the idea of a comprehensive framework for sustainability assessment, real-world applications of the framework are bound to pose many practical challenges. This section describes a tentative application of sustainability assessment, carried out for the fishery system of Nova Scotia, on Canada's Atlantic coast. This work (Charles & Lavers 2000) was designed to provide a 'fishery account' as one component of a genuine progress index (GPI), i.e. a comprehensive measure of social, economic and environmental well-being. This application took place under the auspices of GPI Atlantic, a non-profit research organisation, as part of a project endorsed by the Canadian government through Statistics Canada.

The sustainability assessment focused on the sustainability components discussed above, within each of which a set of indicators was evaluated. However, for various reasons, it was not possible to implement many of the indicators listed in Tables 10.2–10.4. This illustrates the point made above that the framework should be seen not as a rigid set of indicators, but rather as an illustrative approach to be adapted to specific circumstances. The factors driving the need for adaptation in this case included (1) a requirement for the assessment to be applied to the province in its entirety, and (2) a limited time frame and a consequent need to rely on secondary data. The first of these constraints prevented the assessment of community sustainability, owing to an inability to obtain specific data at a local level. The second constraint meant that institutional sustainability was assessed in large part qualitatively, in the absence of suitable survey data, and even in the more quantitative areas of ecological and socio-economic sustainability, it was not possible to locate data needed for many of the indicators discussed earlier.

As a result of this, the list of indicators produced was quite different from those shown in Tables 10.2–10.4, with alternative indicators substituted where necessary, and with some indicators (notably the institutional indicators) developed only qualitatively. The resulting set of indicators used in this example is shown below.

Ecological indicators

- Harvests relative to 'safe' levels based on historical information.
- Biomass of targeted species.
- Level of resource depreciation (declines in the monetary value of the resource).
- Average size of fish (affected by over-harvesting and environmental stress).
- Spatial extent of protected areas (refuges from harvesting).
- Extent of habitat alteration/destruction (e.g. due to impacts on the ocean bottom).
- Marine diversity (measured through trawl surveys and in terms of catches).
- Level of toxic contamination (using various organic and inorganic measures).
- By-catch, discarded/dumped fish (estimated from various sources).

Socioeconomic indicators

- Level of employment relative to that calculated from 'safe' harvests, as above.
- Landed value of fish caught (for comparison with resource depreciation, above).
- Level of exports.
- Resilience (age structure of fishers, extent of licensing for multiple species).
- Concentration of access and wealth (across fleet groups and ports).
- Level of debt and bankruptcies among fishers.
- Safety at sea (measured by rate of injury and death).

Institutional indicators

- Level of resources allocated for science and conservation.
- Priority placed on sustainability in management institutions.
- Cooperation and sharing of power with fishing communities.

The reader is referred to the report itself for detailed results on each of these indicators. It is notable that while several resource collapses occurred within this fishery system over the course of the late twentieth century, the indicators in the sustainability assessment were not all negative. Some (such as resource depreciation and size of fish) were indeed negative, but others (e.g. some aspects of toxic contamination) were positive, and still others (such as socioeconomic resilience) were neutral. This reinforces the key point that rather than seeking an overall aggregation of the results, it is preferable to display the various results and let policy makers and the public determine the balance among indicators, and the consequent actions required.

Based on Charles & Lavers (2000); co-authored with Amanda Lavers

be that a biomass lying below some critical level implies non-viability of the stock, and thus an obvious lack of ecological sustainability. Such indicators must be given special treatment in the analysis. In the above case, values of the biomass indicator below the critical level might be reset to zero, so that the use of multiplicative averaging in calculating the *ecological sustainability index* will imply non-sustainability (an index of zero).

Finally, even if it is felt to be appropriate to determine aggregated indices for the relatively similar indicators within each sustainability component, can this logic be extended to determine an overall index of sustainability for the fishery system as a whole? Such a single index would need to reflect a judgement of the balance among the components. There is an analogy with *cost–benefit analysis*, which is used to assess the feasibility and desirability of proposed economic activities. If it were possible for all benefits and costs to be measured in the same monetary units, then the sets of benefits and of costs could each be summed, a benefit/cost ratio calculated, and a verdict reached on whether the activity is desirable, or at least feasible, based on whether or not the ratio is greater than one, so that benefits exceed costs. However, in reality, benefits and costs cannot all be measured in identical terms. (For example, despite efforts to calculate the value of a human life saved or lost due to a particular activity, there is a widespread reluctance to adopt such thinking.)

It seems that this conclusion should also apply to matters of sustainability assessment. If, as seems to be the case, ecological, socioeconomic, community and institutional sustainability are fundamentally non-commensurable, then the inescapable trade-offs between them should be a strictly *political* task, beyond the scope of quantitative analysis. Sustainability assessment does, however, provide a means to examine the implications of such trade-offs.

10.7 Validation of sustainability indicators

To what extent is a set of quantitative sustainability indicators useful in practice? This question relates to the task of *validation*. Unfortunately, it is not possible, given the nature of sustainability, to prove *a priori* that a given set of indicators will properly *predict* whether or not a given system will be sustainable. The best we can hope for is that the set of indicators being used has proved itself in the past. This implies the need to analyse the performance of the set of indicators across a number of case studies, with suitable contrasts across biophysical, ecological and human dimensions.

The idea is to determine systematically why some systems were sustainable while others were not. There is an intrinsic difficulty with this, however, since non-sustainable systems do not persist. There may well be a lack of suitable time-series data on such systems, thus preventing their full evaluation in an historical analysis. This is comparable to the assessment of species extinction rates, which is confounded by the fact that many species became extinct before ever being studied. The best hope may be to study currently problematic fisheries, where at least one component of sustainability has declined within recent history, and to incorporate temporal information (time-series) where possible, so that a comparison of adjustment dynamics can take place. In any case, it should be noted that there will always be some uncertainty about the utility of sets of indicators, since quantification of sustainability inherently requires projections into the future.

10.8 Methodological challenges for sustainability assessment

10.8.1 *Sustainability versus stability*

A key difficulty in assessing sustainability lies in differentiating between apparent stability and long-term sustainability. It is certainly difficult to determine whether a fishery is 'stable', given natural cycles, environmental influences and random effects, but it is even more of a challenge to assess sustainability. For example, is a fishery system sustainable if the biomass is observed to be fairly stable at a reasonable level, with catches controlled by government? Not necessarily. If a controlling 'heavy hand' is holding the system in equilibrium over a long period of time, that equilibrium is in reality unstable, giving the false illusion of sustainability. Eventually a crisis develops, perhaps caused when either (a) fishers find ways to thwart restrictions, with excessive harvests leading to stock decline (ecological non-sustainability), or (b) massive levels of control are exerted to avoid this, so that government costs climb excessively (institutional non-sustainability). At that point, the equilibrium vanishes. The question is: what will replace it?

Since we cannot conclude that an apparently stable fishery is sustainable, equilibrium is not a sufficient condition for sustainability. For example, fisheries that have been stable for a long period of time, given a certain technology, may eventually collapse as technology changes. Indeed, stability is not even a situation to strive for. More important to sustainability is the system's *resilience* (Chapter 15), i.e. how well it adapts to change and absorbs the perturbations that are certain to occur sometime. Thus, a key element of sustainability assessment is to understand the consequences of change, and to assess whether the essential characteristics of the current system will survive that change.

10.8.2 *Sustainability versus non-sustainability*

It is probably more straightforward to assess a fishery system's *non-sustainability* than its sustainability. A non-sustainable system tends to display high levels of stress on certain aspects of the system, such as a precipitous drop in fish biomass, a decay in the infrastructure within fishery-dependent communities, or an inability of the management institutions to cope with pressures upon them. There may be sufficient resilience in a stressed system to overcome these problems, but in general, the greater the stress, the greater the tendency to non-sustainability. In contrast, if a fishery system has reached a sustainable state, it is unlikely to be under significant stress. This could present itself in terms of a healthy biomass and strong institutions controlling exploitation, or a low but stable biomass with little pressure to exploit the resource. However, as noted above, we must be careful not to assume that the presence of such attributes 'proves' that the system is sustainable.

This implies that it may be more feasible in practice to develop indicators of non-sustainability than of sustainability. In turn, one might expect greater success in locating 'trouble spots' of non-sustainability than in verifying sustainability for a proposed activity. (This is not unlike the case of environmental impact assessment. It is not too hard to point to environmental crises, but considerably more difficult to determine whether as yet untried projects will be environmentally benign.)

10.8.3 *Micro-indicators versus macro-indicators*

It can be argued that while the development of sustainability indicators to date has focused on the 'macro' (national and/or international) level, there needs to be more emphasis on the 'micro' (local, regional, community) level, at which development projects tend to operate. At the latter level, each locale has its own peculiarities, implying the need for both location-specific analysis and the search for common features. Comparability across the range of local experiences is needed, if broadly applicable conclusions, or even methodologies, are to be obtained. It is also important to consider how analyses of sustainability should differ, if at all, between 'macro' and 'micro' levels – say, between national-level plans on the one hand and watershed or community levels on the other.

10.9 Summary

Assessing and predicting the sustainability of natural resource systems is a crucial challenge for society. The sustainability assessment framework in this chapter describes one approach to this challenge, a set of quantitative indicators for assessing the process of sustainable development in fishery systems. However, a caveat is in order: practical 'testing' must take place to demonstrate the validity of any proposed indicators; the feasibility of measuring each indicator must also be determined under varying circumstances.

Therefore, any set of indicators can *only* be proposed on a tentative basis. It is likely that the preferred approach to sustainability assessment will lie in careful selection of relevant indicators for each particular fishery situation, and fine-tuning these over time. Indeed, since the quantification of sustainability, whether at the local level or at the level of the coastline or watershed ecosystem, inherently implies predictions of the future, even the success of efforts towards sustainable development can only be judged in the course of time.

Chapter 11

Uncertainty and the Precautionary Approach

The fishery is an unusual economic activity, in that no one can be certain how much of the key ingredient is available in any given year, what amount of product should be produced that year, or what effect that production will have on the future availability of the fish. Uncertainties are ubiquitous in fisheries. Yet despite an abundance of research attention on the subject, the profound impact of such uncertainties on the sustainability of fisheries seems to have been widely underestimated in the past. However, we have entered a new era in our appreciation of the role uncertainty plays in fishery systems, and in our realisation of the need to allow for uncertainty not only *quantitatively* (for example, in setting catch and fishing effort limits) but also *qualitatively* in determining suitable characteristics of, and approaches to, fishery management (Gulland 1987; Hilborn 1987; Charles 1994).

The theme of this chapter is the idea of *living with uncertainty* in fishery systems. Topics to be reviewed include:

- the various sources of uncertainty and forms in which uncertainty appears;
- the various analytical tools used to examine these uncertainties;
- appropriate management approaches to best operate in an uncertain environment;
- specific ideas of robust, adaptive and precautionary management;
- recent attempts at implementing these approaches, and the implications of their absence.

11.1 Sources of uncertainty in fishery systems

11.1.1 Natural sources

Among non-human sources of uncertainty (Table 11.1), the most obvious one to fishers and those involved in real-world fishery management is probably uncertainty about the size of the fish stock (in biomass or numbers). This is in contrast to the fisheries research literature, where the most widely discussed, and usually the first addressed, source of uncertainty is that inherent in stock–recruitment relationships, both in terms of stochastic fluctuations around the underlying deterministic relationship and uncertainty in the parameters of the relationship itself. Of course, many other sources of uncertainty exist, including those pertaining to:

- age structure;
- natural mortality and multi-species interactions, notably predator–prey effects;

Table 11.1 Sources of uncertainty in fishery systems.

Natural sources	Human sources
Stock size and age structure	Fish prices and market structure
Natural mortality	Operating and opportunity costs
Spatial heterogeneity	Discount rate
Migration	Technological change
Stock–recruitment parameters	Management objectives
Stock–recruitment relationships	Fisher objectives
Multi-species interactions	Fisher response to regulations
Fish–environment interactions	Perceptions of stock status

- spatial complexity: spatial heterogeneity, stock concentrations, migration patterns;
- fish–environment interactions, notably the impact of ocean conditions.

Further complicating the situation is the fact that these sources of uncertainty interact with one another. In particular, stock assessment practitioners face the reality that uncertainties in each of the above areas (e.g. age structure) impact on the level of uncertainty in the stock–recruitment relationship.

11.1.2 Human sources

Sources of uncertainty on the human side of the fishery (Table 11.1) are much less well studied. There has been some effort to examine the effects of random fluctuations in fish price, fishing costs and the discount rate, as well as uncertainties in market structure. However, the more challenging forms of uncertainty, both in terms of analysis and practical importance, are less tractable to analysis and thus less studied; these include:

- technological change (its adoption and its impact);
- management objectives and fisher objectives;
- fisher response to regulations;
- differing perceptions of stock status (as opposed to uncertainties on the natural side).

11.2 A typology of uncertainty

A wide variety of uncertainties arise in fishery systems, on the natural, the human and the management sides. This section reviews the various forms of uncertainty and how these might be categorised, integrating together a classification scheme of Francis & Shotton (1997) and one developed by the author (Charles 1998a).

11.2.1 Example: the stock–recruitment relationship

First, consider an important example where uncertainties arise in fishery systems, namely the fundamental matter in fishery science and stock assessment of the stock–recruitment relationship, i.e. the functional form relating the current fish stock and subsequent recruitment. Suppose that a particular fishery is blessed with a long time series of very reliable and well-behaved data on the fish stock. Then after suitable analysis, we may be quite confident that we have correctly deduced the structure of the stock–recruitment relationship (e.g. perhaps a Beverton–Holt function) and that we have correctly estimated all parameters of the relationship. Yet there will remain fluctuations inherent in that relationship, i.e. random fluctuations ('noise') that are intrinsic to the behaviour of the system. This *randomness* produces what Francis & Shotton (1997) classify as 'process uncertainty', a form of uncertainty that is relatively well studied and often reasonably well understood. Random fluctuations often have a frequency of occurrence that can be determined from past experience.

Stepping back now, even if we remain confident of the form of the stock–recruitment relationship, in the real world its parameters cannot be estimated with certainty. For example, parameter estimates are often made through least-squares statistical fitting to what are typically highly variable sets of data. Thus, such *parameter uncertainty* (imprecise estimates) reflects both 'observation uncertainty', i.e. measurement and sampling error arising in the process of collecting data, and 'model uncertainty' concerning the parameters of the analytical model being used (Francis & Shotton 1997). The existence of observation uncertainty, and the vagaries of human differences in assessing stock status, also lead to *uncertainty in the state of nature,* i.e. in this case, we are uncertain about the biomass of fish in the sea. This is compounded by 'estimation uncertainty', arising in the statistical process of estimating the stock–recruitment relationship. Not only is an estimate inherently uncertain, in addition there may be fundamental errors underlying the statistical method. All this greatly compounds the difficulties involved in stock assessment.

Finally, it was assumed above that the structure of the stock–recruitment relationship was known with certainty. This will not be the case in practice, since we do not know the true nature of the relationship. Specifically, there will be uncertainties with respect to appropriate variables to be included, the set of relevant components and the underlying model structure of the stock–recruitment relationship. Such uncertainties can be grouped as 'model uncertainty', which in turn represents a form of *structural uncertainty* – certainly the most challenging form of uncertainty, and one which will be discussed in detail below.

11.2.2 Randomness/'process uncertainty'

The best studied form of uncertainty involves the random (or 'stochastic') fluctuations inherent in fishery systems. These occur whether or not humans are measuring or managing the fishery, and arise as variations over time in the various quantities in a fishery, with the extent of the variations being probabilistic. The resulting probability distributions can be estimated (to varying degrees of certainty) based on past data and experience. We observe a quantity fluctuating over time, but underlying this is a deterministic (non-fluctuating) value or functional relationship. For example, fluctuations in price may centre on a deterministic mean value (the average price), and the number of fishers choosing to fish on a certain ground may fluctuate around a clear-cut average level. In the case of random fluctuations inherent in stock–recruitment relationships, the function itself fluctuates around an underlying deterministic function. (In the latter case, however, the *cause* of the uncertainty could be random fluctuations in a single quantity, such as the fish survival rate, around its mean value.)

The process uncertainty inherent in random fluctuations is irreducible and implies a lack of full predictability. One does not know what the fish stock will be next year, what the number of fishers will be next week, or what the fish price will be tomorrow (Fig. 11.1). Thus, this process uncertainty is clearly inconvenient from a planning perspective. However, in itself, it does not reflect fundamental ignorance about the fishery system, since the stochastic dynamics revolve around a *specific* probability distribution. On the other hand, in practice this probability distribution must be estimated, to some imperfect degree. Such estimation is based on past data and experience, which are subject to the observation uncertainty that arises due to measurement and sampling error, as noted above. In addition, there may be trends in the system which need to be separated from the randomness and similarly estimated. These various realities add to the difficulties posed by randomness.

From the perspective of fishery management and planning, it is important to note that random fluctuations in one component of the fishery system (such as the fish stock) indirectly

Fig. 11.1   At the fish retailer, prices vary randomly in response to fluctuations in supply and demand. Efforts to predict prices must deal not only with this randomness but also with parameter uncertainty (arising, for example, in demand relationships) and structural uncertainty in our understanding of distribution, markets and consumer preferences.

create fluctuations in other components (such as fleet investment). An analytical example of this is provided below.

11.2.3 Parameter and state uncertainty

While randomness (process uncertainty) is an inconvenience in fisheries, it is even more challenging from both analytical and management perspectives to deal with uncertainties due to the limited nature of data collection procedures and the resulting imprecision in the parameters estimated in fishery models. These types of uncertainty are discussed here, with each of the subheadings below corresponding to an uncertainty category of Francis & Shotton (1997).

Observation uncertainty

This form of uncertainty (Francis & Shotton 1997) reflects the reality that nothing in a fishery can be measured perfectly. As we attempt to observe the system, we face the inherent uncertainty resulting from the random fluctuations discussed above, which complicate the task of understanding underlying system structure and dynamics, as well as uncertainty due to technical aspects (e.g. limited sample sizes in surveys, or limited capabilities to measure certain quantities of interest), procedural considerations (such as differing methodologies for obtaining such measurements) and human behaviour (e.g. the effects of uncertain levels of misreporting).

This directly impacts on *state uncertainty*, i.e. the uncertainty that arises in the magnitude of system variables (such as the fish stock size or the fleet size) which reflect the 'state of nature'. Uncertainties in the states of key system variables play a major role in fisheries, and in particular, imprecise knowledge of fish stock status has had grave consequences in many cases (Charles 1995b, 1998a, b; Hilborn & Walters 1992; and references in Pitcher *et al.* 1998, Flaaten *et al.* 1998).

Model uncertainty

Parameter uncertainty is an aspect of 'model uncertainty' (Francis & Shotton 1997), reflecting the fact that parameter values (e.g. the survival rate of fish at sea, or the maximum demand for fish in the market-place or the rate of labour mobility in a fishing community) are uncertain quantities. This arises as a direct result of observation uncertainty, as well as an indirect effect of random fluctuations (in the absence of which, the parameters would be relatively easy to measure). Parameter uncertainty will be present even within what may be considered a known functional relationship of the fishery system (e.g. the stock–recruitment function, or the demand curve). In other words, if we are willing to view model structure as given, we must still face the problem of parameter uncertainty. Note as well that model uncertainty also contributes to the level of state uncertainty, if (as is typically the case) the state of nature is being deduced from a certain model or functional relationship.

Estimation uncertainty

As noted above, observation uncertainty impacts on our uncertainty about the state of nature, whether this be the size of the fish stock, the number of active fishers in a community, or some other state variable. Also contributing to *state uncertainty* (and to parameter uncertainty) is 'estimation uncertainty' (Francis & Shotton 1997). This arises when, in using data and models to estimate a particular quantity such as a stock level or a fishery parameter, the estimation process itself contributes additional uncertainty to the output (in addition to that arising from observation and model considerations).

Uncertainties in states of nature and in model parameters can be reduced, at least in theory, by learning over time. For example, imprecise estimates of survival rates may become better known as more years of data are collected, while research surveys may help to refine stock size estimates. This process may be passive, for example by simply updating information at each time step as a new observation of the fishery system is obtained, or it may be active, by actively manipulating the system to acquire information, or by actively reducing observation uncertainty. In either case, since the underlying cause of the uncertainties in estimates is the existence of random fluctuations, one cannot predict exactly how the learning process will proceed. Thus, learning is not smooth and deterministic, but rather a process the results of which depend on the specific stochastic outcome that occurs. An example of this is presented below.

11.2.4 Structural uncertainty

Structural uncertainty is the most challenging form of uncertainty, reflecting basic ignorance about the nature of the fishery system, its components, its dynamics and its inherent internal interactions. Structural uncertainty can manifest itself in a number of ways.

Model uncertainty

As noted above, Francis & Shotton (1997) use the term 'model uncertainty' to refer to the technical aspects of parameter uncertainty. They also use the term in reference to a fundamental lack of knowledge concerning the structure of the fishery system, whether relating to the true nature of the stock–recruitment function, discussed above, or in other key aspects of the fishery, such as:

- species structure, i.e. the number of species interacting in the fishery system;
- fleet structure, i.e. the number of fishing vessels that will take part in a fishery opening;
- spatial complexity, i.e. spatial heterogeneity, stock concentrations, migration patterns;
- fish–fish interactions, i.e. multi-species interactions, notably predator–prey effects;
- fish–environment interactions, i.e. interactions between ocean conditions and fish stocks;
- technological change, i.e. the changes that will be adopted and the resulting impact.

While from the specific perspective of choosing a desired 'model' to use in a fishery it is theoretically possible to view uncertainty in model structure as merely parameter uncertainty in a 'super-model' (Francis & Shotton 1997), in reality the fundamental uncertainty involved here goes well beyond the problem of choosing a mathematical model to include challenges arising from very basic ignorance about the system.

Implementation uncertainty

From the perspective of the manager, structural uncertainty may appear as uncertainty regarding the extent to which management measures can be implemented successfully, or specifically, 'the degree of control over the harvest accompanying a particular management decision' (Rosenberg & Brault 1993: p. 244). The very existence of this form of uncertainty is due in large part to the limited nature of communications between managers and fishers, and in particular uncertainties which management faces in understanding and predicting:

- the objectives being pursued by fishers;
- the factors driving fisher decision making;
- the response of fishers to specific regulations;
- the effectiveness of enforcement measures.

Institutional uncertainty

Structural uncertainty may emerge in terms of interactions among those in the management system (O'Boyle 1993), and in particular uncertainties relating to:

- how fishers and others adapt to new management institutions;
- the societal/management objectives being pursued in the fishery.

Structural uncertainty can have a major impact on the outcome of fishery management, particularly since the accompanying lack of understanding can produce *surprises*, i.e. dramatic, unanticipated changes in the system (Holling 1973, 1978). Structural uncertainty also differs in a very practical economic (and financial) sense from randomness, which is at the other end of the spectrum: one can insure against random fluctuations, for which the probabilities of occurrence are known, but this is much more difficult in the face of basic ignorance about the system.

While the effects of structural uncertainty are considerable, this form of uncertainty is little studied, and is challenging to standard modelling approaches (Walters & Hilborn 1978; Sissenwine 1984; Hilborn 1987; Hilborn & Walters 1992). It is also unclear, in many cases, to what extent structural uncertainties are reducible over time, since achieving an understanding of the nature of a fishery system is not something that can be planned for completion within some arbitrary time frame. For example, in most fisheries, there is structural uncertainty in the nature of the underlying stock–recruitment relationship; reducing the level of structural uncertainty requires an unknown level of basic research and statistical data analysis.

11.3 Impacts of uncertainty: risk

If the random fluctuations that actually occur in the future are in an 'unfavourable' direction, the fish stock or the fish price may be lower than expected. If we do not know the true values of the parameters used in our predictive model, or the structure of the fish stock, decisions may be made that turn out (in retrospect) to be wrong. These are manifestations of a basic reality, namely that uncertainty creates *risk*. Here risk can be defined as a certain probability that the outcome will be 'negative' (FAO 1995a), or defined in a decision-theoretic manner to incorporate both the probability *and* the consequence of a negative outcome – so 'risk' increases with both the probability and the consequence of a negative outcome.

It is clearly crucial to understand the risks we face in fishery systems, and to develop the means to deal with those risks and the underlying uncertainties that produce them. This challenge is usually viewed as comprising two distinct tasks (e.g. Pearse & Walters 1992; Francis & Shotton 1997; Lane & Stephenson 1998).

- *Risk assessment* involves technical approaches to analysing uncertainty, measuring risks, and predicting the outcome of given harvesting and management scenarios within an environment of uncertainty.
- *Risk management* involves efforts to manage, reduce or otherwise cope with risks in fishery systems, through both technical (analytical) means designed to derive 'optimal' management plans in the face of uncertainty (perhaps to minimise certain risks, or to balance risk and fishery benefits), and through structural (design) approaches involving the creation and adoption of robust management approaches.

Substantial theoretical work exists on each of these themes, but only risk assessment has been applied in a relatively wide variety of practical settings within nations and international organisations. On the other hand, risk assessment is a rather technical matter, while risk management is perhaps the greatest challenge. Thus, extra attention is now devoted to the latter, as we examine the two tasks of risk assessment and risk management in turn.

11.3.1 Risk assessment

Risk assessment has been the subject of considerable attention among fishery scientists in recent years, particularly in the light of the emerging importance of the precautionary approach, as discussed below (see, for example, the many papers in Smith *et al.* 1993). Risk assessment involves a set of primarily technical methods focused on two goals:

(1)   determining the magnitude of the uncertainties that arise through random fluctuations, imprecise parameter estimates and uncertainty regarding the state of nature;
(2)   assessing quantitatively the 'risk' that certain undesired outcomes will occur, by analysing alternative harvesting or management scenarios from a probabilistic perspective.

The first of these goals is often approached through statistical analysis. In examining random fluctuations, the method may be as straightforward as calculating the standard deviation or coefficient of variation of a certain variable using appropriate time-series data. In assessing parameter uncertainty, statistical approaches are typically more complex, focusing on procedures for parameter estimation and fitting functional relationships, notably with respect to population dynamics (e.g. Deriso 1980; Collie & Sissenwine 1983; Ludwig & Walters 1989; Smith *et al.* 1993; Chen & Paloheimo 1995). With regard to uncertain 'states of nature', uncertainties in resource status are most often shown in stock assessments by displaying statistical confidence limits. (However, in the past, published stock assessments often failed to provide such confidence limits, and indeed, in some cases, such uncertainties were downplayed, this being rationalised on the basis that, faced with such uncertainty, the fishing industry and decision makers would regularly opt for the upper confidence limit, thereby implying a potentially excessive allowable catch.)

The second goal is typically approached in one of two ways, both based on 'simulating' the effects of random fluctuations, parameter uncertainty and uncertain states of nature (Francis & Shotton 1997). The first of these focuses on the immediate needs of stock assessment, the task being typically to provide an assessment of the risk of undesirable outcomes in the following year(s) as a function of harvesting decisions this year (as in various papers in Smith *et al.* 1993). For example, the goal may be to provide the probability that next year's spawning stock biomass will be below a certain threshold, as a function of this year's total allowable catch (TAC).

The second approach involves 'Monte Carlo' modelling (simulation) of the stochastic dynamics of the fishery system to assess the longer-term implications of given harvesting and management scenarios, by determining mean performance and the corresponding variability of selected outputs and/or system variables. Examples of this approach include those from Dudley & Waugh (1980), Hightower & Grossman (1985), Swartzman *et al.* (1987), Charles (1992a) and Criddle (1996). Many such analyses take a 'bioeconomic' perspective, incorporating both biological and economic factors (e.g. Andersen & Sutinen 1984; Clark 1985, 1990; Lane 1989), while others focus on stock assessment and predator–prey models (e.g. Collie & Spencer 1994). Behavioural modelling is a particular approach to simulation that has proved useful in exploring the economic responses of individual fishers to uncertainty, and the aggregate results produced by a set of fishers that is heterogeneous with respect to behaviour under uncertainty, in particular varying levels of risk aversion and risk preference (see Opaluch & Bockstael (1984) for a significant early contribution on this topic).

11.3.2 Risk management

While risk assessment is concerned with quantifying the risks involved in each possible fishery option, whether this relates to the level of next year's harvest or to the nature of a long-term harvesting strategy, risk management concerns decisions about the 'best' course of action in the face of those risks. Francis (1992) has noted, with respect to the orange roughy fishery in New Zealand, that a process of risk management can improve decision making and at the same time serve to improve the understanding of risk among managers and fishers.

Before discussing the various approaches to risk management, it is useful to consider the concept of *adaptive management*. The idea of adaptive management is to account systematically for uncertainty in fishery management by properly utilising, or even seeking out, new information (Holling 1978). This approach has benefited in particular from the methods of

> *Non-adaptive models*
> Average over uncertain parameters
> *Passive adaptive models*
> Update parameters each period but ignore future changes
> *Active adaptive models*
> Incorporate the value of future information in decisions

adaptive control. Fisheries applications of this method were popularised in the 1970s and early 1980s (Walters & Hilborn 1976, 1978; Ludwig & Walters 1982) and are reviewed in two comprehensive books on the subject (Walters 1986; Hilborn & Walters 1992). The key aspects of the method involve two options:

- *passive adaptive* learning, in which parameter estimates are updated as new information becomes available;
- *active adaptive* management, which deliberately attempts to accelerate the learning process by 'probing' the fishery system experimentally.

In discussing risk management, it becomes apparent that both 'passive adaptive' and 'active adaptive' methods are applicable to circumstances in which learning can improve the knowledge base on which decisions are made.

Risk management efforts range from the theoretical to the applied. Beginning with the former end of the spectrum, there has been considerable technical (analytical) research into determining *theoretically optimal* harvesting and management practices under conditions of uncertainty. Such work may focus on maximising *expected* benefits from the fishery, minimising certain risks, or balancing risk and fishery benefits. A common tool for these theoretical studies is stochastic optimisation, based on control theory or dynamic programming, which has been used to derive optimal harvesting policies in a random environment (e.g. Reed 1979; Charles 1983b; Mangel 1985; Clark 1990).

As an illustration of the theoretical approach, consider the key issue in fisheries of deciding on 'optimal' (rent-maximising) fishing fleet investment levels. How does the magnitude of random fluctuations in the fish stock affect the desired investment? It has been shown (Clark *et al.* 1979; Charles 1983b) that the *qualitative* nature of the optimum is not affected by random fluctuations. In all cases, the optimal strategy is to invest until the current fleet capacity matches a target level which is dependent on the fish stock size, a level shown as $h(S)$ in Fig. 11.2. (Similarly, the optimal harvest involves reducing the stock down to a target stock size curve, shown as $s(K)$.) However, the optimisation analysis also shows that the degree of randomness affects the *quantitative* level of investment desired, which may increase or decrease with increasing uncertainty, depending on the bioeconomic parameters. Furthermore, random fluctuations have a major impact on the fishery's dynamics (Fig. 11.2). The existence of fluctuations prevents an equilibrium from being reached, so that *optimal* management of fish harvesting and fleet investment is a continual task of responding to random *shocks*. For example, when an unusually large stock appears, the system shifts strongly to the right in

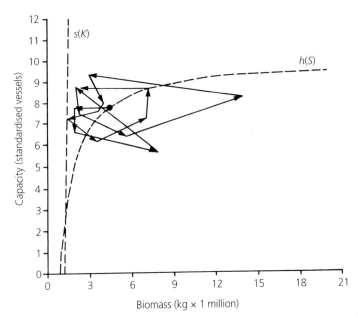

Fig. 11.2   A feedback diagram in which *optimal* decisions are made depending on the particular combination of stock size S and fleet size K. Given a fleet size K, optimal harvesting should reduce the biomass down to the $s(K)$ policy curve (or no harvesting if below the curve). Given a biomass on the horizontal axis, investment should take place up to the $h(S)$ curve (or no investment if above the curve). Following this strategy will produce optimal results *on average* for a fishery with a randomly fluctuating fish stock (i.e. stochastic fish population dynamics). The effects of such random fluctuations are shown through an example of a single 15-year trajectory of the fish-and-fleet system, beginning at a point that would be the fishery's equilibrium if it were not subjected to fluctuations.

Fig. 11.2, inducing new investment, up to $h(S)$, followed by increased harvesting and a resulting tendency of the system to move back to the left (reduced stock, reduced fleet) until another random shock occurs.

Theoretical studies are also undertaken using a Monte Carlo simulation approach. For an application of this approach to examine the impacts of parameter uncertainty, let us continue with the fishery investment analysis above, but now assume that the fishery is subject not only to randomness in the fish population dynamics, but also to imprecise parameter estimates in the stock–recruitment relationship. Suppose that at each time step, we *assume* that the parameter estimates are correct, calculate the 'target' curves $h(S)$ and $s(K)$, described above, and use these in our decision rules. In reality, however, uncertainty in the parameter values means that their estimates are probably incorrect, so that this will tend to lead to undesirable investment and harvest levels. There is a need to 'learn' over time, using new data to improve our knowledge of the parameters. Charles (1992a) developed a 'Bayesian' algorithm to simulate a 'passive adaptive' learning process, embedding the algorithm within a Monte Carlo analysis to examine the expected impacts of the learning process on harvesting and fleet investment. Starting with uncertain or misspecified parameter values, it was shown that on average, such a learning process eventually succeeds in 'finding' the correct parameter values and achieving the desired biomass and investment levels that would apply in a *perfect information* scenario (with no parameter uncertainty). However, since this learning process takes a considerable time, the expected economic benefits obtained (averaged over time) depend inversely on the degree of uncertainty inherent in the initial parameter estimate.

Turning now to the use of risk management in *practical* settings, two approaches may be noted: (a) attention to structural (design) measures, with the goal of developing management tools that are 'robust' to uncertainty, and (b) technical measures, typically based on analytical (often simulation) modelling, aiming to determine preferred management plans under uncertainty. The first of these is discussed later in the chapter, while the focus here is on the second.

The technical approach to risk management, while still uncommon in fisheries, is receiving increasing attention. It focuses on the development of decision rules through which the results of a risk assessment are translated into management actions. Depending on the circumstances, the decision rules may reflect annual *operational* goals (e.g. concerning a risk of year-to-year stock decline) or the pursuit of *long-term* objectives (e.g. concerning multi-year benefits).

In both cases, a precautionary risk-averse decision rule would typically be structured so that, other things being equal, a lower level of harvesting (lower TAC or fishing effort limit) will be chosen when there is greater uncertainty in key variables, such as stock status. Not only does this imply that uncertainty is less likely to produce damaging outcomes, but it also helps the risk management process by creating incentives (a) for scientists to ensure that uncertainty is fully incorporated in their analyses, and (b) for fishers to help to reduce uncertainty, perhaps through industry-supported surveys. The latter incentive to increase information flow arises because, as Cochrane (1999) notes, within a precautionary view of the fishery, 'significant reductions in uncertainty should allow increases in the yield taken from the resource, towards the maximum that can be achieved with perfect knowledge.'

An example of an operational-level risk management is that carried out annually in determining a suitable catch limit (TAC) for the Canadian portion of the George's Bank cod stock, which lies on the Canada–US border on the Atlantic coast of North America. The process begins with an assessment of uncertainty, focusing on that inherent in the virtual population analysis-based stock assessment, which allows calculation of the probability of stock decline under a range of harvesting scenarios. The TAC is then set based on a decision rule that 'chooses' the maximum catch compatible with a probability of stock decline of no more than 20%, and a fishing mortality rate no greater than the maximum which is considered safe (FRCC 1997b). This process, while relatively institutionalised, is somewhat *ad hoc* and continues to be revised as new situations arise.

A rigorous, longer-term form of risk management is that referred to as a management procedure, an approach that typically uses simulation to compare risk and performance measures for alternative management options. Specifically, Butterworth *et al.* (1997: p. 84) say:

'What exactly is a management procedure? The underlying philosophy is that all parties (scientists, industry, managers) should agree upon clearly defined rules before the management game is played. These rules specify exactly how the TAC (or the level of some other regulatory mechanism, such as fishing effort) is to be computed each year and what data are to be collected and used for that purpose.'

Cochrane *et al.* (1998: p. 180) focus on two features that characterise management procedures:

> 'The first is the selection of decision rules on the basis of predicted performance estimated from projections into the future of catches and stock status, based on an operating model ... The second is ensuring the robustness of the selected decision rules within the procedure to plausible changes in the assumptions underlying the operating model.'

The management procedure approach is being explored in South Africa (Cochrane *et al.* 1998) and internationally, in the International Whaling Commission, IWC (Kirkwood 1992). In the latter case, extensive 'robustness trials' were carried out by many fishery analysts, leading to the development of a so-called revised management procedure (International Whaling Commission 1994), in which the decision rule is such that greater certainty (a lower coefficient of variation in the estimates and more frequent estimates) leads to higher TACs, other things being equal. While the management procedure is emerging as a useful tool of risk management, there is little success to date in its implementation. Indeed, in South Africa, management procedures have been adopted but not consistently followed (Francis & Shotton 1997; Cochrane *et al.* 1998), while in the IWC, the approach was adopted but has yet to be applied.

11.4 The challenge of structural uncertainty

Most of the uncertainties addressed with analytical risk assessment and risk management are based on randomness, imprecise parameter estimates and uncertain states of nature. While there have been a few applications of analytical methods to problems of *structural uncertainty*, for example, in the functional form of the stock–recruitment relationship (Walters 1981), in most cases such uncertainties have not proved particularly amenable to quantitative analysis. This is not surprising, since structural uncertainty reflects, by its very nature, a fundamental lack of knowledge about the fishery system, making even the development of relevant models difficult. Thus, structural uncertainty creates extensive risks while being analytically rather intractable.

While it is desirable to see appropriate analytical methodologies developed to overcome this problem, in reality such developments may or may not ever succeed. Thus, it is unlikely that most structural uncertainties can be resolved in a clear-cut analytical manner. Instead, they must be addressed primarily through changes to the practice of fishery management. The design of a management framework within which uncertainty (and particularly the fundamental problems of structural uncertainty) can best be addressed is surely a key element of a strategy for sustainable fisheries. Two key aspects of such a framework can be identified:

- implementation of a *precautionary approach* to management decision making, in order to ensure that we make a proper allowance for uncertainty in our decisions and 'err on the side of conservation';

- redesign of the management system so that its structure and methods are *robust* and *adaptive*, with the overall goal that it should provide acceptable results even if our understanding of the fishery system is incorrect (cf. Ludwig *et al.* 1993; Levin 1993).

The first of these, the precautionary approach, focuses on the underlying philosophical basis, and the specific 'ground rules', of management decisions. This topic has received much attention in recent years, and is explored in the remainder of the chapter. The second aspect noted above probably implies substantial changes ('redesign') to the *practice* of fishery management. Such changes, and their rationale, are discussed in detail in Chapter 15.

11.5 The precautionary approach and the *burden of proof*

Within the fishery management system, all decisions must be made under uncertainty. For example, stock sizes are never known precisely within a given year (much less from one year to the next), and the impacts of fishing methods on the resource or on the environment cannot be predicted exactly. Given this inherent uncertainty, decisions must always be made by balancing risks: for example, the risk of stock and ecological collapse (due to excessive exploitation or environmental damage) versus the risk of lost economic benefits if conservation measures are overly conservative, so that harvests are lower than necessary (Charles 1998a, b; Walters 1998). It is a matter of attitude, indeed of philosophy, as to the precise nature of this desired balance. The fundamental question is: in considering the various risks, does the burden of proof favour exploitation or conservation? If we 'err', as we are bound to do at some point, on which side of the exploitation–conservation spectrum does this occur?

This question came to the fore over the course of the 1990s through efforts to apply the *precautionary approach* to fishery systems (Garcia 1994; FAO 1995a). The fundamental idea of the precautionary approach is to guide fishery scientists and managers into a mode of *erring on the side of caution* in the face of uncertainty. As FAO (1995a: p. 6) noted:

- 'Management according to the precautionary approach exercises prudent foresight to avoid unacceptable or undesirable situations, taking into account that changes in fisheries systems are only slowly reversible, difficult to control, not well understood, and subject to change in the environment and human values.'
- 'Precautionary management involves explicit consideration of undesirable and potentially unacceptable outcomes and provides contingency and other plans to avoid or mitigate such outcomes. Undesirable or unacceptable outcomes include over-exploitation of resources, overdevelopment of harvesting capacity, loss of biodiversity, major physical disturbances of sensitive biotopes, or social or economic dislocations. Undesirable conditions can also arise when a fishery is negatively influenced by other fisheries or other activities and when management fails to take action in the face of shifts in the external conditions affecting, for example, the productivity of the fish stocks.'

In recent years, a major focus of attention in fishery systems, both nationally and internationally, has been on efforts to operationalise the precautionary approach. Much of the attention has focused on 'technical' matters, specifically, how to be precautionary in the task of determining and setting suitable TACs, effort limits and harvest rates. This involves a combination of *risk analysis* and analytical *risk management*. The approaches proposed typically call for adjusting exploitation levels to take uncertainty into account in some reasonable manner. In particular, the methods of risk management discussed above, such as management procedures, provide frameworks in which a precautionary approach can be implemented analytically.

These technical adjustments to the fishery decision-making process constitute an important aspect of the precautionary approach. A key example of this may lie in adjusting harvest levels to reflect the magnitude of uncertainty in fish stock sizes. At least as important as this technical side, however, are adjustments to the assumptions and approaches underlying fishery decision making. For example, structural uncertainty may exist in the form of a range of conflicting views of fish stock status. In the absence of accepted probabilities concerning the 'correctness' of each view, which view is to be adopted? Addressing this question is not an analytical matter, but rather one of determining suitable 'rules' *governing* decision making (as opposed to decision rules *per se*).

11.5.1 The burden of proof

How can we adjust the nature of the decision-making process itself to make it inherently precautionary? This typically involves reversing the *burden of proof* (Charles 1998a, b, c; Walters 1998), which in turn can mean reversing the 'default assumptions' built into scientific analysis and management approaches. Instead of requiring scientists, the public and others to 'prove' that harvesting levels or methods are harmful, the FAO (1995a: p. 6) has noted that 'often, the precautionary approach has been taken as requiring that human actions are assumed to be harmful unless proven otherwise (reversal of the burden of proof).'

At the same time, there is an important related matter, the *standard of proof*, which refers to the extent to which an action must be *proved* to be acceptable. This relates to 'the responsibility for providing the relevant evidence and the criteria to be used to judge that evidence' (FAO 1995a: p. 6). Specifically, 'the standard of proof to be used in decisions regarding the authorisation of fishing activities should be commensurate with the potential risk to the resource, while also taking into account the expected benefits of the activities'. This implies, for example, that 'although the precautionary approach to fisheries may require cessation of fishing activities that have potentially serious adverse impacts, it does not imply that no fishing can take place until all potential impacts have been assessed and found to be negligible' (p. 6).

As examples of possible applications of the precautionary approach and the burden of proof concept, consider the following claims (hypotheses), each of which, given the high levels of uncertainty involved, is plausible yet not conclusively supported by empirical evidence:

(1)   reducing the spawning stock will tend to produce fewer recruits;
(2)   substantially increasing harvests will deplete fish populations;
(3)   fishing gear pulled along the ocean floor will damage the habitat.

Ideally, the validity of each of these would be determined based on a risk-averse judgement of available data in a given circumstance, but if such data are not available, what actions should or should not be taken to address these claims? A typical approach in the past would simply have discounted the claims due to a lack of clear evidence. However, a precautionary approach, *erring on the side of caution*, cannot ignore such claims, unless of course they were *proved* to be incorrect. In this way, the default view of the world is one that favours conservation. The above three hypotheses are examined in sequence below.

11.5.2 *The stock–recruitment relationship*

The stock–recruitment (or spawner–recruitment) relationship is a particularly fundamental part of fishery science. It is based on a common sense idea: the future size of a fish population will depend functionally on how many new fish are born (and eventually 'recruit' to the fishery), which in turn depends on the number of adult fish spawning in the present generation. The essence of this concept is captured in the popular expression: 'no fish, no future'. A key implication is that catching too many fish this year means that too few will have the opportunity to reproduce, which will be detrimental to the future of the fish stock. (For some species, one can also have *too many* spawning fish. This can arise if there is potential, at high densities, for (a) cannibalism, (b) transmission of diseases, or (c) destruction by spawning fish of one another's spawning grounds. The implied functional relationship in this case first rises and then falls with the spawning stock (Hilborn & Walters 1992: p. 261).)

However, given the fluctuations in, and inherent complexities of, ocean and fish stock dynamics, together with uncertainties in the corresponding data available to scientists, it is not at all easy to *prove* that a spawner–recruit relationship exists in a given fishery. Lacking such proof, the logical connections between spawning stocks and the resulting young fish have not always been taken into account in the stock assessment process that determines allowable catch levels. Such a shortcoming, and in particular a lack of attention to the need to maintain spawning potential and the reproductive process, leaves the management process open to the possibility of so-called *recruitment overfishing* (arising from overestimates of the amount of fish that could safely be caught). The latter effect, and the long-term negative socioeconomic consequences it produces, are not inevitable, but they have occurred in many fisheries. (See Hutchings & Myers (1994) for the Atlantic Canadian groundfish case.) This situation arises in particular when the common 'yield per recruit' (YPR) method is used in setting TACs. In contrast, a method based on biomass dynamic models attempts to avoid the problem by incorporating stock–recruitment and other 'density-dependent' effects, albeit not without its own limitations. See Hilborn & Walters (1992) for a discussion of these different approaches.

In retrospect, it seems that a major reason why this situation persisted in the face of structural uncertainty has been a misplaced burden of proof; requiring scientists to 'prove' the common sense nature of spawner–recruitment relationships. A reversal of the burden of proof, through the precautionary approach (Garcia 1994), might involve several steps. First, a positive connection between spawners and subsequent recruitment would be taken as the default assumption (unless demonstrated otherwise). One approach that has been advocated for this is to assume, initially, a density-independent relationship (a straight line through the origin on a graph of recruitment versus stock size) rather than no relationship (a flat line). Second, such assumptions lead to an emphasis on maintaining strong spawning stocks as a basic necessity. Third, also implied is the need to devote more attention to research into determining *healthy* levels for spawning stocks; levels that are more likely to produce strong recruitment. The latter is reflected in ongoing efforts to develop guidelines for 'reference points' to assist in management decision making, based on desired spawning biomass levels per recruit.

11.5.3 Overfishing versus the environment

There is a long history in fishery research of debating and analysing whether fishery collapses are caused by overfishing or environmental change. Yet there also seems to be a tendency to focus blame on non-human causes first. For example, when Canada's Northern cod stock (in NAFO area 2J3KL) collapsed in the early 1990s, the government issued an initial press release that made no mention of human impacts on the resource, nor of problems with the fishery management process. Instead, it stated that 'the devastating decline in the stock of northern cod' was due 'primarily to ecological factors' (Department of Fisheries and Oceans 1992). The burden of proof, it seemed, often appears to be placed on non-human causes.

Yet it seems that while ocean conditions might have acted as a 'trigger' to initiate a stock collapse, the principal underlying cause of the collapse was more likely to have been high levels of resource exploitation. Although the dynamics of collapses, and the connections between fish populations and environmental change, are complex, there does seem to be a pattern to the typical sequence of events, as described below.

(1)   When ocean and environmental conditions are 'good' or 'average' (from the perspective of the fish), fundamentally unsustainable harvest levels may *appear* to be sustainable.

(2)   Inevitably, and quite naturally, ocean conditions will change at some point, so that heavily harvested fish populations become subjected to additional stress, i.e. environmental conditions that inhibit growth and reproduction.

(3)   Faced with intense fishing pressure, and a trigger in the form of an adverse environment, the fishery collapses.

This scenario corresponds to a considerable number of cases experienced around the world, from Canada's Pacific coast herring (*Clupea harengus pallasi*) fishery collapse of the 1960s (Hourston 1978) to the 1972 Peruvian anchovy (*Engraulis ringens*) collapse, triggered by the ocean cooling known as El Niño, but due fundamentally to massive levels of fishery exploitation (Pauly & Tsukayama 1987; Hilborn & Walters 1992: pp. 18–19).

Recent research on the Atlantic Canadian groundfishery also supports this interpretation (e.g. Taggart *et al.* 1994). Indeed, Hutchings & Myers (1994) and Walters & Maguire (1996) indicate that fishing pressure alone is sufficient to explain the collapse of Northern cod; invoking 'bad' ocean conditions as an explanation is simply unnecessary. In any case, a precautionary approach would suggest focusing more on reducing human impacts, rather than on blaming the ocean in the wake of fishery collapses.

Faced with this evidence, what does it tell us about a precautionary approach for the future? First, maintaining a sustainable fishery depends on keeping exploitation levels in line with the productive capability of the fish stocks, which in turn will vary with a complex mix of biophysical conditions. Unless the latter can be predicted closely (which is very unlikely!), a precautionary approach calls for caution in setting exploitation rates, since high rates may turn out to be unsustainable if those biophysical conditions change.

Second, it is worth noting that it is only *prior to and during* a collapse that management actions may be able, if not to avoid the collapse, then at least to minimise its magnitude and the subsequent recovery time. Furthermore, management actions clearly can only hope to affect human activities, not the state of nature. Thus, in the practical application of a precautionary approach, it would seem prudent to *assume* initially (at least in heavily exploited fisheries) that downturns are due to overfishing, rather than environmental change (unless the evidence is strongly to the contrary). This working assumption that human impacts are responsible for stock declines and that conservation actions should be taken to limit exploitation can then form the basis for corresponding conservation actions that 'err on the side of caution'.

11.5.4 Habitat protection

Now consider the controversial matter of trawling and its impacts on the ocean bottom. Many non-trawling fishers express concern about this impact, feeling that it results in a loss of habitat productivity and harm to their livelihood. Some analysts suggest a causal relationship between certain fishery collapses and the emergence of trawling as a dominant activity. However, others point to cases where trawling has taken place over long time periods, apparently without harming fishery productivity.

What is clear is (a) a certain knowledge that trawling has impacts on the habitat, and (b) a high level of uncertainty about the implications (positive or negative) of these impacts. Structural uncertainties in the ocean environment make it virtually impossible to prove (or disprove) any negative impacts on the habitat, the food chain and ocean productivity, and until recently, little research has been carried out to clarify the matter. In the absence of definitive proof, there are various possible responses to the issue of impacts, ranging from a complete or selective ban on trawling (i.e. avoiding the use of potentially damaging technologies in sensitive habitats), to a reduction in such activities, to amelioration efforts to reduce impacts, to maintenance of the status quo (where such impacts remain prevalent).

What response is appropriate in a given situation? How does the precautionary approach apply here? The first step lies in ensuring that the burden of proof is placed on the human activity in question – trawling – to demonstrate its lack of harm, rather than the reverse. Second, subject to this underlying need to 'err on the side of caution', the response chosen should depend on a careful assessment of the balance of evidence. Such a precautionary approach may lead to any one of the actions noted above.

However, it is remarkable that in the past, actions involving the removal or reduction of trawling activity have rarely been chosen. Instead, the chosen response has been to maintain the status quo (no reduction in trawling), despite a lack of strong evidence indicating no negative impacts. This suggests (while not proving!) the lack of regular application of a precautionary approach on this issue. It is less clear why such an approach was not applied. This may have been a simple case of avoiding actions which involved differentiating between users of the various gear types, because these would have political repercussions, given that trawlers are powerful players in world fisheries and catch a large share of the harvest.

As noted, in recent years there have been efforts to improve our knowledge of the subject through considerable research, which has indicated that impacts depend on such factors as the structural complexity of the habitat, and how the rate of fishing-related disturbance compares with natural rates of disturbance. While, not surprisingly, there are no definitive answers, this research may allow more selective decision making, depending on habitat type and environmental factors. With decisions on this issue now being possible from a broader knowledge base, and with more diligent application of the precautionary approach becoming mandatory, habitat impacts may receive different treatment in the future than they did in the past.

11.5.5 Implications of the precautionary approach

As described above, implementing the *philosophy* of the precautionary approach is an important move in many areas. However, putting the precautionary approach into practice properly is not a simple task, but requires new information and new research. For example, the FAO (1999c) has described a number of implications for the work of its Regional Fishery Bodies:

- 'better data are needed;
- uncertainty should be systematically investigated;
- outputs should be identified corresponding to objectives;
- target and limit reference points should be established;
- methods used for assessment need to be revised;
- robustness of management regime to (a) overfishing and (b) environmental change should be assessed;
- contingency plans should be developed.'

11.6 Summary

This chapter has reviewed the various forms of uncertainty in fisheries, the methods available for analysing them and the particular challenge posed by structural uncertainty. In responding to this challenge, emphasis must be placed on *living with uncertainty*, through:

- appropriate risk assessments within the scientific realm;
- fishery management that is precautionary, as well as robust and adaptive (Chapter 15);
- appropriate institutions to implement such a management approach (Chapters 13 and 15).

A shift in fishery thinking towards *living with uncertainty*, while in many cases coming in response to dramatic crises, is very much a positive development. It reflects a realistic view of fishery management, in which analytical tools are used wherever possible to understand and assess uncertainty, but at the same time it is recognised that uncertainty will never fully disappear. That the fishery world is inherently uncertain is cause for some humility. Yet that humility, combined with suitable new management approaches and management institutions, may be exactly what is needed to move us in the right direction, towards truly sustainable fisheries.

Chapter 12

Complexity, Diversity and the Ecosystem Approach

Fishing is not infrequently portrayed in the media as a 'simple' activity, one that evokes a bygone era, one not in keeping with the complex world in which we now live. That sense of 'simplicity' is reinforced by the stereotypical view of a recreational angler, travelling from an urban home to the countryside for the 'simple' life, sitting with a fishing rod beside a quiet pond.

Yet as we have seen throughout this book, viewing the fishery as a system leads to a recognition of the variety of closely interacting, dynamically varying components involved. Many different species of fish inhabit the aquatic ecosystem, living out of sight, their populations changing, sometimes dramatically, from year to year. A spectrum of fishers, including full-timers and part-timers, fixed gear (e.g. hook and line or gill nets) and mobile gear (e.g. trawlers), small-scale (artisanal, usually inshore) and large-scale (industrial, typically offshore), try to find the fish and catch them, using a fleet that changes in number and power over time. Beyond the harvesting sector, the system includes processors, distributors, marketing channels, consumers, government regulators and support structures, as well as coastal communities and human institutions. In the background, but also of great importance, are the social/economic/cultural and the biophysical environments within which the fish and the fishers live. Even the recreational angler mentioned above is part of a system that includes the pond ecosystem, sport fishery outfitters, managers, researchers, transportation infrastructure and so on.

In what sense are these systems *complex*? From an informal perspective, a system might be considered to be complex if we do not understand its structure and functioning very well, as is certainly the case with the fishery. More formally, a variety of precise definitions of *complexity* have been developed by specialists in the subject. For the present discussion, we take an intermediate path, viewing a complex system as one comprised of many components, with many interactions among those components. In this light, the greater the number of species in a system, the greater is the complexity, other things being equal, and a system with a given set of species is more complex the more intricate are the interactions among those species.

In this way, complexity is closely related to *diversity*. The most widely discussed form of diversity is *biodiversity*.

'At the simplest level, biological diversity (biodiversity) is defined as the degree of variation in the numbers and kinds of species and ecosystems within or among regions at a given time. More precisely, biodiversity has been expressed as the collection of genomes, species, and ecosystems occurring in a geographically defined region …'

de Young et al. (1999)

Note that diversity in this sense concerns not only the variation among individual animals (genetic diversity) and populations (species diversity), but also *functional diversity* that pertains to 'what the organisms do and the variety of responses to environmental change, especially the diverse space and time scales to which organisms react to each other and the environment' (Hammer *et al.* 1993: p. 97), the latter form of diversity being driven by variation at the ecosystem level.

The overall concept of diversity is also usefully applied in an analogous manner to the human sub-system in the fishery:

'… human (social and economic) diversity can be viewed as the degree of variation, or "richness", in economic and social entities within a region. This might be reflected in the existence of diverse livelihood/employment opportunities, diverse coastal communities (each with its own special characteristics), or diverse management approaches for conserving and exploiting natural resources.'

de Young et al. (1999)

The recognition that complexity and diversity are dependent to a considerable extent on the number of entities involved in a system, and the extent of interactions amongst them, implies

---

Some sources of complexity in fishery systems

- Multiple and conflicting objectives.
- Multiple species, and ecological (trophic) interactions among them.
- Multiple groups of fishers, interacting with households and communities.
- Multiple fishing fleets, and conflicts among them.
- Multiple gear types, and technical interactions among them.
- Multiple post-harvest stages: from fisher to consumer.
- Multiple ancillary activities, and interactions with the fishery.
- The marine environment and biophysical influences on the fishery.
- The social structure, and sociocultural influences on the fishery.
- The institutional structure, and interactions between fishers and regulators.
- The socioeconomic environment and interactions with the macro-economy.
- The coastal system and interactions among its components.
- Dynamics of fishers, fleets, technologies and resources.
- Dynamics of fishery information and dissemination.
- Dynamic interactions among fish, fishers and environment.
- Objectives and behaviour of fishery participants.
- Uncertainties in each component of the fishery system.

that many factors contribute to the complexity of fishery systems. Some of these are outlined in the box on p. 223. Furthermore, it is important to note that the *resolution* with which a system is examined can influence the complexity we see. For example, what appears in aggregated form as a 'simple' fishery, targeting a single fish stock with a single fleet of vessels, may, when viewed more closely, contain a complex mix of fishers within the fleet, complex spatial interaction among subregions, and so on. The relatively simple 'macro' view in this case belies the complexity at the 'micro' level.

There is a great tendency to view complexity (as well as that other omnipresent feature of fisheries, uncertainty) as a regrettable nuisance. It is *difficult* to manage, or to operate, within a highly uncertain, complex fishery system. Surely the manager would prefer a simpler arrangement, such as the 'laboratory' environment a scientist establishes – a controlled, deterministic ecosystem of predator and prey within a closed box!

Yet it is crucial to recognise a fundamental truth: ultimately, uncertainty and complexity are beneficial. As noted in Chapter 11, uncertainty is in part a manifestation of the fluctuations and perturbations that, while problematic to the manager, also help to build the system's resilience, i.e. its ability to 'bounce back' or absorb unexpected shocks and perturbations without collapsing. Furthermore, this resilience seems to be fundamentally related to the complexity of the system, as well as the diversity inherent within it. As noted by de Young *et al.* (1999), diversity is helpful at each of the genetic, species and community levels:

> 'a population with a wide range of genetic information among individuals is more likely to adapt successfully to changing environmental conditions than is a population with greatly reduced genetic diversity due to strong selection, witness the numerous examples of single-variety agricultural crops suffering from massive outbreaks of diseases. Similarly, a community composed of many species, some of which can fulfil the functional roles of other species, is more likely to persist through periods of considerable environmental change.'

In other words, the two key ingredients that make the fishery hard to deal with in the short term, complexity and uncertainty, are also major factors contributing to long-term sustainability! Even if it were possible, it would be counter-productive to 'solve' the 'problem' of complexity. Just as we must learn to live with uncertainty in fisheries, as discussed in Chapter 11, it is also important to embrace diversity as a positive attribute of fisheries, and adapt to the complexity inherent in fishery systems. This perspective should not be seen to imply that any and all human activities in the fishery, including, for example, destructive fishing methods, are defensible on the basis that they add to the 'diversity' in the human system. Rather, this approach is reflected positively in the considerable worldwide interest in biodiversity conservation: see Charles & Leith (1999) for a review of the literature on this subject, with a focus on aquatic systems.

Given the importance of, and challenges posed by, complexity in fishery systems, this chapter focuses on how a *systems perspective* can be applied to fishery management and policy making, taking into account, as far as possible, the many components of the fishery, as well as the broader natural and human systems, and the complex interactions among all these. (In Chapter 15, complementary approaches to embrace complexity are considered, through *robust management* and '*sustainable fishery*' policy.)

'A systems approach to fisheries has to be built into decision making. This requires understanding of the interaction between the resource, natural environment, fishing activities, technology (including the impact of fishing gear on species and habitats), human behaviour and social and economic factors. In a systems approach, integrated management principles should be used. This means decision making that brings all affected interests and players together in the management process. Scientists of different disciplines should work together to solve fishery problems … Fishing activities must be understood as well as the fish resource.'

FRCC (1997a)

Thus, a systems perspective involves *integrated* approaches both to studying and to managing the fishery. The goal is to incorporate key elements of fishery complexity into our thought processes and decision-making processes. It is crucial to note, however, that taking a holistic view of fisheries will not lead to nirvana – to perfect knowledge of the system. Indeed, embracing complexity implies recognising the *limits* to management, as well as the need for such management. Efforts are needed to counter the persistent *fallacy of controllability* (Chapter 15), i.e. a belief that more can be controlled in the fishery than is possible in practice, given the inherent complexities and uncertainties, and to focus instead on the challenge of developing management and policy measures to maximise the overall sustainability of what is in reality an uncontrollable system.

Several integrated approaches are discussed in the remainder of this chapter. First, we focus in considerable detail on a dominant direction in fishery systems worldwide, the *ecosystem approach*. We then turn to specific integrated approaches in the areas of fishery management, fishery development, fishery research and fishery modelling.

12.1 The ecosystem approach

As discussed in Chapter 2 and subsequently, fish stocks inhabit aquatic ecosystems that clearly have great influence on the fish, and which, from a human perspective, are both inherently complex and highly uncertain. These are by no means new revelations. Furthermore, the role of the ecosystem in creating and maintaining a sustainable fishery, and the need for fishery management to take into account ecosystem interactions and ecological factors, are themes that have long been recognised (Caddy & Sharp 1986), and which have received increasing attention in recent years (e.g. Botsford *et al.* 1997). These themes are at the heart of the *ecosystem approach* to fishery research and management, an approach the various defining characteristics of which will be examined fully below. First, however, we begin with a discussion of one particular element of the approach: the incorporation of multi-species interactions into fishery stock assessment and management.

12.1.1 Multi-species stock assessment and management

In considering the incorporation of ecosystem factors into fishery thinking, among the first pieces of the puzzle which might be included would probably be the interactions that exist among the fishery's target fish stocks as well as between these and the non-fished species with which they interact. The interactions may be ecologically based, involving predators, prey or competitors in the same ecosystem, or may be fishery-related. Examples of the latter include:

- *by-catch*, e.g. discarding 'trash' species in shrimp trawling, or catching juveniles of one species while targeting another species;
- *non-fish by-catch*, e.g. seabirds caught in long-line fisheries, dolphin caught in tuna nets, seals and whales tangled in fishing gear;
- *technical interactions*, e.g. target species caught together in fishing gear, but not in the desired proportions.

To understand these multi-species interactions, and to take them into account in fishery management, it would seem logical that assessment and management approaches are needed which deal with multiple species together (Gulland 1982). Indeed, there have been important research projects looking at multi-species interactions, in some cases within the context of fishery management: see, e.g. Ursin (1982), Daan (1987), and papers in the symposium volumes of Mercer (1982) and Daan and Sissenwine (1991). Such approaches can draw on a considerable body of research, built up over several decades, on ecologically oriented interactions among aquatic species.

Practical examples of multi-species assessment and management can also be found. In tropical and/or developing regions, where fisheries often exploit a large number of intermingling harvested species, and resources for assessment and management are very limited, there may well be a *de facto* focus on multi-species 'stock complexes' rather than on individual species (Pauly *et al.* 1989). In 'northern' nations, incorporation of multi-species approaches has been attempted in certain cases, notably in the North Sea (Daan 1997). There have also been efforts to examine the special nature of multi-species management from an economic perspective (e.g. Wilson 1982).

Overall, however, it seems clear that the practice of stock assessment and fishery management, at least in industrialised nations, has developed largely on the basis of *single-species* approaches. Typically, each fish stock is analysed and managed individually, separate from other species and from the surrounding ecosystem. This almost exclusive focus on single species may be found, for example, in classic stock assessment manuals such as Ricker's (1975) *Computation and Interpretation of Biological Statistics of Fish Populations*. Even today, most stock assessment practice, and almost all fishery management endeavours, are based on single-species methods.

It is interesting to note the ongoing discussion among stock assessment specialists regarding the merits of multi-species methods. For example, the eminent fishery scientist John Gulland noted, in the early 1980s, the 'common criticism by scientists ... of traditional stock assessment work [is] that it is too often concerned only with what happens to a single species ... and that proper account is not taken of what is happening to other species with which the target species interacts' (Gulland 1983: p. 186). With regard to this criticism, he commented that:

> 'In the past, the history of analysis and advice based on single-species approaches ... has been moderately satisfactory.'

but warned that:

> 'This comforting situation is changing with the increasing range of species being exploited by present-day fisheries. There is an increasing risk that analysis and advice based on a single-species approach will be incomplete in some important aspects.'

Gulland (1983: p. 187) elaborated on the latter point by providing several practical reasons to move from a single-species to a multi-species approach.

(1) 'Changes in the fishery on target species can have effects on other species, and to the extent that it is increasingly likely that the non-target species will themselves support significant fisheries, these effects may have important practical impacts.'
(2) 'Another reason for a single-species analysis giving rise to misleading conclusions is a change of interest by the fishermen from one species to another ... For example, statistics of [catch per unit effort] of a given species will not provide a satisfactory index of the abundance of that species if it ceases to be a preferred target of the fishermen.'
(3) 'Also, fisheries on other species can affect the stock of the target species, so that the conclusions from a single-species analysis may become somewhat irrelevant.'

If Gulland's perspective is valid, why is there so little multi-species assessment and management today? One response is that such efforts are very difficult to implement and require considerable resources. Gulland (1983: p. 186) offered a specific concern about embarking on multi-species stock assessment, relating to limitations on personnel and financial resources, and the trade-off between single- and multi-species studies:

> '... any good stock assessment scientist would indeed like to take interspecific interactions into account. [But] too much attention paid to the wider ecological questions can too easily mean inadequate attention being paid to determining the effect of fishing on the target stocks (whether of one or several species). The result can then be poor advice on the effects of changes in fishing strategy (development or management).'

Remarkably, a similar sentiment was expressed years later by the Committee on Fish Stock Assessment Methods (1998: p. 4), which stated:

> 'The committee believes that single-species assessments provide the best approach at present for assessing population parameters and providing short-term forecasting and management advice. Recent interest in bringing ecological and environmental considerations and multi-species interactions into stock assessments should be encouraged, but not at the expense of a reduction in the quality of stock assessments.'

Other difficulties with implementing multi-species approaches include (a) limitations on the data required for such approaches, (b) shortcomings, and a lack of testing, of specific multi-species methodologies, and (c) a possible inertia in some fishery management agencies that could inhibit the development of practical experience with multi-species approaches.

Despite these challenges, it is certainly true that fishery science has long recognised the importance of multi-species interactions, and that considerable research has been carried out on such themes (e.g. Pitcher & Hart 1982). While, for various reasons, stock assessment and fishery management remain focused largely on single-species methods, the situation may be slowly changing. This may be due partly to the emergence of the ecosystem approach, and partly (in terms of fishery management practices) to such major international initiatives as the FAO's *Code of Conduct for Responsible Fisheries* adopted by the United Nations in 1995

Multi-species assessments

While stock assessment has remained largely focused on single species, interest has been expressed for decades in pursuing multi-species approaches (Gulland 1982). Not surprisingly, such approaches tend to be rather complicated and data-intensive, but some success has been achieved (see the discussion on the topic by Hilborn & Walters (1992) among others). Some of the methods available are given below.

- Multi-species virtual population analysis (MSVPA) is an extension of the standard VPA approach that was developed largely through work in the International Council for the Exploration of the Sea (ICES). As Daan (1997: p. 126) notes, 'The primary objective of MSVPA … is to quantify feeding interactions among species in relationship to the interaction between fish stocks and fisheries.' The method has been applied particularly to the North Sea, where it has provided useful insights, although there have been few changes in management measures relative to those suggested by single-species models.
- Analyses of ecosystem structure and function, while not within the usual scope of stock assessment, can be a useful tool in studying fisheries. The best known methodology for this uses the computer software ECOPATH, which first appeared in the early 1980s (Polovina 1984) and is now widely distributed globally (Christensen & Pauly 1996).
- Studies of the status and dynamics of ecological communities within an aquatic environment can be undertaken using diversity indices and similarity indices (e.g. Saila *et al.* 1996).
- Approaches based on aggregating the various species into a small number of logical groupings include the aggregated production model (Ralston & Polovina 1982) and empirical multi-species analysis using Markov transition matrices (Yang 1989; Saila *et al.* 1996).

As interest in an ecosystem approach becomes widespread, it is likely that the development of multi-species assessment methods will also progress as an important component of this effort, although it must be noted that adopting an ecosystem approach is by no means dependent on solving the tricky problems inherent in multi-species assessment!

(FAO 1995b). This seeks to reduce the negative conservation impacts of the technical interactions noted above, and is having an impact on how fisheries operate at the international level as well as within many nations. Specifically, the *Code*:

- 'sets out principles and international standards of behaviour for responsible practices with a view to ensuring the effective conservation, management and development of living aquatic resources, with due respect for the ecosystem and biodiversity;'
- 'covers the capture, processing and trade of fish and fishery products, fishing operations, aquaculture, fisheries research and integration of fisheries into coastal area management.'

Implementing the *Code of Conduct*, as well as the precautionary approach and the ecosystem approach, may well require increasing moves to incorporate multi-species approaches into fishery assessment and management in coming years.

12.1.2 The scope of an ecosystem approach

The previous section focused on the importance of multi-species interactions in fishery systems, whether ecologically oriented or fishery-related. It is notable that such interactions may have an immediate and direct impact on the fishery in question (for example, through harvesting of competitors in an ecosystem), or an indirect impact on other fisheries harvesting interacting species (e.g. if fisheries for cod are affected by the harvesting of herring, a food source of the cod), or impacts on unfished species playing possibly unknown but potentially crucial roles in the ecosystem.

The latter possibility, which relates to concerns for biodiversity conservation, leads us to a broader view of fishery management. Indeed, within such a perspective, it is important to look beyond multi-species interactions, at the impacts of fishing on habitat integrity and quality (e.g. impacts on demersal habitats from trawls, dredges, etc.), and broadly at the structure, dynamics and integrity of the ecosystem as a whole. These themes are part of what is referred to as the *ecosystem approach*, which the FAO (1999c) has referred to as 'a primary policy direction of contemporary importance ... often considered as an element of the precautionary approach'.

The ecosystem approach is also reasonably referred to as *ecosystem-based management*. Another term that is sometimes used is 'ecosystem management'. The latter is perhaps used simply as a form of shorthand for the above terms, but unfortunately has a literal meaning that suggests an objective of 'managing the ecosystem'. In reality, the goal would seem more modest: to ensure that the ecosystem is taken into account when managing human involvement in fishery systems; in other words, to adopt an *ecosystem approach* within our management efforts. (In the following, relevant studies are referred to, whatever term is used therein to describe the ecosystem approach.)

What, then, *is* this ecosystem approach? Larkin (1996: pp. 146–147) suggests that in the context of marine ecosystems, the term:

'... is scientific shorthand for the contemporary appreciation that fisheries management must take greater note of the multi-species interactions in a community of fish species and their dependence on underlying ecosystem dynamics.'

Larkin suggests further that, as applied within the marine environment, an ecosystem approach has three essential components:

(1)   'sustainable yield of products for human consumption and animal foods;
(2)   maintenance of biodiversity;
(3)   protection from the effects of pollution and habitat degradation.' (p. 149)

A Canadian advisory body, the Fisheries Resource Conservation Council (FRCC 1994: p. 118) has used plain language to put the idea of an ecosystem approach to fishery management more broadly:

> 'The various bits and pieces of ecological knowledge must be reflected in a better understanding of the whole system. Thinking in terms of whole ecosystem must become an essential and integral part of day-to-day activities, not just for Science, but within the [fishery management organisation] generally.'

This goal is clear in the mandate of an organisation often referred to for its strong pursuit of the ecosystem approach, the Commission for the Conservation of Antarctic Marine Living Resources (CCAMLR). The CCAMLR is guided by a convention, Article II of which notes such principles as 'maintenance of the ecological relationships between harvested, dependent and related populations of Antarctic marine living resources' and 'prevention of changes or minimisation of the risk of changes in the marine ecosystem which are not potentially reversible over two or three decades' (CCAMLR 1999). The CCAMLR documentation goes on to note:

> 'These stringent principles embody what has been called the ecosystem approach to living resource conservation and set the Convention apart from other marine resource management regimes. Management of fishing must not only aim to conserve the targeted species but take into account the impact of fishing on those animals that prey on and compete with the targeted species. In its broadest interpretation, the Convention requires that management action should take account of the impact of activities on all living organisms in the Antarctic ecosystem or subsystems.'

As a result of discussions on 'further integration of fisheries and environmental protection, conservation and management measures' in the North Sea, European environment ministers (Anonymous 1997: p. 9) called for the adoption of an ecosystem approach, and suggested this be based on three elements:

- 'the identification of processes in, and influences on, the ecosystems which are critical for maintaining their characteristic structure and functioning, productivity and biological diversity;
- taking into account the interaction among the different components in the food webs of the ecosystems (multi-species approach) and other important ecosystem interactions; and
- providing for a chemical, physical and biological environment in these ecosystems consistent with a high level of protection of those critical ecosystem processes.'

The ecosystem approach in the North Sea

Conclusions of the 1998 Workshop on the Ecosystem Approach to the Management and Protection of the North Sea (Anonymous 1998).

(1) It may be difficult or impossible to manage the North Sea towards a desired ecosystem state. We may, however, manage the human activities in an integrated manner to achieve sustainable use and protection of the North Sea.

(2) There is a need for agreed definitions of terms such as 'ecosystem' and 'ecosystem approach'.

(3) Clear objectives for an ecosystem approach to the management and protection of the North Sea must be formulated. There is a need for objectives both at the general level, as overall or integrated objectives, and at the specific level, as more detailed and operational objectives.

(4) The management of the North Sea should be based on the best use of the present scientific knowledge. In particular, there is a potential for more extensive use of existing ecological knowledge.

(5) The present knowledge of the North Sea as an ecosystem does not provide a sufficiently good basis for full implementation of an ecosystem approach to North Sea management. There is, therefore, a need for focused research on the North Sea ecosystem, including climatic, biological and human driving forces of ecosystem variability.

(6) The present monitoring of the North Sea is often insufficient to reveal human impacts on the ecosystem. There is a need for improved, integrated monitoring through coordination and harmonisation of existing national and international monitoring activities, as well as through implementation of new methods and technology.

(7) There is a need for integrated assessments prepared by experts on North Sea fish stocks, environment and socioeconomics.

(8) Stakeholders, along with scientists, managers and politicians, should be involved at different stages of the decision process to promote openness, transparency and responsibility.

Thus, the ecosystem approach is precisely that: an approach. It is not a methodology as such, and in particular should not be confused with 'ecological analysis', but rather it is a method of thinking, of tackling fishery and ecosystem problems. Two major components of the 'big picture' of the fishery system need highlighting.

First, it is important to incorporate aspects of the biophysical environment (Chapter 2). This is already reflected somewhat in the evolution of thinking in fishery science, as described by Mann & Lazier (1996: p. 282):

'Fisheries scientists began to try to define the maximum catch that could be taken on a yearly basis while still permitting the stock to remain vigorous; they called it the maximum sustainable yield. At first these calculations were done with the environment of the fish (e.g. physical and chemical conditions, food supply, predator pressure) held constant. Gradually, fisheries scientists became aware that these assumptions were unrealistic ... Contemporary fisheries management policy, therefore, seeks to control fishing effort against a background of natural fluctuations in stock sizes. For many years, the fluctuations seemed totally unpredictable, but as people began to study oceanographic processes on larger and larger scales, it gradually became apparent that many local changes in fish stocks were related to large-scale processes in the ocean.'

Second, it is crucial to consider the human system as well. Schramm & Hubert (1996: p. 7) note that 'Ecosystem management recognizes that humans – including their societies, technologies, economies, needs and *values* – are part of the ecosystem.' This point is also made strongly by Kay & Schneider (1994: p. 38):

'If we are truly to use an ecosystem approach, and we must if we are to have sustainability, it means changing in a fundamental way how we govern ourselves, how we design and operate our decision-making processes and institutions, and how we approach the business of environmental science and management. This is the real challenge presented by an ecosystem approach.'

Hammer (1998: p. 51) notes the need for specific institutional arrangements for the flow of information:

'To implement an ecosystem approach, there is a need for social mechanisms by which information from the environment may be received, processed and interpreted to build sustainable and resilient socio-ecological systems.'

Finally, Slocombe (1993) has synthesised a useful checklist of key characteristics to be incorporated in an ecosystem approach, focusing on the perspective taken by humans attempting to study and/or manage systems such as the fishery. The ecosystem approach:

- describes parts, systems, environments and their interactions;
- is holistic, comprehensive, transdisciplinary;
- includes people and their activities in the ecosystem;
- describes system dynamics (such as cause-and-effect relationships, self-organisation, homeostasis and feedbacks);
- defines the ecosystem naturally, e.g. bioregionally, rather than arbitrarily;
- looks at different levels or scales of system structure, process and function;
- recognises goals and takes an active management orientation;
- includes actor–system dynamics and institutional factors in the analysis;
- uses an anticipatory, flexible research and planning approach;
- entails an implicit or explicit ethics of quality, well-being and integrity;
- recognises systemic limits to action – defining and seeking sustainability.

Ecosystem integrity

'For an ecosystem, integrity encompasses three major ecosystem organisational facets. Ecosystem health, the ability to maintain normal operations under normal environmental conditions, is the first requisite for ecosystem integrity. But it alone is not sufficient. To have integrity, an ecosystem must also be able to cope with changes (which can be catastrophic) in environmental conditions; that is, it must be able to cope with stress. As well, an ecosystem which has integrity must be able to continue the process of self-organisation on an ongoing basis. It must be able to evolve, develop and proceed with the birth, growth, death and renewal cycle. It is these latter two facets of ecosystem integrity that differentiate it from the notion of ecosystem health.'

Kay & Schneider (1994: p. 37)

12.2 Marine protected areas

While the multi-species assessments, ecosystem analyses and lists of required attributes discussed above are all useful aspects of an ecosystem approach, perhaps the best concrete manifestation of such an approach, reflecting its *integrated*, holistic nature, is the development and implementation of marine protected areas (MPAs, referred to more or less synonymously as marine reserves). There are many different views of what MPAs are about, but the most basic and generally agreed definition is perhaps derived from the name itself: MPAs are areas of the marine environment designated for some form of 'protection'. As discussed below, this may be implemented through complete prohibitions on entry or harvesting, through comprehensive controls on extraction and use, or through regulation of specific designated activities.

12.2.1 The potential benefits of MPAs

The Independent World Commission on the Oceans (1998: p. 200) provides a list of some potential benefits, within both the natural and the human systems, provided by MPAs:

- 'protection of marine species at certain stages of their life cycle;
- protection of fixed, critical, habitats (e.g. coral reefs, estuaries);
- protection of cultural and archaeological sites;
- protection of local and traditional sustainable marine-based lifestyles and communities;
- provision of space to allow shifts in species distributions in response to climate and other environmental changes;
- provision of a refuge for recruits to commercial fisheries;
- provides a framework for resolving multiple stakeholder conflicts;
- provides models for integrated coastal zone management;
- provision of revenue and employment;
- provision of areas for scientific research, education, and recreation.'

Two additional benefits should be noted. First, MPAs can play a role in helping to diversify the coastal economy, e.g. through tourism and conservation work. This can reduce stress on fish stocks and increase community resilience. Second, perhaps the most fundamental role of a marine protected area is a hedge against uncertainty, a form of conservation 'insurance policy' (Dugan & Davis 1993; Rowley 1994; Holland & Brazee 1996).

12.2.2 Types of MPAs

Essentially, three major approaches are taken to this protection, which together define the range of what might be considered an MPA: closed areas, no-take reserves and multiple-use zoned MPAs. Each of these is discussed briefly below.

12.2.3 Closed areas

First, there are specifically targeted closed areas, commonly instituted to prevent fishing on spawning and/or nursery grounds, but also to prohibit other activities, such as oil and gas exploitation. In such areas, the closure may be temporary (seasonal) within a given year, perhaps during a determined spawning period, or it may be permanent. In either case, however, the restriction is typically placed only on a certain activity. For example, fishing for groundfish may be prohibited in a certain closed area known as a cod-spawning ground, but fishing within that area may be permitted for shellfish. Or mineral extraction may be prohibited, even though fishing is allowed to take place.

   An example of a permanent but limited closure is the 'haddock box', a designated area of ocean on the Scotian Shelf off Nova Scotia, Canada. This area was identified by fishers as being important for spawning and as a juvenile area for haddock. Fishers themselves therefore pushed for, and strongly support, a prohibition on targeting the haddock stock within the 'box' (despite a lack of scientific 'proof' of its effectiveness in helping the stocks). The fishers are convinced that protecting the area is both in their own self-interest and the 'right thing to do' for conserving the haddock stock.

   Closed areas have the advantage that their purpose is very apparent and clear-cut. Fishers, for example, can see that the goal of a fishery closure is to protect and conserve a specific stock. On the other hand, this narrowness means that while such closed areas play an important role in fishery conservation, and may well provide indirect ecosystem benefits, they do not provide comprehensive *ecosystem conservation*, and do not reflect a full ecosystem approach.

12.2.4 No-take reserves

Perhaps at the other end of the range of 'protection' are more restrictive MPAs in which no extractive activities are permitted: so-called *no-take reserves*. These reserves may well permit tourism and recreational activities, such as diving and boating, but essentially they are meant to be parts of the ecosystem that are relatively unaffected by human use. The establishment of a single such MPA may reflect a specific objective, such as protection of a fragile or unique habitat, while a system of several no-take areas could serve the broader pursuits of biodiversity conservation and ecosystem well-being. Bill Ballantine (1994), one of the most influential advocates of no-take reserves, has also noted the important social roles of such reserves as tools for public education and as a means to reflect 'heritage and moral values' (p. 210).

Because no-take reserves are (by definition) particularly restrictive in the uses allowed, obtaining societal support can be difficult. For this reason, they tend to be relatively small in size, albeit possibly within the context of a network of such areas, or within a zoned approach (see below). On the other hand, the value of 'large space–time refuges', indeed, the idea of closing most of a fishing area, leaving only a small part open, has been suggested as a way to increase the sustainability of a fishery system (Walters 1998).

Sobel (1993: p. 23) notes that 'among the earlier and better documented reserves are New Zealand's Leigh Marine Reserve (from Cape Rodney to Okahari Point), established in 1977; France's Scandola Nature Reserve established in 1975; and the Philippine's Sumilon Reserve, established in 1974.' (See Russ & Alcala 1994, for a discussion of the latter.) The impacts of no-take reserves are under investigation by various researchers both experimentally (as reviewed by Rowley 1994) and through modelling approaches (e.g. Holland & Brazee 1996).

12.2.5 Zoning the ocean

In contrast to no-take reserves, the major alternative approach is that of larger-scale zoned, multiple-use MPAs. An increasing proportion of MPAs are of this form, designed from a multi-objective perspective to allow multiple uses. Spatial zones are designated within such an MPA, with certain activities permitted within each. For example, there could be an interior no-take area at the 'core' of the MPA, together with harvest limitations throughout the MPA. From a fishery perspective, conservation and stock rebuilding efforts in outer zones of the MPA might be aided by larval migration from no-take areas, and pressure on the fish stocks might be reduced through economic diversification in the neighbouring coastal communities (Chapter 15), supported by tourism development that makes use of the MPA.

King (1995: p. 282) notes that among the possible zones in a multi-use MPA are:

- preservation zones (no access allowed);
- wilderness zones (with access allowed, but no exploitation permitted);
- recreational zones (with only regulated recreational fishing allowed);
- traditional fishing zones (with exclusive fishing rights held by local fishers);
- scientific zones (with exploitation allowed only for scientific purposes);
- experimental zones (to assess the impacts of differing levels of exploitation).

Perhaps the best-known zoned MPA is the Great Barrier Reef Marine Park (GBRMP) in Australia. The Independent World Commission on the Oceans (1998: p. 199) notes that this is the world's largest MPA, at roughly 350 000 km^2, and it is also 'the largest living feature on earth'. It incorporates approximately 2500 individual reefs, with more than 4000 species of mollusc, 1500 species of fishes and hundreds of species of corals and birds. In keeping with the pattern of zoning noted above, the GBRMP is organised so as to provide varying levels of protection, and to provide for a wide range of uses. Another example of a zoned MPA, this time at local community level, is referred to in Fig. 12.1.

12.2.6 Planning and evaluation of MPAs

Some key factors in planning a marine protected area are given below.

- *Scientific design*. There is, at times, a sentiment that MPAs should be put in place quickly, because any effort to protect areas of the ocean is better than no such effort. However, a number of complexities must be considered.
  - First, an MPA of a less than optimal size, put in a poor location, may have some benefit, but not enough to meet expectations. This could lead to a loss of support for the entire concept of MPAs, thereby being counter-productive overall.
  - Second, the placement of an MPA in one location will probably lead to greater fishing activity in other locations. If a particular fish species has a sub-stock structure, this could lead to excessive effort exerted on one stock component.
  - Third, there is considerable uncertainty as to how an MPA will perform within a given environment, due to both (a) the lack of consensus on design criteria for MPAs, and (b) the unavoidable vagaries of ocean conditions and of migrations by aquatic species into and out of the MPA at various life-cycle stages.

Fig. 12.1 An important example of a marine protected area (MPA) based on the 'ocean zoning' approach is the Soufrière Marine Management Area, located next to the community of Soufrière, shown here, within the Caribbean state of St Lucia. This community-designed MPA incorporates zones for fishing, yacht mooring, recreation, no-take reserves and multi-purpose use. Its creation has helped protect local coral reefs while also supporting local employment, notably in the areas of tourism and transportation. (See Chapter 13 for further discussion of the process by which this MPA is managed.)

These issues are by no means resolved, but some insights can be obtained from existing research into the spatial dynamics of fish stocks. This suggests in particular that a key issue to consider is the *rate* at which aquatic species move between the MPA and other areas. The idea can be illustrated from a strictly fishery perspective: if there is little flow out from the MPA, any increases in biomass within the MPA will not translate into fishery benefits, while if there is rapid exchange, fishing pressure outside the MPA will prevent a build-up of the stock inside. In any case, the key point to reiterate in terms of scientific design is that careful consideration is needed before establishing MPAs, and monitoring is needed after the fact.

- *Acceptability*. Experience has shown that the imposition of MPAs without broad consensus is a recipe for failure. It is crucial, therefore, to (a) understand the nature of situations under which acceptance is most likely, and (b) undertake consultation, design, implementation and monitoring of the MPA using participatory processes. Note that the need for acceptability and the need for a scientific basis in establishing MPAs may lead to conflict. For example, it is possible that an 'acceptable' MPA is determined to be small and away from fishing grounds, while a scientifically 'desirable' MPA may be large and centred on the fishing grounds.

- *Evaluation of benefits and costs*. It is important to develop the means to determine the success or failure of an MPA in meeting its objectives. This is a challenging task, since a major objective in many cases is to improve or maintain the structure and functioning of the marine ecosystem (in the spirit of an ecosystem approach), but it is not a simple task to determine whether the objective has been achieved. An adaptive management approach can be useful as a means to assess the MPA continuously as new information becomes available, and indeed to manage the MPA in such a way as to maximise the flow of information. In any case, a 'benefit–cost' framework can assist in undertaking the evaluation. Some of the major classes of such benefits and costs are listed below.

Benefits

- Increased fishery and other direct resource use benefits.
- Increased benefits from non-consumptive use (e.g. eco-tourism).
- Increased 'spin-off' benefits to the coastal economy.
- Non-use value (e.g. increased oxygen production from the sea).
- Existence value (the societal value derived from the existence of the MPA).
- Option value (the value of maintaining a marine ecosystem for future use).

Costs

- Opportunity costs (foregone catches, etc., due to restrictions in the MPA).
- Management costs (additional costs incurred to manage the MPA).

- *Equity issues*. In addition to measuring the benefits and costs of MPAs, it is also important to examine distributional considerations. Who receives the benefits? Who suffers the costs? How are benefits and costs distributed spatially, and at local, regional and national levels? How are benefits and costs distributed over time?

12.3 Integrated approaches in the fishery management system

The ecosystem approach is a key conceptual tool to embrace complexity in fisheries, and a good example of applying *integrated thinking*, in which the system is considered as a whole, incorporating the various components and interactions amongst them. The same kind of thinking can also be fruitfully applied throughout the fishery management system. Here, we consider some aspects of an integrated approach within the areas of fishery management, development, research and modelling.

12.3.1 Fishery management

An integrated approach to fishery management is based on structured decision making, grounded in a set of societal objectives, governed by a set of operating principles and utilising a set of management tools. This approach, which has been referred to as 'fishery management

---

A conservation framework

A systematic approach to formulating objectives, stating guiding principles and defining the means to pursue goals, in line with the stated principles, was described by Canada's Fisheries Resource Conservation Council (FRCC 1997a) as a framework for conservation within the Atlantic groundfishery. The key points are:

Conservation goals
Rebuilding of depleted stocks
Sustainable utilisation
Conservationist practices
Optimisation of benefits

Conservation principles
Understand the resource
Protect resource renewability
Precautionary approach
Systems approach
Consistency
Accountability
Flexibility and responsiveness

Conservation 'tool-kit'
Ensure conservationist harvest rates
Maintain adequate spawning potential
Establish diverse age structure in stocks
Protect genetic diversity
Safeguard ecological processes
Protect critical habitat
Minimise resource waste

science' (Lane & Stephenson 1995), draws on concepts of *systems analysis*, and requires a reasonable understanding of the various pieces of the fishery system puzzle.

In particular, within this approach, it is crucial both to determine and to acknowledge explicitly the objectives being pursued (Chapter 4), since the best management approach cannot be chosen without knowing the criteria for judging success. Furthermore, there will typically be multiple objectives, with management being a balancing act amongst these goals, so suitable analysis is needed to inform decision makers of the full implications of their actions (Charles 1988). It is necessary not only to understand society's goals, but also to understand the objectives driving human behaviour in the fishery. Too often, these objectives are *assumed* rather than assessed. For example, it is often assumed that fishers are solely profit maximisers, yet evidence in some cases shows that fishers do not seek to maximise any combination of conventional measures in the fishery (income, catch, rents, etc.), but rather they *satisfice*, seeking to meet their needs by taking 'enough' from the resource. Such real-world behaviour needs to be understood and taken into account in management.

The principles governing actions in the management system are drawn from a range of sources that could include national legislation, international conventions, local norms, and so on. The precautionary approach, and indeed the ecosystem approach, are two such possibilities. (See the box on p. 238 for an example of a set of principles.)

Finally, it is necessary to determine a policy and management 'tool-kit', in order to meet the stated objectives, subject to relevant constraints and dynamics. Appropriate analysis is needed to determine the optimal choice among fishery policy and management measures. For example, a useful method in this regard is adaptive environmental assessment (e.g. Walters 1986), a participatory process of computer simulation, used to study the effects of proposed management options. Another approach is discussed below under the heading of integrated fishery models.

12.3.2 *Diversity conservation*

A key ingredient in embracing complexity within fishery management is the adoption of measures to conserve (and enhance) diversity, both in the ecosystem and in the human system. Some examples of areas of concern include those listed below.

- Avoid dramatically altering the species mix in an ecosystem as a result of severely over-harvesting individual species, or inadvertently depleting species as a result of harvesting pressure exerted on 'target' species.
- Avoid severely depleting genetically distinct sub-stocks of species caught in a 'mixed' fishery with other 'stronger' stocks that can withstand higher exploitation levels. This is a major concern for salmon stocks, which are caught in a mixed fishery while at sea, before returning to their natal rivers.
- Encourage diversity in fishing methods and fishery objectives; such diversity may be threatened inadvertently (e.g. through the pressure of economic forces) or deliberately (through policy actions). For example, structural adjustment policies in many developing countries led to a major focus on one objective, short-term production for export, with a corresponding move in favour of capital-intensive fishing methods.

- Encourage diversity in employment patterns within fishing communities. This can also be impacted negatively (whether inadvertently or deliberately), for example through policies that promote the transformation of the fishery into an industrial form, and/or penalise fishers who engage in 'occupational pluralism', i.e. the risk-reducing behaviour of maintaining more than one source of livelihood. This theme is discussed in more detail in Chapter 15.

12.3.3 *Fishery development*

A systems approach to fishery development, referred to as integrated fishery development, deals as fully as possible with the complexity of the fishery system. This means that wherever possible, development is carried out in a holistic manner, treating the fishery system as a whole, including harvesting, processing, marketing and fishery management activities, as well as the broader ecosystem/human system of the ocean and coastal areas. As the FAO (1984) noted:

'Development plans should take into account all aspects of the fisheries sector, not only harvesting, processing, marketing, servicing and material supply, but also the development of the infrastructure, technology and human resources to enable developing countries to better exploit their fishery resources, to increase the value added to the economy and to improve employment opportunities.'

The value of a more extensive framework of integrated fishery development lies notably in avoiding mistakes that can occur easily if development efforts are carried out on a piecemeal basis. Consider an example. Development of marketing cooperatives for fishers has been a common policy prescription. The motivation is to enable fishers to bypass 'middlemen' in selling their catch, thereby improving their incomes. This seems a logical move when we focus solely on the harvesting sector, but if we broaden the perspective to examine the post-harvest sector, the picture may change. A system-based analysis suggests that while this measure may be successful if the 'middlemen' so displaced are external to the local fishery system, in many cases women from the local community dominate the marketing sector of the fishery (see, e.g. FAO 1984; CIDA 1993). Thus, these women may be the 'middlemen' being removed, potentially resulting in negative outcomes such as loss of livelihood, increased differences in income levels among local families, greater internal conflict and loss of community sustainability. Development projects organised from the start to look at fishing communities holistically may avoid such unexpected negative impacts.

If a fisheries agency pursues efforts in the post-harvest sector of the fishery, in fisheries planning, and in programme delivery at the community/cooperative/non-governmental organisation level, opportunities will increasingly arise to undertake integrated fisheries development. Within a fully integrated approach, fishery development involves not only the direct fishery system, but also the coastal areas where fishing communities are located. In other words, the development of fisheries is an integral part of coastal development, and vice versa (Fig. 12.2). This approach is reflected in the concept of *integrated coastal development* (e.g. Arrizaga *et al.* 1989). It is undoubtedly challenging to implement, and contrasts with the majority of past efforts in the realm of fishery development which dealt with fairly specific top-

Fig. 12.2 Integrated approaches to fishery management, development and research can take into account not only the fish and the fishing fleets, but also the fishers, their households and communities, and the broader socioeconomic environment. Such a perspective ensures that all relevant interactions – those internal to the fishery system and those involving external factors – are considered before management and development policies are implemented.

ics: stock assessment, fishery management, organisational development, training of fisheries managers and technicians, fisheries information systems and the like.

12.3.4 Fishery research

Fishery research plays a supporting role in the fishery system, particularly with regard to the management and development functions. Over the past several decades, researchers have succeeded in developing a wide spectrum of fishery theory, most of it of a disciplinary nature (biology, economics, etc.). Governments have used some of this theory to develop sophisticated systems of fishery management. Yet serious ecological, economic and social problems remain in many fisheries. This is undoubtedly due in part to failures at the political and institutional levels, but could it be that something has been lacking in the research itself?

There may well be an argument that the wealth of discipline-based research produced over the years, and indeed the basic scientific method of reducing the fishery system to its basic elements and studying each separately, have not managed to deal adequately with the problems of a complex multifaceted fishery. Of course, this form of reductionist scientific training has been widespread, and certainly such research is necessary, but when are all the small details of the system added together to make a whole? Kay & Schneider (1994: p. 33) made an important observation:

'The standard scientific method works well with billiard balls and pendulums, and other very simple systems. However, systems theory suggests that ecosystems are inherently complex, that there may be no simple answers, and that our traditional managerial approaches, which presume a world of simple rules, are wrong-headed and likely to be dangerous. In order for the scientific method to work, an artificial situation of consistent reproducibility must be created. This requires simplification of the situation to the point where it is controllable and predictable. But the very nature of this act removes the complexity that leads to emergence of the new phenomena which make complex systems interesting. If we are going to deal successfully with our biosphere, we are going to have to change how we do science and management. We will have to learn that we don't manage ecosystems, we manage our interaction with them.'

Research implications of Agenda 21

As noted previously, Agenda 21 was the key action document produced by the United Nations Conference on Environment and Development (the 'Rio Conference'). Agenda 21's Chapter 17 deals with '*Protection of the Oceans, All Kinds of Seas, Including Enclosed and Semi-Enclosed Seas, and Coastal Areas and the Protection, Rational Use and Development of Their Living Resources*'. The discussion falls into seven programme areas, the possible research implications of which are examined below.

(a) Integrated management and sustainable development of coastal areas, including exclusive economic zones.

Emphasis is on the need for integrated interdisciplinary approaches, such as the development of socioeconomic and environmental indicators, coordinated development of integrated coastal research systems, studies of coastal communities and relevant multidisciplinary educational programmes.

(b) Marine environmental protection.

The research focus here is on monitoring and data management at the international level, holistic impact assessment and developing socioeconomically feasible control measures.

(c) Sustainable use and conservation of marine living resources of the high seas.

Research topics include elaboration of monitoring, control and surveillance (MCS) methodologies, research on post-harvest activities (reduction of losses, improved value added, etc.), development of modelling methodologies relevant to management, and analysis of connections between the marine environment, environmental changes and high seas fishery harvests.

(d) Sustainable use and conservation of marine living resources under national jurisdiction.

Research topics include (a) analyses of aquaculture development potential, (b) identification of marine ecosystems with high levels of biodiversity and/or critical habitat, (c) production of biodiversity/resource/habitat profiles of exclusive economic zones, (d) development of analytical and predictive tools such as bioeconomic models, (e) transfer of environmentally sound technology, (f) study, scientific assessment and use of appropriate traditional management systems, and (g) development of mechanisms for transferring resource information and improved technology to fishing communities.

(e) Addressing critical uncertainties for the management of the marine environment and climate change.

Research areas include: methods for improved forecasting of marine conditions, development of standard methodologies for assessing and modelling marine and coastal environments, examination of the effects of increased UV on marine ecosystems, systematic observation of coastal habitats, sea level changes and fishery statistics, and other research on the uncertain impact of climate change on the marine system.

(f) Strengthening international cooperation and coordination.

This theme has no research components *per se*, since it deals with information exchanges and inter-governmental/inter-agency cooperation, but there are research implications in analysing, for example, the connections between world trade and environmental conservation.

(g) Sustainable development of small islands.

Research topics include: 'an environmental profile and inventory of natural resources, critical marine habitats and biodiversity', development of 'techniques for determining and monitoring the carrying capacity of small islands under different development assumptions and resource constraints', adaptation of suitable 'coastal area management techniques, such as planning, siting and environmental impact assessments, using Geographical Information Systems ... taking into account the traditional and cultural values of indigenous people', and the use of 'environmentally sound technology for sustainable development within small island developing States'.

There is a growing understanding that to deal properly with the complexity of the fishery system, fishery research must adopt an *integrated* approach. Indeed, while research in all disciplines has focused on the primary harvesting sector (fish dynamics, fleet economics, social interactions amongst fishers, etc.), linkages among the various components of the fishery system call out for attention. The rationale is that a failure to account for any one of these components could lead to important interconnections being missed, resulting in failed management. Indeed, an 'adaptive' perspective would treat fisheries themselves, and fishery management actions, as 'experiments', to be evaluated through suitable research programmes.

This suggests a perspective on fishery research as a *holistic* activity, incorporating ecological, economic and social components of the fishery, as well as interactions with the external human and natural environment. Such an approach is often referred to as a *systems* perspective. In this light, perhaps fishery research could be formulated as follows.

Fishery research

The integrated, adaptive study of fishery systems, their structure and their dynamics ...

- integrated in its perspective;
- multidisciplinary in its scope;
- participatory in its methods.

Clearly, there is a need for a multidisciplinary research approach, to work towards a fuller understanding of interactions between the fishery and the broader ecosystem/human system. This involves crossing traditional borders between natural science, social science and technological studies, and involves participation by resource users, with a focus on meeting the informational needs of those users (Charles 1991a; Durand *et al.* 1991). This approach should be seen as building on, rather than replacing, the usual disciplinary lines of research. As one government advisory body (FRCC 1994: p. 118) has noted:

> 'It is important that scientists study fishing scientifically as a system and strive to better understand the relationship between fish (resource) and fishing (fishing practices, gear technology, capacity analysis, etc.). This must reflect the recognition that fishery science involves more than the natural sciences and that scientific research is a part of the development, implementation and evaluation of fishery management measures and economic policy tools.'

It is also important to support research carried out cooperatively, implying both (a) joint research involving scientists and fishers, and (b) cooperation among fisheries agencies and across jurisdictions. On the latter point, integrated/systems approaches must overcome the lack of a corresponding organisational research structure, since the scientific research arm within a given fishery agency is usually functionally separate from economics and policy, and from operational management and enforcement. A lack of strong connections between these areas may well lead to failures in the design and implementation of management measures (FRCC 1994: p. 118):

> 'It is important that there be better integration and coordination [in the fishery research establishment], between regions, between Science, Operations, Policy and Enforcement Sectors, and as well, between [government scientists] and the fishing industry generally.'

Ecological economics: a science of sustainable development

Ecological economics is a field of study, dating back to the late 1980s, which attempts to integrate economic and ecological thinking, in order to provide an analytical and conceptual basis for *sustainable development* efforts. It is, in a sense, a science of sustainable development, recognising the key roles of economic forces, of ecological constraints and of societal objectives. Indeed, these components were highlighted in an overview of ecological economics (Folke *et al.* 1994), in which three major problem areas for study were noted:

> '(1) assessing and insuring that the scale of human activities are ecologically sustainable, (2) distributing resources and property rights fairly, both within the current generation of humans and between this and future generations, and between humans and other species; and (3) efficiently allocating resources as constrained and defined by 1 and 2 above, and including both marketed and non-marketed resources.'

Ecological economics is a *systems-oriented* framework, emphasising the importance of:

(1)   natural capital (including fish, natural habitats, the oceans, etc.);
(2)   *ecological services* (or 'ecological life support') provided by nature;
(3)   human goals and aspirations, at the individual and community levels;
(4)   human institutions that implement management efforts.

These ideas are nothing new to most fishery researchers. Ecologists, for example, deal regularly with fish stock conservation, which, as fishery economists recognised decades ago, is a matter of maintaining *natural capital*. Similarly, ecologists recognise that the marine ecosystem provides a series of *ecological services*, and economists study *option values* in attempts to measure the value of those services. So fishery management involves balancing conservation (a long-term investment in natural capital) with the benefits of harvesting (conversion of natural to financial capital).

Thus, ecological economics is, in part, an extension of the basic thinking of fishery science and fishery economics to the larger world, particularly to 'macro'/global matters of national and international economies (Charles 1994). However, ecological economics also has much to offer to fisheries analysis, notably an emphasis on the multidimensionality of the *objectives* involved in economic activities, and on the complexities of human institutions. A considerable amount of attention in ecological economics is devoted to examining how state-of-the-art single disciplines may have missed important interactions between the ecosystem on the one hand and the human system – society and economy – on the other.

For example, it is not uncommon in many countries to find that governmental fishery econo-mists frequently deal with the development and implementation of major policy initiatives, often in the absence of substantive research. Perhaps, in introducing new management sys-tems, problems could be avoided if there was more involvement by scientists and social scientists in examining the implications of the system, and possible alternatives. (For a case study of introducing individual catch quotas, see Chapter 14 and the discussion in Angel *et al.* 1994.)

12.3.5 Integrated fishery models

Given the reality of a complex fishery system, managers and researchers alike are faced with the daunting task of understanding, and intervening in, the system. An important tool avail-able to aid in these tasks is that of integrated fishery modelling. The idea of a 'model' is to represent the real world in a form that can be analysed reasonably easily, allowing us to explore the implications of management options prior to implementing them in practice. A model can be expressed verbally (in words), graphically (using graphs), physically (scale models) or, of most interest here, quantitatively (using mathematics or computer languages). The latter type of model allows us to experiment with management options in a 'theoretical laboratory', in a manner analogous to the use of a physical laboratory, such as a wind tunnel to test aircraft before they are flown. The role of modelling in each case is to avoid the costly and time-consuming 'trial and error' process of implementing full-scale but untested approaches. (See Rodrigues 1990 for a collection of papers developing and analysing such models.)

Integrated models focus on the need to understand the interrelationships amongst compo-nents of the fishery system, providing the means:

- to examine the complex relationships amongst ecological, socioeconomic, community and institutional sustainability in fishery systems (Chapter 10), and to assess possible op-tions for enhancing the sustainability of fishery systems;
- to help us understand the past and predict the future, in particular addressing why certain fishery systems have been sustainable and successful, while others have not;
- to aid in the development of fishery policies (Chapter 15), providing quantitative 'laborato-ries' to test and evaluate policies *before* they are implemented in the real world;
- to identify the most crucial data requirements in fishery systems.

Several illustrations are helpful here. For example, with respect to the first point above, a model could describe the various effects (in both natural and human systems) of a policy that seeks to improve the ecological sustainability of a fishery. Such an assessment might also examine the idea of carrying capacities, as described in Chapter 10, for the resource and the corresponding human population (or labour force) utilising the resources.

As an aid in developing policy and management measures, integrated models help us to understand interactions between fish behaviour and fisher behaviour, something which is es-sential to the management and conservation of fishery systems. As Hilborn & Walters (1992) note, 'it is foolish to study only the prey in the predator–prey system … it is equally important to monitor and understand basic processes that determine the dynamics of the predator – the fishermen'.

Furthermore, integrated models allow the study of how fishers respond (or are predicted to respond) to fishery regulations. This is again crucial to successful fishery management; an absence of such an understanding 'has led to management strategies and regulatory schemes that ignore the dynamic responses of fishermen to changes in stock size and to management itself. These responses can dampen or even reverse the intended effects of regulation ...' (Hilborn & Walters 1992). For example, the response to restrictions on the number of vessels is typically increased investment in each individual boat, while attempts to limit capacity expansion by restricting vessel length are thwarted by increases in other dimensions of the fishing vessel (such as width or engine power).

Integrated models of fishery systems may be very simple if they are being used to give a better understanding of the key features of the system. A major class of such models is described below. Alternatively, models may be very complex, seeking to mimic closely the real-world system. Examples of these include a variety of 'ecosystem models' that attempt to examine multi-species fishery systems, such as those of the North Sea. In general, useful models will be those for which (a) actual use of the model requires less detailed knowledge than was needed in the model's creation, and (b) the model can be used directly or easily adapted for application to a range of fishery configurations. Participatory processes for model creation, such as the use of 'adaptive environmental assessment' (Walters 1986) to develop integrated computer simulation models in 'real time' to analyse proposed management options, can be very useful in this regard.

Given their clear potential, it is perhaps remarkable that so little attention has been paid to the use of multidisciplinary integrated models (cf. McGlade 1989). This is probably owing to two major factors. First, there is the challenge of implementing multidisciplinary research, and bringing together specialists from a variety of fields. Second, there is the difficulty faced in linking modelling specialists, who may well be uncomfortable with the human dimensions of fishery systems, and other researchers (especially social scientists) who may be equally uncomfortable with modelling methods. Nevertheless, such models provide one means to help us to a better understanding of the behaviour of fishery systems, i.e. one more tool that can be used in moving 'towards sustainability'.

To illustrate the ideas of integrated fishery modelling, we will now discuss a broad class of such models, namely bioeconomic and biosocioeconomic models. Bioeconomic models are characterised by an integration of the natural and human sides of the fishery equation, combining:

- *biological* aspects relating to the fish resource and its surrounding ecosystem, such as population dynamics, fish growth and fish ecology, and
- *economic* elements that shape human behaviour in fish harvesting, such as fish supply and demand, profit functions and production economics, fisher decision making, and the investment dynamics that drive entry to and exit from the fishery.

Bioeconomic models fall into two broad classes:

- *behavioural models*, designed to explain and predict fishery and fisher dynamics, and providing realistic tools to examine development and/or management scenarios;

- *optimisation models*, oriented towards determining 'optimal' management or development strategies, given a set of specified objectives.

A generic example of an optimisation model, expressing the task of fishery management as pursuing an economic objective subject to a biological constraint, is set out below.

> Maximise the sum (over all future years) of annual benefits
> [to resource owners, harvesters and consumers]
> Subject to the constraint:
> Stock next year = Current stock + Net growth – Harvest

where net growth depends on the average growth of individuals in the population, as well as the specified population dynamics (levels of natural mortality and reproduction), which in turn depend on the environmental carrying capacity of the stock.

The development of integrated bioeconomic analysis of fishery systems is generally seen to date from the middle of the twentieth century (e.g. Schaefer 1957), although earlier work on the subject did take place (e.g. Warming 1911). These efforts progressed further into dynamic analysis in the late 1960s, notably with the predator–prey-style models by Smith (1968) of fish–fisher interactions. Dynamic bioeconomic modelling, particularly with an optimisation focus, emerged as a major force in fishery studies with the work of Colin Clark and co-workers, starting in the early 1970s.

Clark's (1985, 1990) books provide comprehensive syntheses of the subject from a mathematical perspective, while Seijo *et al.* (1998) survey recent developments in the field, with emphasis on simulation approaches. A range of bioeconomic modelling case studies is provided by Rodrigues (1990), and a brief review of actual and potential applications to fisheries of developing countries is given by Padilla & Charles (1994).

Given the biological traditions of fishery science, the emergence of bioeconomics represented important progress towards more integrated analyses. Bioeconomic models provide a unified framework for multidisciplinary research, since they can have considerable intuitive appeal to both biologists and economists, and provide a means to help bridge the gap between these different disciplines. They have been successful as conceptual tools in generating theoretical insights into the dynamic operation and management of fishery systems, and have had, as a result, a substantial influence on fishery thinking. However, there is still much unrealised potential for full applications to real-world fisheries, notably in the development of management plans and policy advice. The number of such applications to date is minimal, although there is considerable potential to utilise the approach using simulation modelling.

One vehicle for doing so is FAO's bioeconomic analytical model (BEAM 4). As described by FAO (1999b), this is a multi-species, multi-fleet spatially disaggregated simulation model involving an 'age-structured cohort-based fish stock assessment model combined with an economic model of both harvesting and processing sectors'. The BEAM 4 model allows for the prediction of 'yield, value and a series of measures of economic performance as a function of fishery management measures such as fishing effort control, closed season, closed areas and minimum mesh size regulation'.

An extension of bioeconomic modelling to incorporate more fully the complexities of the human elements (social, community and behavioural) is that of *biosocioeconomic* modelling. As the name suggests, this integrates the biological, economic and social structure and dynamics of a fishery system within a systematic framework. In particular, biosocioeconomic models may combine relatively well-studied bioeconomic components (such as fish population dynamics and the capital dynamics of the fishing fleets) with less commonly addressed aspects such as the behaviour and dynamics of the fishers and fishing communities, and the range of societal objectives such as conservation, income generation, employment and community stability.

Some examples of biosocioeconomic models are provided in the Appendix; see the empirical simulation modelling work of Krauthamer *et al.* (1987) for a more detailed example. The biosocioeconomic framework serves to highlight the key information needed for an integrated treatment of fishery systems (Charles 1989; Sivasubramaniam 1993).

In particular, to 'fit' biosocioeconomic models, one needs time series of data not always associated with the fishery itself. For example, the determination of appropriate fishery management policies depends on an understanding of fishery labour dynamics, which depend on the decisions of individual fishers, which in turn depend both on internal conditions in the fishery system (such as fish stock dynamics, per capita incomes and employment rates) and the state of the external economy (such as labour conditions, migration rates and population trends). Thus, all of these factors, including those usually viewed as beyond the fishery system, can be important to fishery management.

12.4 Summary

This chapter has covered a diverse range of subject matter. First, the nature of *complexity* and *diversity* in fishery systems was examined, with attention to the various sources of complexity, and particularly to the importance of 'embracing' (rather than attempting to overcome) the complexity and diversity in fisheries. To this end, the need for *integrated* approaches was highlighted. Key to this is the *ecosystem approach*, which was characterised as 'a method of thinking' that attempts to take into account everything from multi-species interactions in a mixed-stock fishery, to complex ecosystem structure and overall *ecosystem health*, to aspects of the human system. A major manifestation of an ecosystem approach was discussed: namely *marine protected areas* (marine reserves) and the various forms these can take. Discussion then turned to several other *integrated* approaches which are important to the fishery management system: fishery management science, diversity conservation, integrated fishery development, integrated and multidisciplinary research and integrated fishery modelling. Each of these approaches has implications for the pursuit of sustainable fisheries, which is discussed in subsequent chapters.

Chapter 13

Fishery Conflicts and
the Co-management Approach

This chapter focuses on two principal themes. First is the ubiquitous phenomenon in fishery systems of conflict among the participants within the fishery, and with those outside the fishery. Second, the chapter examines the *co-management approach*: a key mechanism, indeed a major emerging direction in fisheries, that can aid in resolving some forms of conflict, namely those arising between fishery users and government agencies, and indirectly between users themselves.

13.1 Conflict and sustainability implications

It is hardly surprising that conflict tends to be prevalent in a biosocioeconomic system as complex and as dynamic as a fishery, with its many interactions among natural resources, humans and institutions. Internal fishery disputes arise regularly over the allocation of scarce fish resources, over the division of fishery benefits between fishers and processors, and over management arrangements between fishers and governments. Meanwhile, external conflicts are increasingly common, with competing users, such as aquaculture, forestry, tourism and ocean mining, vying for access to aquatic space and fish habitat. Underlying these more immediate internal and external conflicts are philosophical debates over ownership, control and overall policy direction in the fishery.

An important source of conflict, of the latter 'philosophical' variety, can arise over fundamental perceptions of what constitutes a *sustainable fishery*: if that is what we want the fishery to be, precisely what is it that is to be sustained? At the same time, other forms of conflict can have substantial negative impacts on the sustainability of the fishery system. For example, conflict can be detrimental to sustainability if it leads to over-harvesting in the midst of 'fish wars', as has happened in the Canada–US dispute over salmon on North America's Pacific coast, which lasted through much of the 1990s, and the so-called 'turbot war' between Canada and Europe, off the coast of Newfoundland in 1995.

Furthermore, quite apart from the direct effect of conflict, the immediacy of the need to resolve such conflicts can simply distract attention from the pursuit of long-term sustainability. Management and policy-making processes may become so embroiled in 'fire-fighting' to deal with specific conflicts that the overall fishery picture is neglected. Or even worse, counter-sustainable actions may be taken to resolve conflicts, for example, 'extra fish' might be given out (as higher catch limits) to help resolve allocation disputes.

Finally, faced with a perceived fishery *chaos* produced by the wide diversity of fishery conflicts (and their protagonists), governments and international agencies may be tempted to abandon support for fishery systems, which would harm institutional sustainability.

The discussion in the first part of this chapter highlights the need to understand fishery conflicts and their relation to fishery sustainability. We begin by exploring the roots of conflicts, based on a set of fishery *paradigms*, or world views, that reflect contrasting perspectives and differing mixes of policy objectives. This is followed by an effort to understand the structure of fishery conflict, through the presentation of a *typology* of such conflicts. Finally, the focus shifts to a discussion of possible mechanisms to resolve conflicts.

13.2 Fishery paradigms

As noted in Chapter 4, it is common to view the various objectives pursued in fishery systems as fitting into three broad categories: conservation, social and economic (FAO 1983: p. 20). The relative emphasis placed on each of these in policy making and analysis may be related in turn to the philosophical preferences between three major paradigms or 'world views' (Charles 1992b, c) (Table 13.1).

Often, the complexities of fishery policy debates seem explainable in terms of natural tensions between these three fishery paradigms. They can be arranged in a *paradigm triangle* (Fig. 13.1), an integrated framework within which fishery policy debates may be analysed.

Table 13.1   Policy objectives and fishery paradigms.

Policy objective	Related paradigm
Conservation/resource maintenance	Conservation paradigm
Economic performance/productivity	Rationalisation paradigm
Community welfare/equity	Social/community paradigm

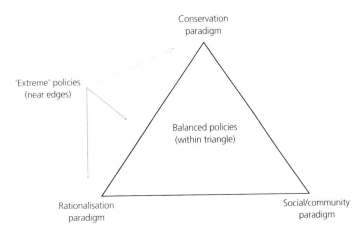

Fig. 13.1   A *paradigm triangle* is an integrated framework within which fishery conflicts and policy debates may be analysed. Fisheries conflicts can be viewed as reflecting tensions between the triangle's three corners, with 'extreme' policy proposals lying relatively close to one of the corners, and attempts at conflict resolution typically aiming at the 'middle ground'.

Fisheries conflicts can be viewed as reflecting tensions between the triangle's three corners, with 'extreme' policy proposals lying relatively close to one of the corners, and attempts at conflict resolution typically aiming at the 'middle ground'.

The paradigm triangle can also be seen as a means to 'locate' the various players in fishery discussions. Most of these players would probably be placed, and indeed would place themselves, somewhere in the interior of the triangle. However, those with relatively extreme views may well be found very near one of the corners of the triangle.

The three paradigms and their characteristics are discussed below (see Charles 1992b, c for further details).

13.2.1   The conservation paradigm

The conservation paradigm is based on recognising (in words if not always in deeds) the obvious dependence of fisher and fishing industry well-being on the state of the fish stocks. The fundamental philosophy of the conservation paradigm revolves around several points.

- *Sustainability* is defined as long-term conservation, so any activity is considered to be sustainable if fish stocks are not over-exploited, regardless of any human-oriented fishery objectives.
- Resource depletion, or even extinction, could potentially be the 'rational' outcome of unmanaged resource industries, so a reliance on unfettered market forces is understood to be inappropriate in the fishery.
- Fishers operate only out of their own self-interest, so it is up to the fishery managers to take care of the fish. In order to save the fish stock (and to benefit future fishers), management must directly control the fleet, restricting the various aspects of fishing time, location, effort and harvest. (Note that this questionable premise has caused considerable problems in fisheries, as discussed later in the chapter.)

This paradigm has been historically dominant (Larkin 1977), leading to a consequent preoccupation with fish stock protection and biologically based management, as well as a dominance of biological and stock assessment studies within the realm of fishery research.

13.2.2   The rationalisation paradigm

In the 1950s, the dominant conservation paradigm was challenged by a rationalisation paradigm focusing on achievement of an economically 'rational' fishery, based on two main concepts.

- Pursuit of a single objective, i.e. maximising the wealth created in the fishery. The latter is typically measured as resource rent (Anderson 1986; Clark 1990), i.e. the economic return to resource owners from the fishery. This is somewhat analogous to the returns that the owners of capital and labour receive for their inputs; profits in the case of boat owners and wages in the case of crew members. The goal in this case is often referred to as maximum economic yield (MEY), in contrast to the maximum sustainable yield (MSY) objective which is traditional in the conservation paradigm; this emphasises increasing economic efficiency, defined narrowly within the harvesting sector.

- A focus on avoiding the so-called *open-access dynamics* that arise in fisheries where fishers are subject to expansionist incentives (because rents are left with fishers as above-normal profits) and the fleet is improperly managed. This leads to rent dissipation and a loss of wealth from the fishery (Chapter 8).

Within the rationalisation paradigm, fisheries that are not economically efficient and do not maximise rents must be *rationalised*. Typically, the measures advocated focus on reducing to an 'optimal' level the number of fishers (seen as profit-maximising firms), and instituting property rights to the fish, notably through individual transferable quotas (Chapter 14).

A major change in fishery thinking over recent decades has been the emergence of the rationalisation paradigm as a dominant force. It has always been pervasive in the literature of theoretical fishery economics, and has traditional support among industrially oriented fishery players, but in recent years it has gained favour among biologists and fishery managers as well, perhaps owing to a perception that by cutting the number of fishers, rationalisation serves conservation goals as well as boosting economic efficiency.

However, it should be noted that ecological sustainability is not implied within the rationalisation paradigm. Indeed, Clark (1973) demonstrated that a *rational* fishery may not be biologically sustainable; the 'optimal' pursuit of maximum rents (MEY) could drive fish stocks to extinction. For example, in the case of whaling, Clark notes that if interest rates (for money in the bank) are higher than biological growth rates (for whales in the sea), returns to financial capital may well be greater than those for natural capital. Hence, rent maximisation may involve 'liquidating' the natural capital and banking the proceeds.

13.2.3 The social/community paradigm

In contrast to the inherent concern for fish conservation and resource rents in the conservation and rationalisation paradigms, public policy debates often revolve more around human concerns. Such concerns for the fishery's social fabric and for the intrinsic value of fishing communities have a base in the social/community paradigm. This paradigm has two key characteristics.

- Emphasis is placed on fishers as members of coastal communities, rather than as components of a fleet (as in the conservation paradigm) or as individualistic fishing firms (as in the language of fishery rationalisation); social and cultural institutions are recognised as potentially playing an important role in fisheries.
- Emphasis is placed on a multi-objective pursuit of socioeconomic goals, contrasting with the rationalisation paradigm in balancing wealth generation (efficiency) and distributional equity. Maintenance of a sustainable yield is seen as important, not because of an inherent concern for fish (as in the conservation paradigm), but rather as a means to preserve the way of life in fishing communities. Fair distribution of the sustainable yield is seen as being as important as its magnitude.

The social/community paradigm tends to be attractive to fisher organisations, to those living in or involved with fishing communities, and to social scientists. Not uncommonly, social/community advocates have sought to protect the 'small' fishers from the impacts of economic forces, and have often opposed policy measures promoted by rationalisation advocates, a fact which explains a large part of the ongoing conflict in fisheries (see, e.g. Marchak *et al.* 1987). However, it is notable that, relative to their role in the fishery itself and in public fishery debates, advocates of the social/community paradigm seem under-represented amongst the staff and in the management initiatives of many government fishery administrations.

13.3 The essence of conflict: efficiency and allocation

As noted earlier, in most fisheries, i.e. those in a state of full exploitation, there appears to be little room available to increase long-term sustainable fishery benefits simply by increasing production. Thus, fishery policy tools are generally limited to (1) increasing the efficiency of harvesting and of management, and (2) making allocation (distributive) decisions, particularly by determining who has the privilege of access to the fish available for capture. It is in these two areas that the conflicts typically arise, with the latter (allocation of harvest shares) generally being at least as critical as optimising the total harvest itself.

The concept of *efficiency* is the source of considerable confusion in policy discussions, both within the fishery and indeed throughout the economy. Essentially, the concept is a simple one: efficient policies are those which give the 'best' results possible (measured in terms of overall well-being or net social benefits) within the means available, or equivalently, those which achieve the desired goals with the least negative effects. Hence, the pursuit of efficiency is desirable, by definition. However, difficulties arise in applying the concept, due to the inherently multifaceted nature of *societal well-being* as a policy goal. There has been a widespread tendency to oversimplify this goal, equating social well-being with wealth (or rent) maximisation, and neglecting other aspects of societal well-being. From an overall policy perspective, it is crucial to note that a blind pursuit of 'efficiency' is meaningless without clearly defining what is meant by the 'well-being' of the relevant players.

In the fishery context, the importance of pursuing efficiency is especially great in fully exploited *zero-sum* fisheries, where an increase in one group's allocation must mean less for others. For example, measures to decrease post-harvest losses have the potential to increase efficiency by simultaneously improving the well-being of all participants. However, in reality, fishery players will probably differ philosophically over objectives to be pursued, and hence over the definition of an *efficient* fishery. For example, should the aim be towards efficiency in generating wealth, in providing employment, in maintaining the sustainability of coastal fishing communities, or in some other measure of well-being?

Given these differences regarding the goals of efficiency-enhancing policies, it is not surprising that such measures lead to fishery conflicts. In practice, such disputes typically revolve around proposals to reallocate limited fish resources to those sectors of the fishery perceived to be most *efficient* (see, for example, the discussion in Anderson 1986). Accordingly, fishery conflicts tend to be dominated by allocation issues (Regier & Grima 1985). Indeed, the allocation of fish harvests is often a key issue faced by fishery managers.

The dual issues in fishery systems of efficiency and allocation, what could be referred to as *maximising the size of the pie* and *allocating pieces of the pie*, are depicted in Fig. 13.2. This chapter will focus principally on an analysis of allocation conflicts (although many such conflicts also involve elements of efficiency, as will be noted where appropriate).

13.4 A typology of fishery conflicts

Clearly, a wide variety of conflicts arise in fisheries. Some arise from fundamental philosophical differences, as reflected in the discussion of paradigms above, as well as arguments related to efficiency and allocation, while others may involve basic matters of allocation. In any case, despite superficial appearances of *chaos*, this wide range of conflicts can be organised into a relatively small number of categories. This section seeks to understand the 'structure' of fishery conflicts through a suitable typology, involving four interrelated categories. The first two of these focus on the structure and operation of the management system, and how it interacts with fishers, while the second two deal with conflict over allocation to resources, whether arising among those inside the fishery, or between fishery players and those outside the system. The four categories are described below.

(1)   *Fishery jurisdiction.* This category lies at the *policy and planning* level (Chapter 4), and deals with fundamental, often philosophical, conflicts over fishery objectives, who 'owns' the fishery, who controls access to it, what should be the optimal form of fishery management, and what should be the role played by governments in the fishery system.

(2)   *Management mechanisms.* This category includes conflicts at the *fishery management* level (Chapter 5) concerning shorter-term issues arising in the development and implementation of fishery management plans, typically involving fisher/government conflict over harvest levels, consultative processes and fishery enforcement.

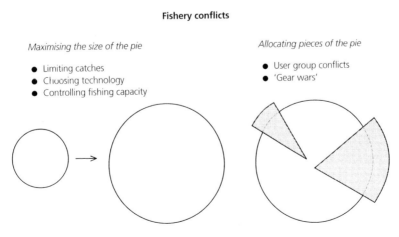

Fishery conflicts

Maximising the size of the pie
- Limiting catches
- Choosing technology
- Controlling fishing capacity

Allocating pieces of the pie
- User group conflicts
- 'Gear wars'

Fig. 13.2   Graphic representation of the dual issues in fishery systems: efficiency (maximising the benefits available) and allocation (determining who receives what portion of the benefits).

(3) *Internal allocation.* These conflicts arise among the direct participants in the fishery system, i.e. the fishers and processors, and relate to differing perceptions of appropriate allocations of fishery access and use rights between the various user groups and gear types, as well as among fishers, processors and other players.

(4) *External allocation.* This category includes conflicts arising between internal fishery players and those outside (or on the edge of) the fishery system, which could include foreign fleets, aquaculturists, non-fish industries (e.g. tourism and forestry) and indeed the public at large.

These four principal conflict categories are meant to include most major forms of conflict, but are not necessarily mutually exclusive. Any given fishery dispute appears to fit under at least one of the headings, although certainly some will fall under more than one. (Examples of the latter may be cases in which apparently straightforward allocation disputes have roots in philosophical conflicts over jurisdiction.) Furthermore, most fishery players are involved in a wide variety of conflicts simultaneously.

Within each of the four conflict categories are key sub-themes; three sub-themes are identified for each category in the conflict typology in Table 13.2.

Each of these conflict sub-themes is described briefly below. For each category, the sub-themes are arranged from that which seems closest to the direct fishery participants (the fishers), through to that which is most 'global' in nature. Thus, under fishery jurisdiction, for example, 'property rights' issues bear directly on the fishers, while 'intergovernmental conflicts' take place, by their nature, at a more remote governmental level. Similarly, under external allocation, 'domestic versus foreign' conflicts take place solely among fishers (whether within or external to the system), while those concerning 'competing ocean uses' involve non-fishery participants as well.

13.5 Fishery jurisdiction

13.5.1 *Property rights*

Debates over property rights involve major long-term philosophical questions concerning legal, historical and/or *de facto* ownership, access and control in the fishery. A particular focus lies in conflict over the relative desirability of fishery property options, such as open access, centralised management, territorial use rights in fishing (TURFs), community-based common property management, market-based individual quotas and privatisation. Such subjects will be considered in detail in Chapter 14.

Table 13.2 Typology of fishery conflicts.

Fishery jurisdiction	Management mechanisms	Internal allocation	External allocation
Property rights	Management plans	'Gear wars'	Domestic vs. foreign
Role of government	Enforcement conflicts	User-group conflicts	Fishers vs. aquaculture
Intergovernmental conflicts	Fisher/government interactions	Fishers vs. processors	Competing ocean uses

13.5.2   The role of government

A major and fundamental conflict is rapidly evolving between the centralised government regulation of harvesting activity and challenges by more decentralised alternatives, including community- and market-based management options, as well as the development of *co-management*, as discussed later in this chapter.

13.5.3   Intergovernmental conflicts

Despite new challenges to the traditionally dominant role of government in fishery management, in most cases there remains a large government presence, and intergovernmental conflict is common both between nations (as in trans-boundary fisheries) and between jurisdictions within a nation (i.e. states, provinces or municipalities).

13.6   Management mechanisms

13.6.1   Fishery management plans

The development of periodic management plans for determining allowable harvest levels, harvest allocation, fishing times and/or fishing gear represents such a major recurrent source of fisher/government conflict as to deserve a category on its own, although in fact these conflicts often reflect fisher concerns over internal and external allocation issues, as discussed below.

13.6.2   Enforcement conflicts

Fisher/government conflicts over enforcement arise in two major forms. The most common situation involves complaints of excessive government enforcement imposed on a particular user group, and the reverse sense based on complaints by one set of users that enforcement is overly lenient when applied to competing users (as is common between inshore/artisanal and offshore/industrial sectors).

13.6.3   Fisher/government interactions

An omnipresent source of conflict is the perception on the part of fishers that government managers and scientists ignore the knowledge and ideas of the fishers. Even in fisheries with elaborate and expensive consultative processes, such conflicts are likely to remain as long as the fishers are excluded from the actual decision-making processes.

13.7   Internal allocation

13.7.1   'Gear wars' conflicts

Conflicts arising within the commercial sector of the fishery are generally focused on allocations between vessel categories. Typically, this involves differences in fishing gear, but differences in scale (as in traditional artisanal/industrial fisheries conflicts) may also fit here, for example, in cases where such disputes involve technological interaction on the fishing grounds.

13.7.2   User-group conflicts

Major disputes, both short- and long-term, arise between the various broad classes of fishery users, generally representing quite different segments of society (such as artisanal vs. industrial fishers, or commercial vs. recreational fishers). An example of a user-group conflict is depicted in Fig. 13.3.

13.7.3   Fishers vs. processors (allocation of economic benefits)

A common feature of many commercial fisheries is the strained relationship between fishers, particularly those of the small-scale variety, and fish processors. On the one hand, as discussed within the user-group conflict heading above, vertical integration by processors could increase conflicts between small-scale and industrial fleets. On the other hand, most fisher–processor disputes tend to be of a typical labour–management form, involving bargaining over prices to be paid for landings, wages to be received, and other economic conditions involving fishers and fish plant workers. This conflict produces a clear dichotomy between the fisher-oriented social/community paradigm on the one hand, and rationalisation advocates on the other, the latter tending to side with the processors as profit-maximising companies that limit payments to fishers in order to be efficient.

Fig. 13.3   On the Pacific coast of North America, an ongoing 'user group' conflict arises in the salmon fishery, over sharing of the available harvest between commercial fishers and the recreational sector – of which this sport-fishing boat is a representative.

Example: internal allocation conflict in the Atlantic Canadian groundfishery

(1)   The classic user-group conflict in Atlantic fisheries revolves around the differing modes and scales of fishery production, in particular the allocation of groundfish harvests between inshore and offshore commercial fishers (Kirby 1983; Lamson & Hanson 1984). The large-scale capital-intensive offshore fishery is viewed variously as representing the 'industrial', 'corporate' or 'efficient' component, while the small-scale inshore fishers constitute the 'traditional', 'artisanal', 'community-based' component. As in other parts of the world, conflicts between these two sectors are based on technological and biological impacts (such as the effects of offshore fishing on spawning grounds), as well as the crucial issue of harvest allocation.

(2)   A *gear wars* conflict now rivals the inshore–offshore conflict for prominence. This concerns harvest and fishing space allocation between so-called fixed and mobile gear sectors, in particular long-liners and hand-liners in the former group, and otter trawlers in the latter. This conflict does not follow 'inshore versus offshore' lines, but pits relatively heavily capitalised, highly mobile trawlers against the fixed-gear fishers, with the latter arguing for a larger share of the harvest on the basis of greater 'environmental sensitivity' relative to bottom-dragging trawlers, and a greater economic contribution to the coastal communities.

(3)   *Fisher versus processor* conflicts constitute the third form of internal allocation disputes. Atlantic fishers, especially those in remote communities, have traditionally relied on processors, 'fish merchants' and other middlemen to buy their catches, at prices set by the latter. A variety of attempts have been made to overcome this 'price-fixing' by increasing fisher bargaining strength through the formation of unions or cooperatives (Clement 1984; Calhoun 1991). However, fishers and processors continue to engage in annual conflicts over the prices to be paid for the catch, a debate which amounts essentially to one of determining how fishery benefits are to be shared between the primary and secondary sectors of the industry.

13.8 External allocation conflicts

As described earlier in this chapter, conflicts among fishers over fishery resource allocation arise in most fisheries, and these range from user-group conflicts and *gear wars* to fundamental disagreements about the direction of fishery policy. However, conflict over allocation goes well beyond the fishery itself; fishers seem to find themselves in continual battles against competing economic forces. Here, we briefly discuss three such 'external' conflicts which arise in many locations and involve foreign fleets from distant-water fishing nations), aquaculture development and interactions with other coastal industries.

13.8.1 Domestic vs. foreign fisheries

A range of conflicts exist between coastal states and domestic fishers on the one hand, and distant-water fishing nations and their fleets on the other. These include problems of illegal fishing within the coastal state's exclusive economic zone (EEZ), legal but unmanaged fishing in areas outside the EEZ (as on Canada's Grand Banks), destructive high seas fishing (as with drift nets), opposition of domestic fishers to bilateral fishing agreements, and so on.

The so-called 'turbot war' of 1995 provides a clear illustration of a conflict that related to catch allocation, but also had overtones of conflict between differing perceptions of the balance between fishery regulation and the *freedom of the seas*. This dispute, between Canada and the European Union (notably Spain), revolved around a dispute over the catch allocations of Northwest Atlantic turbot (Greenland Halibut) to be received by Canada and the European Union through the Northwest Atlantic Fisheries Organisation (NAFO). Using a form of 'opting-out' clause, the EU unilaterally declared a higher catch target than that assigned by NAFO, and several European Union vessels set out to fish just outside Canada's 200-mile limit. Canada argued that this was a threat to conservation, and arrested a Spanish vessel outside the Canadian EEZ for violating the coastal state's regulations. The European Union responded by calling this action 'piracy', and argued that it contravened their freedom to fish the high seas.

After much diplomatic tension, the matter was eventually resolved with a compromise on catch allocations and agreement on the management of future harvests. The event drew broad attention to the problem of conflict over fishery resources outside any nation's EEZ, and the overriding need for fishery conservation and regulation. Perhaps most importantly, the dispute led to the 1995 United Nations Conference on Straddling Fish Stocks and Highly Migratory Fish Stocks, and the subsequent 'Agreement for the Implementation of the Provisions of the United Nations Convention on the Law of the Sea of 10 December 1982 relating to the Conservation and Management of Straddling Fish Stocks and Highly Migratory Fish Stocks'. Despite this diplomatic action, a court case continued against Canada concerning the actions it took outside its 200-mile limit. The matter was eventually dismissed by the court at The Hague in 1998.

13.8.2 Fishers vs. fish farming (aquaculture)

Aquacultural activity is rapidly expanding within many jurisdictions. While globally much of the aquaculture production comes from ponds and other discrete water bodies, there is also an expansion of coastal and lake (reservoir) aquaculture. These forms typically take place within an aquatic environment where fisheries also operate. While the potential exists for a positive relationship between these two fish-based sectors, for example, if aquaculture helps to expand job opportunities for fishers (e.g. Newkirk 1995), in reality, conflict has been more common than symbiosis. This can be traced to a variety of factors, such as poor control and planning of aquaculture development, a lack of training and risk-alleviation programmes for new entrants (e.g. from the fishery), and a fundamental difference between the nature of capture fisheries and that of fish culturing operations. The latter leads to a natural reluctance among fishers to switch between these activities, and a corresponding dominance of non-fishers in coastal aquaculture. This situation adds to the conflict between the two sectors, with a variety of concerns frequently being raised by fishers regarding aquaculture development. Some of these are listed below.

(1)   Competition between fishers and aquaculturists over the use of ocean space (for farms versus fishing locations, e.g. between lobster fishers and fish farmers in Nova Scotia).
(2)   Competition between wild and cultured fish for available aquatic space and environmental resources.
(3)   The introduction of new fish diseases by aquaculturists, thereby harming wild stocks.
(4)   Contamination of the ocean by antibiotics and other chemicals used in fish farming.
(5)   Destruction of fish habitat (e.g. loss of mangrove areas to shrimp ponds in Ecuador).
(6)   Extraction of wild larvae for use in aquaculture (e.g. shrimp larvae used in Thai farms).
(7)   Depression of market prices for wild fish as the supply of cultured fish increases worldwide (e.g. salmon) and issues of market access.

In some cases, as in parts of India, fishers have led vociferous opposition to aquaculture development. Interestingly, there are also conflicts between aquaculture and tourism, when the former is thought to adversely affect the aesthetic attractiveness of an area for tourism, and between aquaculture and shipping (e.g. the operation of ferries in British Columbia can be affected by the location of salmon farms).

   From the point of view of the conservationist paradigm, the principal concern in this conflict is the protection of wild fish stocks, which might lead to a favouring of capture fisheries over fish farming. (However, a 'techno-fix' perspective expresses faith in scientific efforts to overcome any problems with aquaculture and thus avoid damage to wild stocks through either diseases or habitat damage.) The most dominant dichotomy here lies between the rationalisation and the social/community paradigms. The former tends to favour the private property nature of fish farming and the entrepreneurial spirit of fish farmers in developing their businesses, while the latter prefers traditional common-property capture fisheries, and opposes disruption of fisher's lives.

13.8.3   The fishery vs. competing industries

Fishing activity utilises coastal zones (and lakeside zones, among others) that are not only crucial fish habitats, but are also desired by other economic activities, particularly for the ocean space itself and for the natural resources other than fish that reside in these areas (see, e.g. Caddy & Griffiths 1995). Thus, many fisheries face ongoing conflict with, for example:

- forestry (e.g. industrial logging on North America's Pacific coast, which has an impact on salmon habitat in rivers, and the small-scale harvesting of tropical mangroves, leading to a loss of habitat for fish);
- oil, gas and offshore mining (e.g. drilling and extraction activities, as in Canada, Indonesia, Norway and elsewhere);
- agriculture (e.g. where pesticides, other agrochemical run-offs and sedimentation affect mangroves and/or coral reefs, as in Costa Rica's Gulf of Nicoya and Caribbean coast);

- shipping (particularly involving oil spills);
- tourism (e.g. where hotel construction harms mangroves, as in various Caribbean islands, or in the opposite direction, where fish processing activities, and in particular the smells they produce, are seen as a problem for tourism operators).

These conflicts incorporate the most *global* of disputes, those involving the nature of, and the priorities for, use of the oceans and other aquatic systems. Indeed, non-fishery industries can be considerably more important to the local and/or national economy than fishing. Hence, from the 'most profitable use' rationalisation perspective, in such cases there is reason to favour industries other than fishing. However, the conservationist and social/community paradigms focus on potential damage to the fishery system, including both fish and fishers, and are thus generally united in favouring protection of the fishery.

This united stand against environmental damage is reflected in the efforts of fishers to oppose at-sea oil and gas drilling in various locations around the world. For example, a coalition of small- and large-scale fishers, processors, environmentalists and coastal communities, known as 'NoRigs', worked together twice, in the late 1980s and again in the late 1990s, to oppose multinational oil companies who were seeking to exploit oil and gas on the Canadian side of George's Bank, a rich fishing ground on the Atlantic coast of North America. The fishing sector won the resource use battle on both occasions, with governments agreeing to moratoria on drilling.

13.9 Resolving conflicts

So far in this chapter, the focus has been on *understanding* the nature of conflict in fishery systems through an integrated analysis based on a typology of conflicts and a set of paradigms reflecting the philosophical basis of the conflicts. However, it is one thing to understand the nature of conflicts and quite another to develop mechanisms for resolving such conflicts. We turn now to the latter theme.

As we have seen, conflict arises at many different levels in the fishery: between nations, between the various users of ocean space, between different classes of domestic fishery resource users (e.g. recreational versus commercial), between sectors of the commercial fishery (e.g. industrial versus artisanal), and so on. Some fisheries seem to be dominated by conflict at all these levels. Others, however, appear to operate remarkably well without the same level of conflict.

It would appear that fisheries which are relatively conflict-free have achieved a 'policy balance' through a multifaceted approach, with attention being paid jointly to fishery 'health' from ecological, socioeconomic and community-level perspectives. This implies the simultaneous pursuit of sustainability in the natural resource base, the fishing industries and the fishing communities (as discussed in the sustainability framework of Chapter 10), with a focus not on extreme policy proposals, but rather on more palatable intermediate options. Furthermore, the process of *sustainability impact assessment*, also discussed in Chapter 10, may give a more holistic framework within which to seek suitable fishery policies.

At the international level, conflict resolution between nations has been aided by key mechanisms within the United Nations Convention on the Law of the Sea and the more recent so-called Straddling Stocks Agreement (Borgese 1995, 1998) – discussed above. With respect to multiple domestic uses of coastal, marine and freshwater areas, the approach of integrated management (e.g. Clark 1996) has much to offer in providing approaches to resolve conflicts. In the following discussion, the focus is more narrowly on the fishery system within a given jurisdiction, and specifically on a key institutional arrangement that can be useful in alleviating some of the long-standing conflict.

13.10   The co-management approach

A famous example of fishery conflict, and its resolution, is that of the Lofoten Islands cod fishery, in northwest Norway. Over the course of many centuries, this fishery has attracted a large number of participants, leading to ongoing conflicts, and in particular 'gear wars'. Various remedies were tried in the nineteenth century, but the government was unable to succeed in its effort to 'control' the fishers. Rather than continuing to pursue such a path, the government introduced, as a 'last resort', the Lofoten Act of 1890 (Jentoft 1985, 1989). This legislation devolved the powers of the central government over the fishery to local institutions in which fishers played a large role in setting and enforcing regulations. Such a system, now referred to as *co-management*, 'may be the earliest such arrangement in Europe' (NRTEE 1998: p. 54).

In this section, we explore aspects of the co-management approach. While not purporting to remove all conflict in fisheries, co-management does appear to deal successfully with certain major manifestations of conflict. Before focusing in detail on the approach, however, it is helpful to give a brief review of some aspects of the history of fishery management that led up to the present-day situation.

13.10.1 The evolution of participation in fishery management

Most civilisations, and notably those of indigenous populations around the world, have developed, over time, mechanisms and institutions for the management of fishery systems. If any fishery attempted to operate under *laissez-faire* conditions, i.e. avoiding management in favour of the freedom of the seas, based on a sense of limitless marine resources, such a situation would have remained viable only until pressure on the fish stocks grew, leading to stock depletion and fishery collapse. Thus to survive, any fishery requires a management system.

However, processes of colonisation in many nations destroyed traditional management arrangements. Furthermore, over the past century, the industrialisation and commercialisation of fisheries globally has led to the neglect of traditional local means of management (see, for example, discussions in Johannes 1978 and Berkes 1999). In the vacuum created by these losses of traditional management, serious fishery declines have been commonplace. In response to these, a new centralised model of management emerged, focused on governments having the role and responsibility to manage the fish in the sea (e.g. within a nation's exclusive economic zone), these fish being recognised as a public resource. Governments became the dominant players in the management process. Intense regulatory efforts were undertaken to control fishers, fishing fleets and processors, with the various aims of fish stock conservation, allocation of the benefits from the fishery and resolving use conflicts. In many developed nations, this led to the development of extensive regulatory structures.

Government-led regulatory frameworks have often been top-down in nature, with much power being held by the fishery bureaucracy. Underlying this approach lies a sense of fishers as selfish profit-maximisers, and regulators as protectors of the resource. This naturally produced an 'us versus them' view of the world, in which there appeared to be a complete divergence between the goals of exploiters and managers. (This questionable view of fishers is also reflected in the classic *Tragedy of the Commons* argument of Hardin (1968), which is discussed further in Chapter 14.)

Unfortunately, this approach to management did not work out as planned. In fact, many such efforts proved to be futile, or indeed counter-productive. Fishers, being excluded from the formulation of the rules, had no incentive to follow them. Indeed, not only was there little social or peer pressure to follow the imposed regulations, but the desire actually increased to 'beat the system'. Enforcement was costly but ineffective; it was essentially impossible to prevent illegal fishing and over-harvesting at sea. Ironically, the imposed view of fishers as exploiters became self-fulfilling, leading to a lack of effective management, with overfishing and depleted stocks being the result.

To overcome this 'us versus them' attitude, many fishery regulatory bodies shifted to a consultative model, in which government discusses management with the industry prior to implementation. However, in many if not most cases, this model of government *consultation* with fishers, usually carried out without a sharing of the decision-making power, fared only somewhat better. While efforts to improve compliance with fishery regulations through extensive consultative processes gained some nominal success, this was at an exceptionally high cost, with very capital-intensive monitoring, control and surveillance. Despite these efforts, fishers did not 'buy into' what were still government-imposed regulations.

(In the Atlantic Canadian groundfishery, for example, there is evidence that extensive dumping and high-grading of fish, as well as under- and misreporting of catches, took place in the 1980s, leading to stock collapses in the 1990s (Angel *et al.* 1994). There is also a percep-

---

A view from Asia

'In many Asian countries fisheries policies and regulations were designed using a top-down approach and most of these regulations were by-products of colonial legacies. Their legitimacy was frequently questioned by stakeholders in the fisheries. It was for these reasons that fisheries management often failed to achieve its desired objectives. The unwillingness on the part of fisheries administrators to include fishermen's interests in the design and formulation of fisheries regulations and policies partly explains why fisheries management failed badly in many areas. The traditional approach towards fisheries management requires a serious second look. The interests of the stakeholders in fisheries cannot be taken for granted. A shift in the management paradigms among the policy makers must take place. The new fisheries objectives must focus on more pressing intergenerational equity issues, and its implementation must include a more participatory approach, taking into consideration both the government and fishermens' view.'

Abdullah & Kuperan (1997)

tion among inshore fishers that the consultative mechanism, which operated on a centralised basis, favoured the larger-scale fishery players, who were equally centralised in their own organisation.)

Fishery management in commercial and industrial settings is evolving through what is essentially a return to participatory modes (which, it should be noted, have remained in place in many small-scale and/or indigenous systems). In order to be effective, management must not only *regulate the fishers*, but must also *involve the fishers*. This is because management must have the support of those affected, if only for the purely pragmatic reason of ensuring a 'buy in' to conservation and management actions. Without the confidence of fishers, the system may well disintegrate over time, as we have seen in management shortcomings that led to fishery collapses in the past. To win this confidence, fishery management must be seen to be both efficient (working well) and equitable (working fairly). These attributes in turn require the involvement in management of fishers, and potentially of others in the fishery, through the increasingly popular mode of operation referred to as *co-management*, a subject to which we now turn.

13.10.2 The concept of co-management

The essential idea of co-management is the sharing of decision making and management functions between government and stakeholders in the fishery. More formally, co-management can be defined as the creation and implementation of suitable management arrangements through which a set of agreed stakeholders, i.e. fishers and their organisations, work jointly with government to develop and enforce fishery regulations and management measures. These ideas of co-management more or less reflect the definitions that have emerged in the wealth of published literature on the subject (e.g. Kearney 1984; Pinkerton 1989). There follow some examples, generally mutually compatible, of how others have defined co-management.

- '... an arrangement where responsibility for resource management is shared between the government and user groups.' (Sen & Nielsen 1996: p. 406)
- '... the collaborative and participatory process of regulatory decision-making among representatives of user groups, government agencies and research institutions.' (Jentoft *et al.* 1998: pp. 423–424)
- '... government agencies and fishermen, through their cooperative organizations, share responsibilities for management functions ... A part of the regulatory power is transferred from the government to fishermen's organizations. Fishermen's organizations therefore not only participate in the decision-making process, but also have the authority to make and implement regulatory decisions on their own.' (Kuperan & Abdullah 1994: p. 310)
- '... various degrees of delegation of management responsibility and authority between the local level (resource user/community) and the state level (national, provincial/state, municipal) ... a middle course between state level concerns in fisheries management for efficiency and equity, and local level concerns for self-governance, self-regulation and active participation.' (Pomeroy 1995: p. 150)
- '... a partnership arrangement using the capacities and interests of the local fishers and community, complemented by the ability of the government to provide enabling legisla-

tion, enforcement and conflict resolution, and other assistance.' (Pomeroy & Berkes 1997: p. 465)
- '... systems that enable a sharing of decision-making power, responsibility and risk between government and stakeholders, including but not limited to resource users, environmental interests, experts, and wealth generators.' (NRTEE 1998: p. xvi)

In co-management arrangements, the stakeholders:

- have a stake not only in the fishery, but also in the management of the fishery;
- share decision-making power with government;
- share responsibility for ensuring the fishery's sustainability.

Therefore, those involved in co-management arrangements have both rights and responsibilities. The rights in this case are *management rights*, i.e. the right to be involved in the design and implementation of management measures (discussed further in Chapter 14).

There are two underlying motivations for co-management initiatives. First, there is the desire to reduce conflict between stakeholders (particularly fishery users) and government, as well as between stakeholders themselves. By clearly defining rights and responsibilities, by providing an institutional forum for discussion among the decision makers, and by ensuring that management simultaneously serves the public interest *and* operates inclusively, co-management can encourage fishers to support the process and provide a vehicle to help in resolving conflicts.

Second, and closely related to the above, is the need in the fishery for a conservation ethic, i.e. a fundamental conservation perspective within the 'belief system' of fishery participants. The manner by which the management system is structured, and who is able to participate, can affect the presence or absence of this conservation ethic among fishers since, for fishers to want to conserve the fish, they must 'buy into' management. A co-management approach is a means to accomplish this, bringing fishers and others into the decision-making process, and at the same time ensuring that they share responsibility with government for sustainability in the fishery.

13.10.3 Participants in co-management

A substantial list can be drawn up of possible participants in the fishery management system, including:

- governments;
- fishers;
- other fishery sector players (e.g. processors);
- community organisations, decision makers and/or residents;
- interested organisations (e.g. environmental non-governmental organisations);
- the general public.

In the light of the discussion above, the first two of these are essentially compulsory participants in a fishery co-management process. Others may be completely excluded from management, or may be able to participate only on certain topics. Here, we look at three major organisational modes for co-management:

(1)   a fisher–government model;
(2)   a community-based (fisher–community–government) model;
(3)   a 'multi-party' model involving multiple stakeholders and/or public involvement.

13.10.4 Fisher–government (sector-based) co-management

The most straightforward means to rectify the problems of past management related to its top-down exclusion of fisher involvement is to add the fishers into the process. With this rationale, most co-management initiatives, at least in locations such as Europe and North America, focus on a model of joint fisher–government management. This can also be referred to as 'sector-based' co-management, or 'functional representation' (Jentoft & McCay 1995), be-cause it is typically organised on the basis not of all fishers within a certain jurisdiction, but rather on the basis of fishery sectors, i.e. specified groupings or segments of fishers, which could be defined on the basis of:

- species fished (e.g. the herring fleet);
- vessel size (e.g. vessels of a certain length);
- gear type (e.g. otter trawlers);
- producer organisation or cooperative;
- some other desired grouping.

In a sector-based approach, fishers within a given sector share with government the right and responsibility to establish a management plan for the sector. In such cases, fishers are defined essentially on the basis of their membership of a sector (analogous to an 'interest group'). In some cases, this reflects the structure of the fishery within which management has tradition-ally operated, in which case such groupings may be reasonably cohesive. On the other hand, if fishers identify with the geographical community in which they live, this approach has the disadvantage of institutionalising divisions among fishers living within the same community. Jentoft & Mikalsen (1994) suggest that in the Norwegian context, an approach based on gear groupings has been less than successful in bringing small-scale fishers into the management process and in promoting flows of local knowledge into fishery decision making. Neverthe-less, this model remains a dominant form of co-management in many jurisdictions.

13.10.5 Community-based co-management

The main alternative to sector-based fisher–government co-management is a geographical structuring of the management system, usually referred to as a community-based approach. The defining feature of this model is the focus on the geographical unit, which could be a specific 'community', in the common usage of that term (e.g. a municipality), or could be a defined ecosystem, coastal zone or political jurisdiction, albeit one that is usually on a spatial scale more in keeping with a municipality than a province or state. In this model, co-management units are organised geographically, rather than dividing stakeholders according

Co-management in the Dutch flatfish fishery

The beam-trawler fleet of the Netherlands is the principal operator in the North Sea flatfish fishery, controlling 45% and 75% of the total allowable catches (TACs) for plaice and sole, respectively (Langstraat 1999). In the late 1970s, individual quotas for sole and plaice were allocated to vessels; these were at first non-transferable, but in 1985 they became transferable as individual transferable quotas (ITQs) (Dubbink & van Vliet 1996; Langstraat 1999). However, this regime led to serious problems. There was significant investment in larger, more powerful vessels, increasing pressure on national TACs, and 'an increasing number of fishermen started to overfish their individual quotas …' (Langstraat 1999: p. 74). In response, the government implemented vessel capacity and fishing effort controls, which had some success in reducing the extent to which quotas were exceeded and reducing the Dutch fleet size (although many vessels continued fishing under British flags), but this led to a sense that the government was too enmeshed in micro-management while the fishers were left with too little control and too little flexibility. Overall, this '*de facto* command-and-control system' (Dubbink & van Vliet 1996: p. 506) led to 'increasingly poor relations between fishers and government' (Sen & Nielsen 1996: p. 412).

In 1993, co-management was introduced into the fishery on the basis of recommendations from a steering group of government and industry members (Langstraat 1999). Sen & Nielsen (1996: p. 412) note that in the new system:

'responsibility to manage individual fishermen's quotas has been devolved to groups of fishermen who pool their individual quotas … These groups are responsible for implementing and enforcing regulations, imposing sanctions and organising intra-group quota exchanges. The government retains responsibility for controlling the national quota …'

As in any co-management system, the 'Biesheuval Groups', as they are called, have certain obligations, notably to develop fishing plans and to provide requested information to the authorities. The groups also have strong incentives to enforce their fishing plans, since a group can lose its right to operate if it overfishes its quota. Within a group, fishermen must formally agree to abide by the rules of the group. Indeed, 'the members are obliged to transfer their right to manage their ITQs to the executive board of the Group' but at the same time, 'individual members have the right to use their individual quotas under the conditions agreed upon in the fishing plan' (Langstraat 1999: p. 76). These obligations seem to be well received; according to Dubbink & van Vliet (1996: p. 508), fishermen 'appreciate the increase in flexibility they have at present' and 'for the first time since the start of the quota regulations in 1976, there seems to be administrative and political stability in and around the sector'.

to their sector of the fishery, as in the model above. Participation in such co-management could be as in the sector-based model, i.e. restricted to allow only fisher and government involvement, or it may also allow for participation in fishery management by other stakeholders and institutions. The possibilities may include processors, ancillary fishery industries, local government, community organisations and the general public.

Community-based management is a major force in reshaping resource management in developing nations. It is also common in fishery management among indigenous/native peoples in developed nations, and is receiving increasing attention as a potential tool in some small-scale commercial fisheries as well. However, confusion can arise over what is meant by community-based management, and to what extent it is the same as, part of, or disjoint from co-management. Some view community-based management as being carried out in the absence of government, i.e. complete self-regulation (Sen & Nielsen 1996: p. 406), and therefore do not consider it a form of co-management. Others view the two as intrinsically related. For example, Pomeroy & Berkes (1997: p. 467) state that 'A certain degree of community-based resource management is a central element of co-management.' Here, we avoid this discussion by simply treating 'community-based' as an adjective describing the arrangement of co-management on a geographical basis.

Such a perspective is important, because an expanding body of research (e.g. Pinkerton & Weinstein 1995) indicates that when co-management includes those living in the geographic vicinity of the fish stocks (fishers and other residents of local communities), as well as their local institutions, conservation and socioeconomic benefits can be enhanced. Improved conservation benefits arise from both the collective stewardship role of coastal communities and the implicit moral suasion, i.e. the built-in social pressure that community institutions exert on fishers to comply with regulations in support of conservation. Socioeconomic benefits may be obtained if, for example, compliance among fishers is increased, thereby decreasing enforcement costs and increasing the efficiency of management.

Community-based co-management has succeeded in many situations where the coastal communities and their residents have strong ties to, and dependence on, the fishery. In small-scale fisheries, notably in developing countries, connections between the community and the fishery are more often recognised and allowed for in management. In contrast, in many developed countries, coastal communities have tended to be excluded from fishery decision making. (Indeed, in some cases, the term *community* has been distorted from its common-sense geographically based meaning to refer merely to a 'sector' of the fishery, as in the 'community of gill net vessels'.) Pinkerton (1994: p. 2367) put the matter succinctly:

'While it would be a mistake to consider the community-based aspect of co-management an automatic panacea for the array of fisheries management problems, it would also be equally unfortunate not to take advantage of the management benefits available under community or mixed community–government arrangements …'

What, then, might lead to the selection and implementation of a community-based approach to co-management? Whether it is a desired mode of operation, and whether it is implemented, would seem to depend on two major factors.

- The approach must be intrinsically feasible. This feasibility depends on a variety of factors, including the nature of the resource base, the sociocultural environment, the fundamental format of the fishery (small-scale *community-based* versus *industrial*), the nature of social cohesion in the community, and the strength of community institutions (see, e.g. Jentoft *et al.* 1998). The latter is crucial, since a suitable management institution, integrated within the fishing community, is needed as a forum for interaction among stakeholders. If one does not exist, it must be created.

Co-management of Japan's coastal fisheries

Japan's coastal fisheries have a long tradition of management on the basis of geographically defined units of fishers. This tradition dates back for centuries, in particular to the defining of fishing territories in the Tokugawa period of 1603–1868 (Kalland 1996). However, throughout the 1800s, the system was feudally based. The Fisheries Law of 1901 adapted and codified the feudal territorial system by transforming what had been village 'guilds', through which fishers operated, into fisheries cooperative associations (FCAs). These FCAs were charged with 'carrying out fisheries management via the granting of fisheries rights and the issue of licences, as they had traditionally' (Ruddle 1989a: p. 274). The FCAs form the foundation of Japan's fishery co-management structure, which continues to the present day. As Kalland (1996: p. 78) notes:

> 'Although the responsibility for both fishing rights and licences today ultimately rests with the national government, represented by the Ministry of Agriculture, Forestry and Fisheries, one of the main features of the Japanese management regime is the delegation of as much of the responsibility as possible to the local FCA. Through the Fisheries Cooperative Association Law of 1948 ... and the Fisheries Law of 1949, most of the day-to-day administrative tasks are delegated to the FCA.'

Specifically, fishing rights in adjacent coastal waters are granted to the FCA by the local fisheries office, and in turn 'the FCA then allocates these use rights to eligible individual fishermen that comprise its membership' (Ruddle 1989b: p. 171). Note that in this regard, only eligible fishers can be part of an FCA and take part in co-management. This is therefore a case of community-based co-management in which non-fishers do not play a substantial role. Nevertheless, Ruddle (1989a: p. 280) has highlighted the importance of the FCA not only to fishery management, but also to the welfare of the local communities:

> 'In present day Japan, the FCA is a vitally important intermediate organisation that links the central and prefectural governments with the individual fisherman. Although comprising the fundamental unit of governmental fisheries administration, and being the key organisation in the implementation of official fisheries projects, an FCA belongs entirely to the local community of fishermen. The FCA lies at the hub of modern Japanese fishing communities, in which it constitutes the focus of social and economic activities.'

In summary, the Japanese system of co-management represents a particularly long-standing and effective example of community-based fishery management.

• Community-based co-management must be perceived positively by the government, which typically maintains *de facto* control over fishery policy and therefore holds the power to determine the *form* that co-management is to take. While the option of community-based management should be evaluated objectively, in reality it is more likely that such an approach will be considered only if the government (a) sees coastal communities as relevant to fishery decision making, (b) has some familiarity with the community-based option, which, in many developed regions, is rather little known relative to the sector-based approach, (c) has the will to tackle new challenges that arise with community-based approaches, such as devising suitable geographical *community* boundaries, and (d) perhaps most fundamentally, is willing to share decision-making power at a decentralised level, and to develop workable, equitable mechanisms to bring fishers and other community members together with government officials.

13.10.6 Multi-party co-management and the public role

What, if anything, is the role in fishery management of the general public and of those in economic sectors other than the fishery in fishery management? There are two key points to note in this regard.

(1)   First, within the fishery management system *per se*, there is a need for different treatment of strategic management on the one hand, and the tactical and operational levels on the other (Chapter 5). In practice, the general public (as well as non-governmental organisations and others) have largely been excluded from fishery management deliberations at the tactical and operational levels. This may be justified on the basis that short-term choices, such as this year's mesh size restrictions for gill nets, are unlikely to require public involvement if made within a context compatible with society's preferences. At such levels, it may be that co-management is restricted to fishers and government, or if the process is public, the special (perhaps dominant) role of fishers may be recognised. However, at the strategic level, policy decisions concerning 'optimal' use of public resources, long-term conservation, and the generation and allocation of fishery benefits are logically a matter of public interest. Such decisions need to be made within a broad, inclusive process of co-management in which there is scope for the involvement of communities and the public, while still ensuring active participation by the resource *users* who must, in the end, 'buy into' the process (Fig. 13.4).

(2)   Second, we must note that many challenges arising in a coastal or watershed setting, such as those involving conflict resolution and decision making concerning environmental issues or the planning of multiple use in marine areas, extend far beyond the fishery. In such cases, a form of 'multi-party co-management' may be desirable (Pinkerton 1994). Such an approach, which typically requires great attention to procedural issues and institutional development, is fundamental to achieving success in dealing with a range of multiple-use situations, particularly those involving 'integrated coastal zone management' (ICZM), including the establishment of marine protected areas (Chapter 12). An example from the Caribbean is described in the box below.

Multi-party co-management: Soufrière, St Lucia

Soufrière is a community located on the southwest coast of St Lucia, an island nation in the eastern Caribbean. Fisheries are important to the community, but as Renard & Koester (1995) have noted, 'Due to its isolation and a rugged topography, the local economy has been relatively depressed for several decades, and tourism is now seen as an opportunity to bring new benefits to the community ...' Indeed, tourism has grown rapidly, and this, combined with the general development of Soufrière town and accompanying environmental impacts, 'provoked severe conflicts between the various resource user groups and had major impacts on fishermen and their activities. The most acute of these conflicts were between divers and fishermen (over reef areas) and between seine fishermen and yachts anchoring in fishing zones' (Renard & Koester 1995: p. 5).

These conflicts led to participatory planning and resolution efforts in the early 1990s involving all the stakeholders, i.e. fishers, tourism operators and recreational users, other harbour users, and other community members. This process in turn led to the creation of the Soufrière Marine Management Authority (SMMA) in 1995. The Authority 'represents the first attempt to establish, on a large and permanent scale, in a context of multiple use, a system of collaborative management for coastal and marine resources in St. Lucia' (Soufrière Marine Management Authority 1996: p. 1).

At the heart of the new coastal management arrangements in Soufrière is the multi-use zoning of the aquatic part of the coastal zone. There are five types of management zones: marine reserves (with no extractive activity allowed), fishing priority areas (in which fishers have precedence), multiple use areas (notably some reefs), recreational areas and yacht mooring sites (Soufrière Regional Development Foundation 1994). Another important element provided by the Authority is the opportunity for 'diversification of the activities of fishers and other people in the community, with the development of small-scale sea transportation', using local boats as water taxis for visitors and residents (Renard & Koester 1995: p. 7).

The Soufrière co-management system works through several related institutions and organisational structures (Soufrière Marine Management Authority 1996: p. 5):

'The operation of the SMMA is based on the involvement and cooperation of a number of groups and institutions. Its main governing body is the Technical Advisory Committee (TAC), chaired by the Department of Fisheries, which brings together all the main stakeholders. The day to day operation of the SMMA is carried out under the auspices of the Soufrière Foundation ... Key to the success of the SMMA is the effective and consistent participation of the people who are representing the various sectors ... Another important issue is that of organisation at the community level, and particularly among users of the resources of the SMMA.'

While resource use conflicts and challenges remain in Soufrière, there is now in place a clear institutional arrangement for dealing with problems through the multi-party co-management process within the SMMA.

Fig. 13.4 The general public, and environmental organisations in particular, are taking an increasingly active interest in policy development concerning fisheries and the ocean environment. This role has become important in many co-management situations, for example in Europe as well as in the international arena.

13.10.7 Levels of co-management

The second major theme concerning participation in co-management is the balance among the players, specifically, the proportion of responsibility and power held by government, as opposed to stakeholders. If this share were 100%, we would have a completely *top-down* scenario of centralised management, while if the proportion were 0%, this would represent total user control (self-regulation), with all management activities being determined and carried out by those outside government.

Pinkerton (1994: p. 2367) has suggested that levels of co-management can be envisioned along an axis, and specifically that 'it is useful to conceptualise co-management as eight potential points on a ten-point continuum between state management (at point 1) and self-management (at point 10).' In this sense, neither of the latter extremes would qualify as co-management, but the eight intervening 'points' would represent varying co-management levels. Recent studies have come to agree on a related framework for discussing the range of options available, a so-called 'ladder of co-management' (Sen & Nielsen 1996; Pomeroy & Berkes 1997). The steps on the ladder, ranging from centralised management to self-regulation, are shown in Fig. 13.5).

- *Instructive*. Government is in control, through centralised management. It utilises channels of communication with users and communities to inform ('instruct') them about decisions already made and actions to be taken. This has been the 'traditional' top-down mode of operation prevalent in 'modern' fishery management, and is often blamed for management failures.

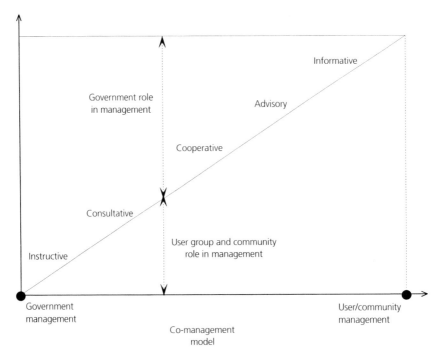

Fig. 13.5   The *ladder of co-management* (e.g. Sen & Nielsen 1996; Pomeroy & Berkes 1997) is a framework for discussing the range of options available in co-management. These range along the horizontal axis from management by government ('instructive') to management by the users and/or communities ('informative').

- *Consultative.* Government makes the decisions in the fishery after gathering opinions and suggestions through consultations with users (and possibly the relevant fishing-oriented communities). This mode often developed as an improvement on the instructive mode when the latter emerged as a failure. The complaint about this mode, however, is often that those being consulted may have their say, but have no actual power over what decisions are eventually made.
- *Cooperative.* Government and users, as well as communities in some cases, engage as partners (possibly, but not necessarily, as 'equals') in management decision making. Note that a cooperative approach may well be used for some aspects of management (particularly operational functions such as the setting of annual harvesting plans) but not others (e.g. setting user-group allocations).
- *Advisory.* Users essentially make the decisions and communicate these to government (i.e. 'advising' the government on what decisions have been made), but government evaluates and accepts these decisions only if they satisfy certain overarching criteria (see the example in the box on p. 272).
- *Informative.* Decision-making authority lies with user groups, perhaps reflecting historical realities, or is specifically delegated by government. Government is merely 'informed' of any decisions. This represents fully decentralised management, bordering on self-regulation. The examples described in this section of the Dutch flatfish fishery, the Japanese coastal fisheries and the Soufrière marine management area might all be considered as falling within this level of co-management.

The advisory level of co-management: Kattegat, Denmark

Examples given earlier in this section illustrate the level of co-management closest to self-regulation, namely the informative level. If it is the stakeholders who develop management plans but, even technically, those stakeholders must seek approval from government before implementation, this indicates a cooperative or advisory level of co-management. The Danish example discussed below illustrates such a case.

'The days-at-sea regulation in Kattegat is an experiment aimed at solving some of the problems created by the quota system – discard and mis- or non-reported landings within the sole/nephrops fisheries. By using days-at-sea effort regulation, fishermen were allowed to land all catch of sole and nephrops, with discard and mis-reported landings of these species being diminished. In the working group managing the days-at-sea regulation in Kattegat, co-management exists as a cooperative process between government (MAF) and user-groups.' (Nielsen & Vedsmand 1997: p. 280)

'Representatives from the government, the fishermen's association and a research institute meet monthly … and make decisions concerning the number of days at sea to be allocated and to review progress. These decisions are communicated to the Danish Regulation Advisory Board … However, all proposals put forward by the Working Group have, so far, been accepted. This co-management arrangement is therefore considered *advisory*.' (Sen & Nielsen 1996: p. 412)

Which of these various levels of co-management will occur in a given situation? Which *should* occur? These questions relate to such considerations as:

- the history, tradition and experience with local self-management (e.g. Dyer & McGoodwin 1994) and the current interest of stakeholders in taking on management responsibilities (and rights), a matter that relates in part to the overall acceptability among fishery participants of the co-management system;
- the capacity of fishers and communities to undertake co-management, something that involves (a) the organisational and financial capability, as well as the social cohesion, necessary to create and maintain strong institutional arrangements for management, and (b) the devolution (or at least availability) at the local level of technical expertise that is typically existing in centralised fishery management structures;
- the willingness of those currently holding power in the fishery (typically the government and those close to government) to share decision-making power with previously excluded fishers and perhaps local communities (and/or the political capability of the latter groups to campaign for and acquire the necessary power);
- the optimal balance, as measured by society, between the potential benefits involved in devolving decision-making power to stakeholders (such as greater manageability and enforceability of regulations), on the one hand, and meeting legislated requirements for resource conservation on the other.

It is also important to reiterate that two other factors may be key determinants of the level of co-management arrived at: (a) the inherent nature of the fish stocks and the fishery, and (b) the level of management itself (strategic, tactical or operational) for which a co-management system is being considered. Considerable research is underway worldwide (e.g. Jentoft & McCay 1995; Sen & Nielsen 1996) to examine and compare cases of co-management, in order to deduce which cases are conducive to, or tend to lead to, the various levels of co-management.

13.11 Summary

This chapter has focused on two principal themes. First was an examination of the nature and structure of fishery conflicts, through an integrated analysis involving (1) a typology of conflicts and (2) a set of fishery paradigms that reflect the philosophical basis of the conflicts. As was seen, conflict arises at many different levels in the fishery: between nations, between various users of ocean space (fishery and non-fishery), between different classes of domestic fishery resource users (e.g. recreational versus commercial), between sectors of the commercial fishery (e.g. industrial versus artisanal), and so on. These various sources of conflict fit within the typology presented in this chapter, and are often driven by the basic differences between the paradigms discussed at the start of the chapter.

Second was an exploration of the possibilities for resolving conflicts. Some note was made of mechanisms for doing so at the international level (between nations) and at the level of domestic coastal and/or ocean management (between economic sectors). Emphasis in this section was placed on implementation of a co-management approach within the fishery system, particularly to deal with conflicts between government and fishers/communities.

The move towards co-management is taking place in many fisheries, most often spurred on by the mutual interest of governments and fishers to transfer some management responsibilities from the former to the latter. Implementation of co-management raises a series of key strategic questions, discussed in this chapter. Under what circumstances should co-management systems be organised by the *sector* of the fishery, regardless of location (e.g. 'those fishers who use gill nets') or by geographical community? Under what circumstances should the players in co-management include non-fishers as well as fishers? What is the desirable, and feasible, balance of power between government and stakeholders?

Also of importance, given that fishery management is increasingly being viewed as connected with integrated coastal and/or ocean management, is the question of scale, and of interconnections. How does co-management differ in structure and function when applied at a fishery-specific level, versus application to multi-user integrated coastal and/or ocean management? Can a single institution deal at both levels?

There is a long-term research element to these questions, in the sense that a comparative analysis of co-management experiences globally can shed some light on the answers. There is also a shorter-term practical element, in that strategies for addressing these issues are needed immediately, given that the implementation of co-management systems in real-world situations is something that is happening on a regular basis already.

Chapter 14
Rights in Fishery Systems[*]

The theme of *rights* has become a popular one in discussions of fishery management, with a multitude of articles, books and conferences appearing on the subject. This chapter explores the various forms of *rights* in fishery systems, how these rights appear naturally or are introduced, and some practical issues that arise in dealing with those rights. We begin by looking at the rationale for and nature of fishery *use rights*, which allow the holder to access and harvest the resource, perhaps subject to specific conditions imposed by society on fishing effort, time spent fishing, etc. The concept of *management rights* is also examined. These relate to participation in fishery management decision making (as referred to in Chapter 13). Use rights and management rights are then discussed in the broader context of *property rights*, i.e. the range of rights that a holder may have, relative to other individuals, with respect to a certain item of value (such as a stock of fish in the sea or a permit to catch a certain amount of that fish). The various types and characteristics of property rights, as well as the range of property regimes, are examined, along with their implications for fishery management, based on the distinction between *open access* and *common property*, and between resource use rights and resource ownership. The second half of the chapter explores use rights in greater detail, providing an overview of the various forms of such rights, their advantages and disadvantages, and issues relating to implementation, such as initial allocation, institutional arrangements and transferability of rights.

Note that property rights, including use rights and management rights, arise in a multitude of contexts well beyond the fishery. For example, consider the owner of a unit in a condominium, i.e. an apartment block in which each unit is privately owned. That individual, as owner of the unit, has an extensive 'bundle of rights' over that specific unit, but also holds use rights with respect to the common grounds surrounding the building, and facilities such as elevators in the building. There is also a management right, allowing involvement in collective management decisions for the building as a whole, such as the setting of rules about acceptable noise levels, or decisions about maintenance. Similar arguments arise in a wide variety of communities; for example, a family may hold full property rights within its household, as well as sharing collectively in use and management rights over community property.

*The first part of the chapter has benefited greatly from the comments of Dr Melanie Wiber (University of New Brunswick, Canada). The second half of this chapter is drawn in part from a paper, Townsend & Charles (1997), written jointly with Dr Ralph Townsend (University of Maine, USA). I am grateful for the insights of both these colleagues, but take sole responsibility for any errors or interpretations within this chapter.

Alternatively, in such a situation, the cultural context may be such that collective (group) rights predominate, as in some indigenous/native societies.

Let us take a more detailed look at the various forms of rights in fishery systems, beginning with the historic situation.

14.1 Open access

Consider the 'high seas' – ocean spaces located outside any single nation's jurisdiction. Until quite recently, high seas fisheries took place without limits to access: anyone could exploit high seas resources. Such situations are referred to as *open access*. This term is actually used in two somewhat distinct senses: to refer to a fishery that is totally unregulated, in which both the fleet and the catch taken by the fleet are uncontrolled, or referring solely to matters of access, so that the output of the fishery (e.g. catch and size of fish) may be regulated, but not the inputs (e.g. the number of boats).

It has become accepted wisdom that the first of these senses of open access, i.e. a limited resource exploited by a completely unlimited fleet, will probably lead to disastrous conservation and economic problems. This conclusion is reinforced in fishery theory, particularly for fisheries operated individualistically without social constraints. For example, bioeconomic analyses (e.g. Clark 1990) show how, even in the case of a sole owner, it may be 'rational' from a private financial perspective to drive a fish stock to extinction in the pursuit of profits.

In practical terms, experience with collapses and depletion in fisheries worldwide supports the view that unregulated '*laissez-faire*' (i.e. free enterprise) exploitation of marine resources is among the greatest threats to the long-term sustainability of fishery systems. Indeed, the threat posed by such open-access fisheries was a major factor leading to efforts to regulate fisheries on the high seas through the United Nations Conference on Straddling Fish Stocks and Highly Migratory Fish Stocks and the subsequent 'Agreement for the Implementation of the Provisions of the United Nations Convention on the Law of the Sea of 10 December 1982 relating to the Conservation and Management of Straddling Fish Stocks and Highly Migratory Fish Stocks', adopted on 4 August 1995 (for details, see United Nations 1999).

The situation is somewhat different in a fishery that is 'open-access' in the second sense above, i.e. with no controls over the fleet but with regulated harvesting levels. In such a case, the fish stocks may not necessarily collapse (if regulations work), but the fleet may overcapitalise, driven by a typical *rush for the fish*: a free-for-all in which those who catch the most fish first are the 'winners'. With more inputs used than are necessary to catch the available fish, profits and rents are dissipated; the resource may be safeguarded, but not necessarily the economic health of the fishery.

Some analysts, however, have noted the possibility that some form of 'controlled open access' could have beneficial aspects. For example, consider a fishery that is open-access within a limited community, so that access is unrestricted for the local users, but at the same time outsiders cannot participate. In such a case, the community will typically limit the overall level of exploitation and set rules on how community members accessing the fishery must operate. Such controlled open access may be desirable in some circumstances: perhaps to reflect local culture and a sense of equity (community fairness) with respect to job opportunities, or perhaps for other economic reasons (e.g. Panayotou 1980; Sinclair 1983). For example, in a study of New England fisheries, Townsend (1985) noted that local-level open access, i.e. the *right to fish*, may be valued as part of community culture, with its very existence potentially enhancing stability by improving employer–employee relations. This might occur on a cultural level, or it may help to increase the bargaining power of workers (by ensuring a readily available fall-back job in the fishery), with the latter leading to higher overall wage levels in the local economy.

While variations of open access may have some value, open-access fisheries generally do not have a good name. Both internationally and within national jurisdictions, there are strong beliefs that in the interests of sustainability, not everyone can have access to the fishery. Thus, while specific decisions about access restrictions are properly made through informed analysis (cf. Rettig & Ginter 1978; Gimbel 1994), the overall need for, and desirability of, restricting access are usually accepted as basic premises in fishery management.

14.2 Use rights

Where open access is the critical problem, the solution would seem to lie in limiting access. If this is done, only a limited number of individuals will have the right to go fishing, i.e. to access the fishery. Those entitled to do so are said to hold *use rights*. In some cases, it will be the resource 'owners' themselves who hold use rights, while in many other cases (Symes 1998b: p. 4), the use rights will be in the form of *usufruct*, which might be defined as 'the legal right to use another's property and enjoy the advantages of it without injuring or destroying it' (World Book 1999).

The need for use rights is by no means a new insight. Informal and traditional use rights have existed for centuries in a wide variety of fishery jurisdictions. Even in cases where direct government regulation of fisheries is dominant, use rights alternatives are being implemented with increasing frequency. Use rights options range widely; for example, each of the following approaches to fishery management involve use rights (to be examined more fully later in this chapter).

- Customary marine tenure (CMT) and territorial use rights in fishing (TURFs) have long been applied by indigenous peoples in determining the usage of local resources.
- Limited entry, providing a limited number of individuals with the right to access the fishery, was the initial approach to use rights in modern state management of fisheries.
- Quantitative use rights approaches are increasingly being adopted, based on allocations of rights either (a) to fish at a certain level of effort, perhaps using specified bundles of gear for certain time periods, or (b) to catch a specified amount of one or more species of fish, with allocations through community institutions or market-based approaches, e.g. community quotas or individual transferable quotas.

Note that there are two levels of use rights. The first two approaches above concern *access rights*, which permit the holder to take part in a fishery as a whole (limited entry) or to fish in a particular location (TURFs). The quantitative use rights above are forms of *withdrawal (harvest) rights*, which allow a certain level of resource usage. Figure 14.1 shows some pos-

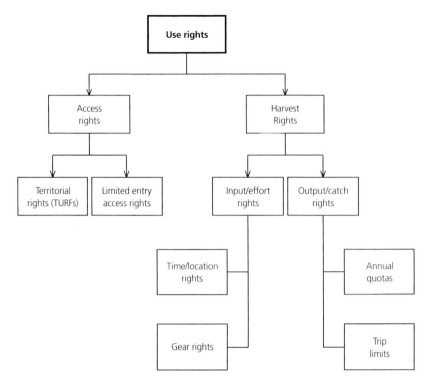

Fig. 14.1   A wide variety of use rights arrangements are possible in fishery systems. These can focus on inputs or on outputs, on quantitative or qualitative measures (e.g. quotas or spatial rights), on individual or group rights.

sible arrangements, for each of input rights and output rights, with a more detailed discussion of these options provided later in the chapter.

Use rights can be seen as an institutional mechanism through which fishers, fisher organisations and/or fishing communities hold clear-cut rights and some security of tenure over access to a fishing area, use of an allowable set of inputs, or harvest of a quantity of fish. If use rights are well established, fishers will have secure tenure and it will be clear who can access the fishery resources and how much fishing each is allowed to do. This can encourage fishers to adopt a *conservation ethic* and move towards fishery sustainability, since conservation measures to protect 'the future' are compatible with their own long-term interests. (However, use rights do not guarantee sustainability. Indeed, ecological sustainability could be diminished if, for example, the economic incentives created by the use right lead to anti-conservationist behaviour at the individual fisher level, such as 'high-grading', which is the discarding of lower-valued fish to maximise individual profits.)

14.3 Management rights

This idea that clearly defined fisher rights over resource *use* may lead to a more sustainable fishery can also be applied to resource *management*. Indeed, the earlier discussion of open access focused not only on the need to limit a fishing fleet's access to fishery resources, but also the need for conservation measures and other forms of management. As was discussed in

Chapter 13, such management has generally been unsuccessful when practised in a top-down manner, and has led to the emergence of new co-management arrangements.

Within the context of fishery rights, discussions of co-management can be expressed in terms of *management rights*. Who has the right to be involved in fishery management decisions? Specifically, to what extent should management rights be held by the government, the resource harvesters, the coastal communities and the general public, and to what extent do the holders of management rights vary across the levels of management: operational, tactical and strategic?

On this point, it is logical to envision management rights in a single fishery system as varying with the level of management itself. Consider, for example, the *operational* level of management decision making, i.e. determining details of fishing times, allowable hook or mesh sizes, and the like. Experience with top-down management points to the pragmatic rationale for ensuring that fishers have management rights at the operational level in order to improve acceptance and compliance with at-sea regulations. However, communities and the general public may have less of a role to play in such operational decisions. In contrast, debates over *strategic* fishery management, which concern the overall objectives and structure of the fishery, are typically matters of public interest. The general public, and fishing communities in particular, are clearly legitimate stakeholders, deserving the right to be involved in setting the broad objectives for use of the fish. (Indeed, in many cases, the public can be considered the 'resource owners' (as will be discussed below), presumably with the right to benefit from that ownership.)

14.4 Property rights

The discussion above has highlighted two main points. First, *use rights*, designating the right to access the fishery and harvest fish resources, provide an important means to avoid the problems of *open access* in fisheries, in which an uncontrolled scramble to access resources often leads to fish stock depletion. Second, *management rights*, designating the right to involvement in management decision making, lie at the heart of co-management initiatives and may be held to varying degrees by a range of stakeholders in the fishery. These access, withdrawal and management rights are specific types of what are called *property rights*.

The term 'property rights' is widely used in present-day discussions within fisheries and many other natural resource sectors, yet its meaning is not always clear. Indeed, a variety of perspectives exist on the topic, notably from legal theorists, economists and anthropologists. For example, the fishery economics perspective is reflected by Pearse (1980), Neher *et al.* (1989), Hannesson (1996) and the historical/conceptual reviews of Scott & Johnson (1985) and Scott (1988). A social science view is found in the works of Ostrom (1990) and Von Benda-Beckmann (1995), as well as the collections by, McCay & Acheson (1987), Berkes *et al.* (1989) and Bromley (1992). Conceptual and practical aspects of property rights issues in fisheries (with emphasis on Europe) are found in the collection of papers in Crean & Symes (1996) and Symes (1998a), while a related discussion across a wide range of resource and environmental applications is provided by Hanna & Munasinghe (1995a, b).

While it can be argued that most of the debates over rights in fisheries relate in fact to the use rights and management rights outlined above, it is nevertheless helpful to have a sense of the broader, albeit rather theoretical, picture of property rights. With this goal in mind, what then are 'property rights'? Our everyday usage of the term 'property' may lead us to envision the term as pertaining to ownership of objects such as houses, clothes, etc. However, as noted by Von Benda-Beckmann (1995: p. 311), analysts of property rights 'emphasise that property is a sanctioned social relationship *between persons* with respect to material and immaterial objects seen as having value and not between persons and things'. From this perspective, one can imagine a triangle (Von Benda-Beckmann 1995: p. 312):

> 'property relationships … consist of three basic components: (1) the social units (regarded as) having the capacity to hold property relationships, (2) the units of the natural and social environment that are seen as being the object of a property relationship, and (3) the rights, duties, privileges, and possibilities that express what property holders may or must (not) do with property objects.'

The triangle can be summarised as involving property holders, property objects and property relationships. Note that within this perspective, two fundamental issues relating to who holds the rights concern (a) the choice of 'property regimes', such as state property or private property, and (b) the balance between individual and group (collective) rights.

The structure underlying property rights may be stated in simple terms as involving (1) a typology of property rights, (2) a set of characteristics of the rights, and (c) a range of rights holders. These are discussed in turn below. An important and long-standing insight into property rights holds that in any given situation, the specific rights, and their characteristics, may be seen as a 'bundle of rights'. This metaphor highlights the idea that 'bundles' vary from case to case (with a differing set of 'sticks' in each bundle), so that a key challenge is to understand what types of rights, and what rights characteristics, are present within a given bundle.

14.4.1 Types of rights

A major consideration in this regard is the point that 'in all societies there is some differentiation … between rights to control, regulate, supervise, represent in outside relations, and allocate property on the one hand, and rights to use and exploit economically property rights on the other' (Von Benda-Beckmann, 1995: p. 314). The latter group of rights corresponds to what are referred to here as 'use rights', or by Ostrom & Schlager (1996) as *operational-level* rights. The former group includes management rights, as discussed above, as well as other *collective choice* rights that provide the 'authority to devise future operational-level rights …' (Ostrom & Schlager 1996: p. 131). According to Ostrom & Schlager (1996), there is a dichotomy between use rights and collective choice rights:

- *Use (operational-level) rights:*
  - *access* rights authorise entry into the fishery, or a component of the fishery, or a specific fishing ground;
  - *withdrawal* (harvest) rights typically involve the right to engage in a specific level of fishing effort or to take a specific catch.

- *Collective choice rights:*
  - *management* rights authorise the holders to participate in management and governance of the fishery;
  - *exclusion* rights provide the authority to determine the qualifications necessary to access the fishery and 'withdraw' resources – thus holders of exclusion rights are mandated to allocate use rights;
  - *alienation* rights authorise the transfer or sale of the other collective-choice rights.

From the perspective of the above rights typology, the comprehensiveness of a given rights bundle depends on the extent to which the rights holder holds the five types of rights: access, harvest, management, exclusion and alienation. Ostrom & Schlager (1996) suggest that the last of these rights is the key characteristic of *ownership*, the idea being that you are not an 'owner' unless you have the right to sell off all your rights, but other authors note a variety of additional considerations involved in ownership, as discussed below.

It is important to note that discussion of a rights bundle is relative to a specific 'property object'. For example, the property object may be a fishing boat, the fish in the sea or a permit to go fishing. A fisher may hold use rights to all three of these property objects: the right to use the boat, to harvest the fish and to make use of the permit to fish. The fisher may also have an exclusion right to prevent others from using the boat and the permit, but not hold an exclusion right over access to the fish stocks. Thus, we must consider specific cases one at a time.

Consider first the property relationships involving a holder of *real property*, present in a tangible material form as a physical asset. For example, the owner of a fishing vessel has the right to use the boat, sell the boat and control all access to and use of the boat, although he or she is perhaps subject to some societal restrictions. The vessel owner holds all five forms of rights listed above (at least to some extent). Precisely the same point can be made about the fish caught by a fisher and in the hold of the fishing vessel: the fisher holds an extensive set of rights over that fish.

But what of the fish swimming in the sea? In most national jurisdictions, it is the state (collectively, the people of the relevant nation) which holds collective-choice property rights over the fish, i.e. the right to decide who are to be the holders of operational rights of access and harvest (who kills those fish), as well as matters of management, exclusion and alienation rights. In this sense, the state 'owns' the fish. (This is certainly not always clear-cut. For example, a nation and an aboriginal community may dispute the holding of collective-choice rights over fish in the sea.)

What about use rights? Within this framework, such rights, as manifested in a fishing licence or an individual harvest quota or a permit to fish a certain number of lobster traps, are forms of property rights, even though they do not involve physical assets (such as fishing boats or landed fish). Use rights, held by the fisher, clearly have value and, in some cases, can be bought and sold. However, by definition, use rights *per se* do not include collective-choice rights (just as the holders of collective-choice rights do not, as such, hold any corresponding access rights).

Use rights are sometimes referred to as *quasi-property rights* to distinguish them from rights involved in ownership of, say, a fishing vessel. In fishery management discussions, where the critical issues often concern matters of access and harvesting, it is often helpful to focus on use rights rather than on property rights more generally. For example, this may help to clarify discussions of individual transferable quotas, ITQs, which are but one of several forms of use rights; they do not incorporate full property rights, despite often being presented as a means to 'introduce property into the fishery'. In fact, the 'property' nature of use rights such as individual quotas has been the subject of considerable debate, and remains unclear (see, for example, the discussion by Wiber & Kearney 1996).

At the same time, debates are emerging over devolution to harvesters and others of collective-choice rights to the fish within a commercial fishery. To what extent should this occur? Who should be permitted to hold such rights? What institutions are suitable for such arrangements? These questions are likely to receive increasing attention in the years ahead. (For example, with respect to institutional arrangements, Townsend (1995) has suggested that forms of corporate governance may facilitate exchange of ownership rights across a range of potential participants.)

14.4.2 Characteristics of rights

The second aspect in examining property rights concerns the extent to which a set of property characteristics is present in a given situation. This provides another means to depict a 'bundle of rights', i.e. as a set of characteristics of property present in the given situation. Bromley (1989: pp. 187–190), following Honoré (1961), notes that 'full, or liberal, ownership' (which can be interpreted as 'complete property rights') involves the presence of 11 characteristics:

- *the right to possess* – 'in the absence of this, there is no ownership';
- *the right to use* – related to use rights;
- *the right to manage* – related to management rights;
- *the right to the income* – i.e. to receive the income accruing from owning property;
- *the right to the capital* – i.e. to alienate, consume or destroy it;
- *the right to security* – notably 'immunity from arbitrary appropriation';
- *transmissibility* – the ability to transfer the right to a successor;
- *absence of term* – 'full ownership runs into perpetuity';
- *the prohibition of harmful use* – 'ownership does not include the ability to harm others';
- *liability to execution* – 'the liability of the owner's interest to be used to settle debts';
- *the right to residuary character* – 'to govern situations in which ownership rights lapse'.

Bromley emphasises that the more of these 11 characteristics are present in a given situation, the more complete the ownership, and accordingly the more valuable is that ownership. However, he notes that it is not necessary to have all characteristics in place in order to have ownership. If one or more is missing, there can still be 'ownership', but in a restricted sense. In this sense, the 11 characteristics represent potential 'sticks' that may or may not be included in a given 'bundle of rights'.

In contrast to the relatively comprehensive listing above, an alternative view of property characteristics is provided by Scott (1988), who lists five characteristics:

- duration – a measure of the length of time over which the right applies;
- exclusivity – a measure of the independence of the right from others;
- quality of title – related to the enforceability of the right;

- transferability – the capability to sell or rent the right to others;
- divisibility – the capability to divide the right into smaller pieces.

Scott argues that in assessing property rights arrangements, the relevant matter is not the absolute presence or absence of these characteristics, but rather their relative presence (e.g. on a scale from one to ten). Thus, the greater the quantitative presence of each, the stronger the property nature of the right, other things being equal.

14.4.3 Property regimes

Finally, we turn to the third approach to classifying property rights: in this case, not in terms of what form they take or what characteristics they have, but rather with respect to who are the *right-holders* and what is the corresponding *property rights regime*. Typically, four such 'regimes' are differentiated (e.g. Berkes & Farvar 1989; Hanna & Munasinghe 1995a).

Non-property

Traditionally, 'high seas' fish stocks, i.e. those addressed above under the heading 'open access', were no one's property. The fish were there for the taking, like unclaimed coins dropped and left on the sidewalk. No one could claim ownership and exclude others. This represented a lack of property rights, a case of 'non-property'. As time has passed, fewer and fewer of the world's fishery resources have been exploited in the absence of property rights. For non-fish examples of a lack of property rights, one might consider the mineral resources at the bottom of the ocean, or indeed the planet's air and water, which historically have been used in an open-access manner.

Private property

As noted earlier, whenever a fisher catches a fish, once it is brought out of the water into the vessel, that fish becomes the private property of that fisher. Even when fish are still swimming in the water, they may be private property. In some nations, the fish in a river that passes through private land can be the private property of the land owner. The same could be said for fish in a lake located entirely on someone's private land. In such cases, only the owner of the resource has the right to decide the use of the resource, subject possibly to societal constraints, such as those that may be imposed to preserve biodiversity. (Of course, the private owner may decide to allow others access to the resource. This access could be free, if the owner is in a good mood, or more likely obtained through paying a fee to the owner for use of the resource. This is a common phenomenon with regard to recreational fishing, for example.)

Note that a distinction can be drawn between private property held by individuals (that referred to in the examples above) and the 'corporate property' held by corporations of various forms, from large businesses to families and households. From an anthropological perspective, this is a component of the important distinction between 'individually held property (in which the individual stands against all others in his/her social world), and various kinds of group-held property (in which several or many individuals stand against single, or multiples of non-property holders' (Melanie Wiber, personal communication, 1999). The idea is that relative to individual property, corporate property (a) is held by a group of (several or many) individuals, often with unequal levels of access, (b) typically involves a somewhat different bundle of rights (e.g. with the 'absence of term' characteristic, i.e. assumed perpetuity), and (c) is legally treated in a different manner from individual property (for example, allowing a 'corporate veil' behind which the corporation operates).

State property

In many nations, fish in the oceans are considered the property of the nation's citizens, and are controlled on their behalf by the government. Legally, the government (the 'state') owns the resource (albeit on behalf of its citizens), and hence this is referred to as a case of 'state property'. The perception of 'state property' is very different in the natural resource context from its use in daily discourse. Whereas state-owned corporations can be sold off in their entirety, natural resources owned by the state (forests, fish in the ocean, minerals and oil in the ground) typically cannot be privatised without legislation, or perhaps even constitutional change. Certainly, these resources can be extracted, thereby becoming private property, but in their original location, they remain state property.

Common property

While historically, the above forms of property have received most attention in studies of resources within the Western world, *common property* is pervasive worldwide, and in recent years has become a major topic of property rights theory (e.g. McCay & Acheson 1987; Berkes *et al.* 1989; Ostrom 1990; Bromley 1992). Common property resources, such as fish swimming in the sea, are those held 'in common' by a certain identifiable group of people, and for which typically (a) there is inherent *subtractability*, in that use by any one user detracts from the welfare of others, and (b) exclusion of potential users is challenging (Berkes *et al.* 1989).

Groups holding common property have collective-choice rights and can seek to establish rules for access, use and management. Examples of such groups include the citizens of a specific local jurisdiction (community) or the members of a native tribe, but not a private individual or company. Note that if the group were defined very widely, as all the citizens of a nation, this would be a case of state property. Indeed in common usage, the fishery resources of a nation are often referred to as common property, rather than state property, although in common property theory, a smaller, more cohesive group is usually envisioned. Nevertheless, as pointed out by Berkes (1989), it is possible to have a blend of property rights systems, with a fish stock formally being state property, but the actual use of the fish being governed by local communal rights.

We have seen, therefore, that in addition to the range of property characteristics, property rights vary both by type (access, withdrawal, management, exclusion and alienation rights) and by regime (non-property, state property, common property and private property, where the latter can be seen as including individual and corporate forms). A simple depiction of the structure of property rights in these two dimensions is shown in Fig. 14.2.

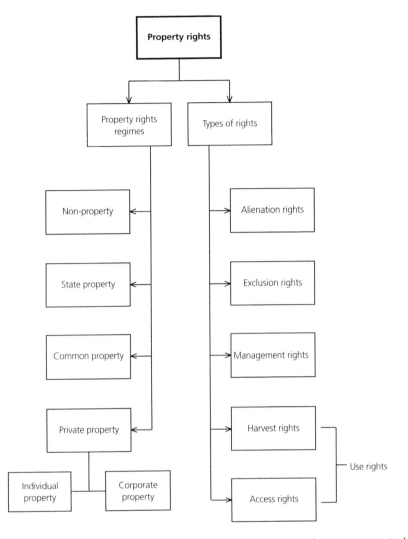

Fig. 14.2 Property rights can be classified by regime (non-property, private, state and common property) or by type (access, harvest, management, exclusion and alienation rights).

14.5 Property rights and fishery management

The preceding overview of property rights theory provides a context within which we can examine use rights and management rights specifically. In this section, two particular aspects involving the links between property rights ideas and fishery management are discussed, before we turn in the following section to a practical review of use rights options in fishery systems.

14.5.1 Resource access vs. resource ownership

Considerable controversy surrounds fishery rights. At the heart of this may well lie the matter of ownership: typically, the difference between ownership of the resource and ownership of the use rights to the resource. As noted earlier, use rights are an important ingredient for sustainability in fisheries, but in some cases, use rights systems (such as individual quotas) are promoted by suggesting that the fishers holding these rights will in fact 'own' fish in the sea, just as they own their fishing boats. In reality, however, in discussions of use rights such as individual quotas, Symes (1998b: p. 4) notes that 'what therefore is being allocated is a right – whether limited in time or granted in perpetuity – to harvest part of the annual yield of the fishery; the stocks themselves are not subject to appropriation'. In other words, while a fisher may hold the right to (attempt to) catch a certain amount of fish, that fisher does not own the fish *per se* until those fish are actually caught. Despite this reality, the suggestion of private ownership over fish in the sea, meant to appeal to fishers, has caused concern among many of them, as well as the general public, who are opposed to *privatising* 'common property' or state property fishery resources.

While debates over 'resource ownership' as such do arise in contexts such as that of aboriginal title to fishery resources, it is important to note that much of the debate around 'ill-defined' property rights in commercial fishery systems refers to *use rights* (and in some cases, management rights as well). In this regard, it is useful to compare the fishery with other natural resource sectors, where the difference between resource access (in the form of use rights) and resource ownership is perhaps clearer. Consider the case of forestry, for example. In jurisdictions with significant areas of government-owned forest, it can be standard practice for industrial harvesting companies to hold leases on specified areas of that forest, giving those companies the *right* to *use* the corresponding resources. This is often subject to conditions, e.g. that reforestation accompanies resource extraction to ensure sustainability. Similarly, in the oil and gas sector, the focus is more on the use right to a particular oil field. The use right *per se* may be 'owned', but ownership of the resources *in situ* is not in question.

This leads us to focus the discussion on a differentiation between the bundle of rights held by fishers, typically a combination of varying levels of access, usage and management rights, and that held by the resource owners, generally some management rights (perhaps shared with the fishers), together with exclusion and alienation rights. In most fisheries, the latter two types of rights in particular are held collectively, within a regime of common property or state property, by the citizens of the jurisdiction (e.g. the nation, the community or tribe), who are effectively the resource owners. There is typically widespread support for keeping such property rights publicly held, so that even while governments engage in co-management with harvesters (Chapter 13), the nation's citizenry (or the government on its behalf) remains the ultimate decision maker concerning strategies to maximise overall societal benefits from the fishery.

In the above framework of contrasting rights bundles, management rights may well represent the link between holders of use rights and holders of exclusion/alienation rights. This explains in part the focus on co-management arrangements, since these define that link between what may otherwise be disjoint groups of rights holders. Certainly, there may well be complex interactions between these; for example, (a) between a coastal nation and the group of licensed fishers or the local fishing community, (b) between the coastal nation and distant-water fishing nations (and their foreign fleets), or (c) between a community or tribe and those of its members designated as having access to the fishery.

Various questions arise. To what extent do use rights simply evolve naturally, and to what extent are they allocated or delegated deliberately by the 'resource owner'? If the latter occurs, to what extent is this process based on economic, social or political factors? More generally, is the establishment of use rights primarily an administrative tool or an instrument of policy? Specifically, are use rights used as a tool to improve compliance and conservationist behaviour? (This reflects the idea, discussed earlier, that if fishers have greater security of tenure (use rights) and a stake in resource management (management rights), their behaviour at sea may be more compatible with obtaining optimal sustainable benefits for society.) While in theory, and in keeping with Ostrom & Schlager (1996), those holding collective-choice rights may be able to determine the initial and ongoing allocation of use rights, in practice all of the above matters are likely to depend on the particular fishery being studied.

14.5.2 *Open access vs. common property*

It is difficult to think of two concepts that have been more widely confused, leading to more extensive practical problems, than the cases of 'open access' and 'common property'. Recall that an *open-access* fishery is one without restrictions on access and use (i.e. unassigned access and use rights), while a *common property* regime, pertaining, for example, to fish swimming in the sea, is one in which the collective-choice property rights are held jointly by the members of a certain identifiable group. Thus, 'open access' and 'common property' are very much distinct concepts, one concerning use rights and the other collective-choice rights. Yet the confusion between them has led to common property regimes often being misinterpreted as open access, and thereby seen in an unjustifiably negative light.

The confusion can be traced back at least to Gordon (1954), but became conventional wisdom in fishery economics and beyond through a famous paper by Garrett Hardin (1968) on the *Tragedy of the Commons*. Hardin envisioned a situation in which too many users placed too much livestock to graze on a community 'common', which eventually became severely depleted. Clearly, this was a bad outcome, but some reflection indicates that the problem was not one of common property (land held in common by a community), but rather a lack of control over *access* to the common. In other words, resource depletion arose not due to the property rights regime *per se*, but rather the failure to specify appropriate use rights (and underlying this, the lack of a suitable management approach).

The point here is not that over-exploitation cannot occur in common property fisheries. Indeed, the possibility, perhaps the likelihood, of this occurring in *unregulated* common property situations has been pointed out in fishery theory for decades (Scott 1955). Furthermore, some of the major problems arising in open-access fisheries have occurred within common property regimes (as well as within non-property, such as high seas fisheries) but, as pointed out strongly by many researchers (e.g. Berkes *et al.* 1989), the truly unfortunate outcome from Hardin's paper was the *equating* of 'commons' with 'tragedy', based on a confusion between common property and open access.

This theme has been explored in recent years within a new and growing field, *common property theory*. This area of study examines the circumstances under which common property systems do or do not develop strong institutions to conserve resources and ensure sustainability. Literally hundreds of cases have been examined, in a regular series of conferences of the International Association for the Study of Common Property and in a range of studies on the subject; see, e.g. the National Research Council (1986) and references given earlier in this chapter on property rights theory and the concept of common property.

The key point is that limitations on access and use can be made under any property rights regime: private property, state property and common property. Such limitations are a key strength of most traditional systems of communal property, as, for example, in many aboriginal fisheries and in South Pacific fisheries. Access limitations have also become widespread in state property commercial fisheries, where *limited entry* systems have been implemented to restrict access. Thus, common property systems, those with the right institutions, can be quite sustainable. Finally, even in the classic case of non-property noted above (the high seas), the 1995 agreement that flowed from the United Nations Conference on Straddling Fish Stocks and Highly Migratory Fish Stocks provides the tools to control usage and delegate responsibility for high seas stocks, thereby prescribing use rights without ownership *per se*.

Open access and its inherent problems can, however, also occur under any property rights regime. Certainly, it is typical to find that open access pertains to cases of non-property, as has been the case historically with high seas fisheries, in which the resources have lacked clear ownership and (prior to the 1995 agreement noted in the previous paragraph) use rights have been unassigned. Indeed, open access and non-property are sometimes expressed almost as synonyms, but in reality, open access can occur even in cases of private property. This may occur purposefully; for example, in the case of a lake lying entirely within private land, the landowner might (for some reason) extend use rights to anyone wishing to fish in the lake. In contrast, open access to private property may arise as a form of 'enforcement problem' when the property holder is unable to prevent others from using the property, for example, harvesters illegally fishing in a private lake, or foreign fleets fishing in a nation's (unmonitored) coastal waters. In such cases, at least at the local level, there is often a mismatch between what the property holder and the resource user perceive to be the 'bundles of rights' held by each.

14.6 Operational aspects of fishery use rights

We turn now from a discussion of rights that has focused on the strategic level of management and policy to an examination of the more practical, operational aspects of fishery rights, particularly emphasising the understanding and evaluation of use rights approaches in fishery systems. This begins with reviews of each major use rights approach. Following this is a discussion of various issues involved in implementing use rights systems, and finally an overview of factors involved in choosing an appropriate use rights system.

14.7 Forms of use rights

The wide variety of use rights utilised in fisheries of the world (Townsend & Charles 1997) can be classified in three principal categories, each of which is described in detail below:

- territorial use rights;
- input (effort) rights, including limited entry;

- output (harvest) rights.

Note that each form of use rights can operate at the scale of the fisher, the fisher organisation or the fishing community. In other words, the rights can be allocated individually or collectively. This point is explored in more detail later.

14.8 Customary marine tenure/territorial use rights

Lawson (1984: p. 80), in examining alternative fishery management approaches, noted:

> 'The most effective method of control exists where it is possible geographically and physically to delineate a territory in a way in which all fishing which takes place within it can be monitored and controlled, and which can, if possible, be supervised by the fishing community itself or by its elected leaders.'

This statement places the focus of fishery management on spatial aspects of management, which include the standard management tool of 'closed areas', in which an entire fleet is limited in its allowable fishing locations (Chapter 5), as well as the rights-based approaches of *territorial use rights in fishing* (TURFs) and *customary marine tenure* (CMT). The latter system assigns rights to individuals and/or groups to fish in certain locations, generally, although not necessarily, based on long-standing tradition ('customary usage'). As Davis (1984: pp. 145–146) points out, in reference to a specific community-based system of marine use rights:

> 'The individual fisherman's right to exploit the resource zone is based on a form of usufruct or use-right … Claims of ownership and control of property is centred in the community, and individual use-rights are derived from membership in the community.'

A classic reference on TURFs is that of Christy (1982: p. 1), who noted that 'As more and more study is given to the culture and organisation of fishing communities, there are indications that some forms of TURFs are more pervasive than previously thought to be the case, in both modern and traditional marine fisheries.' This point is echoed by Gimbel (1994: p. 16), who notes that among marine resource users, 'formal and informal beliefs about territoriality and exclusive ownership exist and have existed for centuries'. Under suitable conditions, such TURFs can add an important, socially supported element to fishery management.

The wide range of examples of TURFs around the globe has been noted by Christy (1982), Ruddle *et al.* (1992) and Lawson (1984). Lawson describes several territorial use rights systems, including: lagoon fisheries in the Ivory Coast, beach seine net fisheries along the West African coast, collection of shellfish and seaweed on a village basis in South Korea and Japan, and controls over 'strangers' and 'fishermen from outside' by fishing communities in Sri Lanka. TURF systems have a particularly long history in traditional, small-scale and indigenous fisheries. For example, in the Atlantic region of Canada, the Mi'Kmaq (aboriginal) people have developed a social process for determining control over fishing territory:

'In the centuries before the arrival of the first Europeans, the Mi'Kmaq ... governed themselves through councils based on consensus in accordance with the laws of nature. District Chiefs were responsible ... for confirming and reassigning hunting/harvesting territories.'

Native Council of Nova Scotia (1994)

A well-known example of territorial use rights in a more commercial fishery context lies in the lobster fisheries on the northeastern coast of North America. In particular, the work of Acheson (1975) was fundamental in showing how fishers in some areas in Maine have been able to maintain extra-legal control on entry (exclusion rights). A common ingredient in these systems is the local solution of usage issues. For example, Brownstein & Tremblay (1994) report on the case of a small community in Nova Scotia, Canada, faced with a lobster poaching problem in the late 1800s. The problem was resolved by the local church minister, who decreed marine use rights based on an extension of property lines out to sea. In addition, if a fisher was unable to obtain a reasonable harvest from his or her area in a given year, the fisher would be given temporary access to a fishing 'common', a reserve area designed to enhance equity in the fishery. The most notable aspect of this case study is that its workable management system has been maintained by the community to this day.

While many examples of CMT and TURF systems exist (see also Chapter 15), such systems tend to be poorly understood by many governments. Very often, little effort is made to incorporate, or even to examine, these systems as options within the management system. This is unfortunate, since there may well be value in nurturing such approaches, where they exist, and in examining the potential for their development in other fisheries (Ruddle *et al.* 1992).

This is not to suggest that CMT/TURF systems are suitable in all cases. Some TURFs may be easily implemented and regulated within the framework of existing social institutions, while others may involve potentially high transaction costs to maintain a specialised institution for the TURF. Even in the latter case, the costs may be outweighed by the inherent value of the institution involved, but alternatively, excessive time and money may be required by participants to make the arrangement work. The point is that the options need to be examined, since CMT and TURFs may provide efficient means of fishery management, depending on the specifics of the fishery system.

Territorial use rights in Chile

'Today, artisanal fisheries management measures in Chile consider the allocation of Territorial Use Rights in Fisheries (TURFs) among fishing communities traditionally exploiting benthonic resources such as Chilean abalone (*Concholepas concholepas*), sea urchins (*Loxechinus albus*) and macha clams (*Mesodesma donacium*), among others. Chilean fisheries legislation, the General Fisheries and Aquaculture Law (GFAL) enacted in 1991, allows the establishment of areas especially reserved for the use of specific artisanal fishing communities, through their legally constituted organizations (e.g. artisanal fishermen's associations and fishermen's cooperatives, among others).'

Gonzalel (1996)

Fortunately, there are some moves to maintain or re-establish CMT and TURF systems. For example, in the fisheries of Oceania (Johannes 1978), long-standing CMT/TURF systems declined as fisheries 'developed' and became commercialised, but there are initiatives in some Pacific nations (such as the Solomon Islands, Fiji and Samoa) to re-establish them. A strong example of long-standing TURF systems is in coastal Japan (e.g. Ruddle 1989a, and as discussed in Chapter 13), where traditional institutions are incorporated in modern resource management.

14.9 Limited entry: licensing as an input right

Limited entry is a common management tool in which the government issues a limited number of licences to fish. This creates a use right, i.e. the right to participate in the fishery. Limited entry bars new fishing boats and/or fishers, with the aim of controlling potential fishing effort (fleet capacity). If limited entry is successful, this limit on effort helps to conserve the resource and also generates higher incomes for the licence holders.

Limited entry has produced some economic benefits. In a number of fisheries, it slowed the rate of expansion of fishing capacity during the 1960s, 1970s and 1980s, when large increases in fish prices would otherwise have attracted significant new entry. Limited licence holders often earned substantial profits, which were reflected in the high value of licences. A good example of reasonable (although certainly not total) success in limited entry programmes is the Alaskan fishery, where the state government actually reduced active licences, and licence values in many fisheries remained high (Twomley 1994).

However, limited entry has also experienced problems. Perhaps the key shortcoming is that limited entry, by itself, does not overcome the *rush for the fish* in which each fisher seeks to catch the fish first, before the competitors get it. This creates an incentive for any given fisher to expand his or her vessel's fishing capacity (including the physical capacity and the technology) beyond the level that would be 'efficient' for the fishery as a whole: a problem known as *capital stuffing*. As a result, while limited entry prevents entry into the fishery, and this may at first control effort, capacity often expands over time *within* the existing fleet. The salmon fishery of British Columbia, Canada, is often cited as a limited entry programme that, while initially successful in creating economic benefits, succumbed to gradual increases in the fleet's fishing power. Various discussions of this point and reviews of experiences with limited entry may be found in Rettig & Ginter (1978), Townsend (1990) and Gimbel (1994).

There is a tendency these days to conclude that the limited entry approach has been a failure, on the economic grounds described above as well as the collapse of various fisheries operating under limited entry. However, this conclusion would be too simplistic, for several reasons.

(1)   First, most limited entry programmes began as moratoria on entry, *after* the number of participants in the fishery had already become too large. Only in rare cases has government effectively reduced the number of licences either by revoking or by buying back licences. Thus, limited entry is a management measure that has been instituted typically after 'the chickens have flown the coop'.

(2)   Second, the introduction of limited entry often coincided with the emergence of a destructive 'us versus them' climate in the fishery (Chapter 13), in which fishery managers saw themselves as 'conservers', focused on controlling fishers as 'exploiters'. This climate was accompanied by imposed top-down regulations and a not surprising lack of acceptance of such regulations on the part of fishers. Regulations therefore became largely unenforceable, and extensive illegal fishing resulted. This situation, rather than any drawback with limited entry, was probably the key factor in producing conservation problems and fishery failures. Certainly, it is not clear that replacing limited entry with more quantitative rights will make a difference in this regard. For example, in the Atlantic Canadian groundfishery, otter trawl fishers operating under what is essentially an ITQ scheme (see below) do not seem any more or less conservationist in their actions, and in the viewpoints they express, than their counterparts in the limited entry fixed-gear fishery. In fact, most sectors of this fishery seem to be influenced by similar entrenched attitudes (Charles 1995b).

(3)   Third, and perhaps most importantly, limited entry alone should not be expected to be the one and only measure to solve management problems. It is now generally understood that fisheries management cannot rely totally on limited entry, but requires a 'management portfolio' that includes tools such as quantitative allocations of inputs or allowable catches, within a suitable form of regulatory institution – as discussed below and in Chapter 15.

Overall, then, access rights inherent in a limited entry programme may certainly play a useful role in fishery management, within a suitable portfolio of management measures.

14.10   Effort rights (quantitative input rights)

In the common situation of an over-capitalised fishery, in which the number of fishing licences, and the average individual vessel capacity (catching power) are excessive, a natural solution would be to limit the total amount of inputs. The range of possible inputs that could be controlled, as discussed in Chapter 5, include time fished, vessel size, amount of gear and gear attributes. When such inputs are controlled by aggregate limits for the entire fishery or fishing fleet, such as an allowable *total boat days at sea* (days fished multiplied by the number of boats), this constitutes an aggregate management measure. However, the assigning of allowable levels to individual fishers (such as a specific amount of fishing time and gear) then produces individual input rights.

The key problem for an input rights programme is the incentive that will exist among fishers to thwart the input controls by locating other uncontrolled inputs to expand. Certainly, fishing activity involves a wide range of such inputs: the various dimensions of the vessel itself, a range of fishing gear and fish-finding technology, and the intensity of use of all those inputs. Whichever of the inputs are actually controlled, fishers have an inherent incentive to manipulate both those and the unregulated inputs to achieve greater output from the allowed inputs. For example, if input rights were implemented over a simple index such as boat tonnage, the vessel owner may expand use of uncontrolled inputs with larger engines, more electronics, more crew, vessel redesign to increase hold capacity, and more days fished per year (e.g. reducing time at the dock, using rotating crews, fishing in poorer weather, etc.).

This implies the need for a multidimensional approach to input rights, by implementing rights over not one but a range of inputs. For example, in the lobster fishery of Atlantic Canada, standard access rights (limited entry to control overall participation) were supplemented with quantitative rights limiting the number of lobster pots per fisher, which was a relatively effective control for many decades (Pringle & Burke 1993). In recent years, however, more frequent hauling and baiting of the traps has improved the effectiveness of each trap and increased *effective effort* even while *nominal effort* (the number of traps) appears constant. If not dealt with, this could lead to problems of over-exploitation (FRCC 1995), but now there is a capability to overcome the problem by adding another dimension, the number of 'trap hauls', to the assigned input rights. As another example, in trawling fisheries, input rights over just one input, such as hold tonnage, vessel horsepower or days fished, may be inadequate, but rights over a set of these inputs may be effective.

In addition to the above tendency to over-expand inputs, there is also a natural process of technological improvement that gradually increases the fishing effectiveness of any set of inputs over time. If the monitoring of this effect, and subsequent adjustments, are not carried out, this can create a problem in that the conservation impact of fishing by a given fleet (even one with a constant number of boats) may be underestimated, leading to over-exploitation.

However, it is possible for an individual input programme to adjust for such improvements in fishing efficiency. This may involve reducing the aggregate level of allowable inputs over time, to reflect the rate of efficiency increase. If all vessels increase fishing power per unit of input at a comparable rate over time, the adjustment might be done by periodically 'taxing' the inputs, and thereby reducing input allocations by a certain amount each year. On the other hand, if fishing power increases are highly variable across vessels, it may be fairer to place the onus on vessel owners to ensure, and demonstrate, that efficiency increases resulting from the construction of new, presumably more efficient vessels (or gear additions) are compensated for by adjustments elsewhere in the fleet, so that overall catching power does not increase. A reduction of the level of input rights may also be done voluntarily on a collective basis. Some trap fisheries (e.g. for lobster in part of Atlantic Canada) have reduced the maximum number of traps allowed per fisher, with the aim of both conservation and cost reduction, and so improving the welfare of fishers.

Thus, input allocations can be a viable approach to rights-based management if care is taken in the definition of the rights, with a multidimensional approach to creating a portfolio of input rights, and if a plan to deal with fishing efficiency improvements is incorporated. It is also notable that input rights may be politically more acceptable to some groups, particularly if better fishers, who can earn more income from a given set of inputs, feel that their skill is better rewarded under input rights than under quotas.

14.11   Harvest quotas (quantitative output rights)

A total allowable catch (TAC) is a conservation control but not a use right, since setting a TAC makes no statement about the rights to catch the fish. The situation changes, however, if that TAC is shared out among sectors of the fishery, individual fishers or communities, in which case these shares of the TAC represent quantitative output rights: collective or individual use rights over the corresponding 'shares'.

Several variations on this may occur. The right may be held collectively by a *sector* of the fishery, with allocations made for small boats or large boats, for hook-and-line fishers or net fishers, etc. Alternatively, rights may be assigned to communities as *community quotas*, so that not only is collective control exerted, decisions on the use of the quota can explicitly reflect community values and objectives.

Finally, harvest rights may be allocated to individual fishers. This may be in one of two forms. Within a fishing season, *trip limits* provide fishers with the right to a certain level of catch on each fishing trip. This right may be combined with a right to a certain total number of trips per year, thereby ensuring that the TAC is not exceeded. On an annual basis, *individual quotas* (IQs) are fisher-specific rights to harvest annually a certain portion of the fish resource (a fraction of the TAC). As described in Chapter 5, IQs appear in two main forms.

- *Individual transferable quotas* (ITQs) are harvest rights that can be permanently bought and sold among fishers in a 'quota market' (Clark *et al.* 1988).
- *Individual non-transferable quotas* (INTQs) are rights that are not permanently transferable. (The impact of transferability is discussed later in this chapter.)

In contrast to effort/input rights systems, which have received relatively little research attention or active promotion within fishery management, IQs (and especially ITQs) have been strongly promoted for many years, notably by economists, managers and large-scale fishery participants. Partly as a result of this, they are becoming increasingly common in industrial fisheries. The best known cases are those in New Zealand and Iceland (Clark *et al.* 1988; Arnason 1996; Dewees 1998), but examples of *quota fisheries* are also found in Australia, Canada, the United States, the Netherlands, South Africa, Chile, Peru and elsewhere. These have been met with varying degrees of acceptance (see boxes below).

If individual harvest rights are considered to be certain within a fishing season, the fisher can plan fishing activity as desired, and this can (a) potentially provide a better match to available markets, and (b) avoid the 'race for the fish', so that individual harvests can be taken at a lower cost, with less incentive for the over-capitalisation that can occur with limited entry and input allocation programmes. This benefit exists with both trip limits and individual quotas, but more so with the latter since the fisher is able to plan fishing activity over the course of a

---

ITQs in New Zealand

'New Zealand has introduced a range of fisheries into individual transferable quota (ITQ) management since 1982 when explicit mechanisms to introduce transferable property rights were first used in New Zealand's deep-water fisheries. Predating this, there were individual quota arrangements in the Bluff oyster fishery and the Ellesmere eel fishery. A comprehensive regime for quota management and individual transferable quota was introduced into New Zealand fisheries law in 1986.'

Major (1994)

ITQs in Chile

'In the late 1980s, strong controversies arose on the prevailing regulations for Chilean fisheries. This was related to the rapid growth of the industrial fishing sector between the mid 1970s and the late 1980s. The controversies led to a reform of the Chilean fisheries law ... A significant innovation attempted to introduce Individual Transferable Quotas (ITQs) to the most important marine fisheries in the country. But this faced strong opposition from a significant sector of incumbent firms. As a result, *ad hoc* restrictions on ITQs were legislated. After 5 years since the enactment of the new fisheries law in 1991, ITQs have been used in only two relatively small fisheries, while the most important industrial fishing grounds remain under (annually renewable) closed entry regulation and periodic biological closures.'

Pena-Torres (1997)

full year, rather than on a trip-by-trip basis. IQ advocates argue that such incentives lead to (1) a reduction in fishery inputs (fleet size and number of fishers), (2) increased rents in the fishery, and (3) increased product value, either through more attention to quality or through the development of higher-valued product forms (e.g. fresh fish instead of frozen). Such benefits have been addressed widely in the literature (see, for example, the papers in Neher *et al.* (1989), OECD (1993) and Pikitch *et al.* (1997)), although some economic analysis questions the extent to which benefits are achieved in reality (e.g. Squires & Kirkley 1996). There is also likely to be a set of social concerns arising in these systems, and these will be discussed later in the chapter.

With respect to conservation implications, just as systems of input rights face inherent incentives to thwart input controls (discussed above), conservation problems arise in using output rights (individual quotas or trip limits) due to (1) the incentives inherent in quota fisheries to maximise the value of the fisher's quota, combined with (2) difficulties in enforcement faced under such programmes. A number of potential problems have been noted with individual quota systems (e.g. Copes 1986, 1994, 1995, 1997a). Three areas are discussed here, each of which can be seen as a challenge to fishery enforcement.

First, the inherent incentive for fishers to under-report catches, given restrictions on catch through a TAC, is made worse with individual quotas. In a competitive fishery, if one fisher reports his or her catch and that leads to a closure of the fishery, the 'pain' of such a closure is spread among all fishers. However, with individual harvest rights, every unreported fish is one less deduction from the fisher's own quota (and thus one more for catching on the next trip) or one less fish for which quota must be purchased, at considerable cost, from other fishers.

Second, an individual quota system seems to increase the incentives to dump, discard and high-grade fish in order to maximise the value of the catch obtained from the corresponding quota (e.g. Angel *et al.* 1994; Arnason 1994). This may be seen by first considering a competitive system, in which all fishers have equal access to the allowable harvest. In such a case, profit maximisation creates an incentive to land all that is caught and marketable, assuming a vessel's hold capacity is not limiting. In contrast, allocating quotas individually creates an incentive on the part of the fisher to maximise the profit from every unit of quota. This 'personalises' the benefits of dumping, discarding and high-grading. For example, dumping of small fish or unwanted species allows the fisher to continue seeking out more desirable fish, thereby increasing profits. Similarly, if there is a significant price differential between large fish and small fish, then high-grading (dumping of small fish) directly increases the income obtained from a given quota. This will be a profitable action if the difference in value between a kilogram of large fish and of small fish exceeds the cost of searching for and catching the larger one. (Note that this argument about incentives applies primarily to individual quotas, rather than TACs or fleet sector quotas, since in the latter cases, if the same fisher were to dump fish, the allowable catch thereby 'saved' would have to be shared by the entire fleet, and would therefore keep the fishery open only marginally longer. Thus, in the latter case, dumping would imply sacrificing a moderately valuable catch, while receiving only slight benefits.)

Note that anti-conservationist incentives may be reduced if quotas are transferable, since fishers can simply buy more quota from others to account for extra harvests. However, a fisher may not find it financially desirable to do so in practical situations. For example, consider a system of individual transferable quotas (ITQs) in a mixed-stock fishery for two species, say haddock and cod. Since the two species are unavoidably caught together, a point will be reached (particularly toward the end of the fishing season) at which fishers have nearly exhausted quotas for the low-abundance species (haddock, say). The demand for haddock quota will rise (as fishers attempt to continue fishing for cod), so the cost of that quota may well rise to a point at which the most profitable course of action is to dump haddock, rather than buy quota for it.

Given these inherent incentives for anti-conservationist practices at sea, possible remedies tend to be technological in nature, such as around-the-clock observer coverage and/or video surveillance on fishing vessels, and satellite monitoring using 'black boxes'. These technologies are currently expensive and/or incomplete, although it is possible that they may become cheaper and more reliable over time.

A final aspect of conservation concern arises largely with industrial and commercial fisheries. It is common for fishers to go into debt to buy their boats, and then feel the need to go fishing to pay the debt. In transferable rights systems (such as ITQs, as well as some limited entry licensing approaches), fishers may go further into debt to purchase what may be expensive rights (quota) from others, and thus need to fish even harder to make a reasonable income and pay off the debt. This can result in financial hardship among fishers, and a resulting pressure to increase the TAC that may be difficult for decision makers to resist.

14.12   Issues in the implementation of use rights

Whatever the chosen form of use rights in a given fishery system, several key issues arise in their implementation. The three issues examined below are:

- the initial allocation of rights;
- the appropriate institutional approach (market-based versus community-based);
- the transferability of the rights.

14.13   Initial allocation of rights

To establish rights-based fishery management, the rights must be allocated. This is the most contentious part of the process, since those who feel short-changed by the allocations may well oppose the entire programme. There is no 'right' way to allocate rights. The theory developed for quantitative rights schemes notes that maximisation of economic efficiency would call for the auctioning of rights (Moloney & Pearse 1979). On the other hand, the political reality has been that rights are allocated to fishers on the basis of historical participation, i.e. the past participation and/or *catch history* of the fishers involved. However, this can be problematic.

What is the best way to define historical participation? Consider the case of individual catch quotas. If only recent catch histories are used in the allocation, then long-time fishers who exited the fishery temporarily are penalised. Indeed, if those fishers cut back on fishing due to declining fish stocks (i.e. if the period over which the 'history' is calculated was one of overfishing and stock depletion) then those who receive the lowest quotas are those who contributed least to the overfishing. On the other hand, those who bent the rules (or fished illegally without being caught), e.g. using fine-mesh liners in the nets to increase catches, or dumping less desirable fish to fill their holds with higher-valued species, are rewarded, into perpetuity, with a larger quota. While such a situation is unfair, the alternative of not using recent history will lead to objections from recent entrants, who may well be numerically large, technologically advanced and politically important. In all cases, there is a strong possibility that social conflict will arise.

In an effort to deal with these problems, hybrid schemes for initial allocations may be used. For example, there is some experience with formulas in which part of the total is allocated based on catch history and the remainder is allocated equally among fishers. In any case, the idea is generally to divide up the total in a manner that seeks to minimise conflict, and to accompany this with some form of appeals process to manage the variety of special cases that are bound to arise. While such processes sometimes resolve conflicts over initial allocations of rights, the debates involved at this stage can be so extensive that they delay implementation of rights systems for years, thereby delaying the realisation of potential benefits as well.

14.14   Institutions and rights: market-based and community-based rights

A key issue in fishery policy debates about rights concerns the mechanism by which the holding of rights is itself managed. In many cases, this revolves around the choice between two institutional arrangements for determining who are to be the fishery participants: a market-based approach, and one based on multi-objective planning, often at the community level.

14.14.1 The market

An increased reliance on market forces (and adjustment mechanisms to achieve this, such as privatisation and deregulation) has become a popular direction among many governments and international financial institutions. This leads to a market-based approach to fishery policy, in which strategic-level decisions about who is to participate in the fishery and who is to receive allocations of allowable catch or effort are determined through the buying and selling of rights in the market-place, as typified by ITQ systems.

Who it is that buys, or sells, the rights will depend on the situation at hand; it is an empirical question. It may be, as economic theory suggests, that more efficient players buy out less efficient ones, or it could be that the buyers are those with better access to financial capital (which may be a particularly important concern in developing countries), or there may be some other factor that dominates. Furthermore, while market-based rights are typically discussed in the context of individual fishers, there is nothing conceptually to prevent an entity operating at a collective (corporate or community) level from buying or selling on a fishing rights market. However, the actual bundle of rights resulting from such a transaction may differ depending on whether the buyer is an individual, a corporation or a community; perhaps, for example, because there are differing regulatory constraints on these various rights-holders. (This relates particularly to the distinction, discussed earlier in the chapter, between individual private property and corporate property.)

Broadly speaking, a market-based rights system can be expected to display the various advantages and disadvantages of the overall market system, and inspire similar debates to those arising with respect to market mechanisms elsewhere in the economy. For example, depending on one's perspective and the case at hand, markets may (or may not) be the most cost-efficient institutional arrangement to handle transactions between fishers, and may (or may not) increase flexibility in fisher operations. Given a widespread familiarity with markets in many economies, these may be relatively easily implemented fishery rights systems, but they may have financial impacts on the pursuit of new policy directions (if, for example, those with market-based rights must be compensated if policies are contrary to the self-interest of rights-holders).

14.14.2 Multi-objective planning and community-based rights

In contrast, a planned approach assigns use rights in a more deliberate manner (whether permanently or periodically) through a decision-making process that (1) recognises multiple societal goals, (2) is carried out by institutions operating at a suitable scale, whether community-based, regional or national, and (3) involves rights specified through a combination of legislation and governmental decisions, on the one hand, and tradition and informal arrangements on the other. These rights may operate at the individual fisher level, but (as with the market approach) could alternatively be allocated at the group (collective) level, with allocations being made through the relevant institutions.

Such arrangements have a lengthy history in real-world situations. For example, they frequently arise in the context of cooperatives, marketing boards and indigenous/native communities. The popularisation of *common property theory* has led to increased study of such examples, and theoretical developments of these themes (McCay & Acheson 1987; Berkes 1989; Ostrom 1990; Bromley 1992; Pinkerton & Weinstein 1995).

This approach can be especially important at the local level, with fishing communities potentially playing a role in regulating fisheries in which they have a 'community self-interest'. This institutional environment draws on the group dynamics in such communities to create a collective incentive to ensure that the resource is managed wisely. In particular, this can involve:

- efficiently managing the allocation of catches and fishery access (and also helping to pre-vent the *rush for the fish* noted above);
- increasing management efficiency by bringing fishers and fishing communities fully into the management process, encouraging self-regulation or 'co-management' jointly by fishers and government, and implementing local enforcement tools.

An example of this process was described by Acheson (1975: p. 187), who discussed the role of traditional fishing rights in the lobstering communities of Maine. He noted that 'to go lobster fishing at all, one needs to be accepted by the men fishing out of one harbor; and once one has gained admission to a "harbor gang", one is ordinarily allowed to go fishing only in the traditional territory of that harbor'. Acheson notes that this informal system is maintained through 'political' inter-harbour competition, since the territorial system 'contains no "legal" or jural elements'.

Davis (1984: p. 146) presents a similar situation in his discussion of implicit property rights within the small-boat fishery of Port Lameron Harbour, Nova Scotia. In such com-munities, 'claims of ownership and control of property are centred in the community, and in-dividual use-rights are derived from membership in the community'. Davis suggests that this community-based approach is 'basically foreign to the system of owner/non-owner property relations' inherent in a market economy, and hence the existence of these informal property rights is not well utilised in the development of government policy.

The key research question raised is to determine why community or fisher self-regulation works in some jurisdictions and not in others, and whether incentives can be formulated to improve the success rate of this potentially cost-effective management tool.

14.14.3 *Community quotas*

An important application of the latter approach is that of 'community quotas', i.e. fishing quotas (portions of the TAC) allocated by government to communities rather than to individu-als or companies. While suffering some of the inherent flaws of any quota-based scheme, community quotas defined on a geographical basis tend to bring people together in a com-mon purpose, rather than focusing on individualism. Fishers in the community manage them-selves, perhaps also with the involvement of their community. The fishers create fishery man-agement plans and divide up the quota (or other form of rights) to suit their specific local situation and to maximise overall benefits, rather than leaving it to the market to make choices for them. As well as providing many of the benefits of individual quotas, this approach may also enhance community sustainability (Chapter 10), allowing each community to decide for itself how to utilise its quota. For example, one community may decide to allocate its quota in a rent-maximising auction, while another may prefer distributing the quota so as to achieve a mix of social objectives, such as community stability, employment and equity. A strong example of this approach is Alaska's system of Community Development Quotas (CDQs) (see, for example, the Committee to Review the Community Development Quota Program 1999). Some other possible forms of community quotas are discussed by Hatcher (1997), by several authors in Loucks *et al.* (1998) and in the case study of Canada's Atlantic groundfish-ery in Chapter 9.

14.15 Transferability of rights

Closely related to the debate over the extent to which the market will be used to deal with the allocation of fishery rights is the key policy issue of whether or not to make rights transferable. Several approaches to transferability are possible.

(1)   With *complete non-transferability*, the licence, allocation or right can only be used by the holder, and ceases to be valid when that fisher leaves the fishery.
(2)   Transferability only *within the fishery sector or community* in which the fisher is located will provide greater stability within the sector or community with respect to numbers of licences, although the localised concentration of holdings is still possible.
(3)   There could be *differing classes of licences*, some transferable and some not, with defined policy measures determining which fishers have which form of licence. (For example, in the lobster fishery of Atlantic Canada, two classes of licence were issued: transferable 'A' permits for full-time lobster fishers, and non-transferable 'B' permits for part-time lobster fishers.)
(4)   *Non-divisible transfer* of fishing licences, and any attached input allocations or quota rights, may be allowed between fishers, but only if such transfers are done as a complete package.
(5)   *Divisible transfer* of input or quota rights is the ultimate version of uncontrolled transferability, with such rights being freely transferable from one vessel to another.

A variety of issues need to be considered in determining the level of transferability to be allowed.

14.15.1 Efficiency

Transferability is often promoted as a means to improve economic efficiency, using an argument such as the following. To be economically efficient, the participants in a fishing fleet are those most profitable in their harvesting of a given fish quota. In theory, a market-based system (such as an ITQ system), with divisibility and transferability of input or output rights, leads to increasing efficiency, as vessel owners who can do the most financially with quota buy up that quota from others, like any commodity on the market. The idea is that the more 'efficient' vessel owners remain in the fishery, while others sell their quota and leave, in a 'survival of the fittest' process leading to increasing overall efficiency of the fishery.

   This efficiency argument for transferability is made so often, and so emphatically, that it is useful to highlight some serious potential qualifications. First, while economic theory calls for efficiency to increase through the market process, there is no guarantee that this will happen. For example, there could instead be strategic buying of quota to gain greater control of the fishery, not unlike the manner by which financial mergers can take place elsewhere in the economy. This still concentrates ownership and reduces participation in the fishery, but the impact on efficiency is unclear.

   Second, and more fundamentally, it is important to recall a theme discussed in Chapter 13, namely that while efficiency (*obtaining optimal benefits for a given set of inputs*, or 'doing the most with what we have') is desirable, it needs to be viewed at a level appropriate to the problem at hand. Unfortunately, this point is often neglected in considering the relation between efficiency and use rights transferability. Specifically, the focus of the efficiency argument above is at the individual vessel level (generating for the vessel owner the greatest monetary value of fish at the least cost), while what is probably important from a policy perspective is the broader level of the fishery system. There is no reason to believe that efficiency at a vessel level corresponds to that at a system-wide level, where one takes into account (a) all players rather than just the owner of an individual vessel, and (b) all related monetary and non-monetary benefits from catching a fish, not only net revenue to the boat owner. These related benefits will depend on the specific situation, but would typically include benefits to crew members and onshore workers, as well as broadly to the onshore economy and relevant coastal communities. Such considerations are generally omitted from a vessel-level analysis, but must be taken into account in assessing the desirability of rights transferability.

Third, a system-level assessment may adopt a conservation-oriented view of efficiency, in which the key input in the fishery is the available catch. Then efficiency may mean *producing the greatest net benefits per fish caught*, and using every available fish in the allowable catch optimally. From this perspective, an efficiency analysis of the fishery system may again look quite different from one that deals with the harvesting operation alone. For example, a common prediction is that transferable rights (such as ITQs) will increase the technical efficiency of harvesting, at the same time as it reduces participation in the fishery. Assessing efficiency for the fishery system as a whole requires an analysis of the impacts of such a move in terms of net total benefits per fish in the coastal economy. What the actual results will be in a given situation is an empirical question, but it is important to note that if we take a broad perspective on efficiency, the impact of transferability is not clear *a priori*. Instead, any conclusion should focus on which measures best contribute to an efficient fishery system.

14.15.2 Fisher mobility

Transferability allows a fisher to exit the fishery when the revenue to be gained from the sale of the licence or quantitative right exceeds the expected benefits of remaining in the fishery. This provides maximum flexibility for the fisher, but at the expense of less stability in fishing communities if there are no restrictions on keeping rights within the community. Conversely, non-transferable systems provide better stability, but face difficulties in the reduction of fishing power over time (capacity reduction), since the mobility of the fishers is reduced. An incentive exists to keep a non-transferable right in use as long as possible in order to maximise its benefit. If licences apply to the vessel, incentives are created to use the boat beyond its technological life, thus creating safety problems. If a non-transferable licence applies to the fisher, the incentive is to continue fishing into old age, particularly if this provides employment for younger family members. Finally, if there is any prospect of a later decision to allow transferability, the participants may retain their licences as long as possible in the hope of a financial windfall. Thus, attrition may be a haphazard process; rules may be needed to smooth the dynamics of rights holdings over time.

14.15.3 Social cohesion

Transferability can have a major impact on social well-being (Copes 1995, 1997b; Bogason

1998). First, since rights are often held solely by the vessel owner, the selling of those rights may well leave the most vulnerable in the fishery, the crew members, without jobs and without compensation. Second, transferability can lead to a loss of social cohesion within the communities as a whole. In particular, if fishery policy allows transferability of fishing rights, this may lead some community members to sell rights to individuals outside the community. This results in a shift from community-level fishing to greater involvement by 'outsiders', with the potential to (a) reduce morale among those remaining in the local fisher, and (b) reduce the effectiveness of community pressure in preventing overfishing and illegal activities. Non-transferable rights, on the other hand, can help to stabilise the local economy by ensuring that a certain portion of the rights resides in the local communities. It should be noted, however, that in the case of non-transferable quota systems, there remains an inherent pressure to shift to transferability, with all the implications of that change, and it seems that this has occurred in most examples to date.

14.15.4 Concentration of rights

Transferability generally leads to a concentration of fishing rights. There are legitimate social and economic reasons to be concerned about the concentration of rights ownership, in particular those related to the potential loss of community welfare and socioeconomic sustainability if inequities are created. For small-scale fisheries that are closely linked to local social structures, such drawbacks of concentration may outweigh any advantages of transferability. Non-transferability, on the other hand, is often seen as a way to maintain the traditional organisation of the fleets and fisher communities, to prevent consolidation of fishing rights (particularly in the hands of processors or dealers) and possibly also to maintain employment of the crew.

Rather than prohibit transferability, some management plans attempt to limit the maximum amount of rights that can be owned by any person or firm. If enforced properly, this objective of restricting concentration can largely be achieved through an owner–operator requirement. However, in other cases, such limitations are difficult to achieve. Attempts to limit total rights ownership may be evaded through legal contracts and through nominal ownership by family, relatives and employees. In the scallop fishery of southwest Nova Scotia, for example, the courts have upheld the validity of such contracts, creating essentially a dual ownership system. This reinforces the caution needed even in instituting a non-transferable quota system, for example, since *de facto* transferability may result.

14.16   Choosing a use rights system

Use rights are important ingredients in fisheries management, and thus not surprisingly, rights-based fishing has become a 'hot' topic of discussion worldwide. However, it is important to remember that use rights schemes are implemented as a means to maximise the benefits accruing from the fishery system. Thus, the specific choice of use rights to maximise benefits is a context-sensitive empirical question, dependent on three key considerations (e.g. Berkes *et al.* 1989; Charles 1994, 1998c):

- society's objectives with respect to the fishery system;
- the structure, history and traditions of the fishery in question;
- the social, cultural and economic environment relevant to the fishery.

We should be sceptical of any claim that one form of use rights is somehow inherently superior to others. Such claims may reflect philosophical or ideological beliefs more than any systematic empirical evidence. The truth is that just as people come in different shapes and sizes (so one size of clothing does not fit all), similarly, the enormous biological, economic and social complexity of fisheries implies that one policy answer does not apply everywhere. The appropriate role for the government fishery managers and planners is not to impose a favourite system, but rather to seek out, on behalf of resource owners and together with those affected (such as vessel owners, crew members, aspiring fishers and other stakeholders), a form of use rights that will work in practice, based on understanding the history of the fishery, the philosophy of those involved in it, and the nature of the resource (Fig. 14.3).

For the analyst, perhaps the key questions are these: which types of use rights systems are suited to which types of fisheries? How do we choose between alternatives in specific fisheries? There is a need for a thoughtful, empirical comparison and assessment of use rights options, with the goal of predicting the circumstances under which the various options might be preferred. For example, when have output rights (such as ITQs) proved more effective than input (effort) rights, and vice versa? When have TURFs (territorial use rights) proved desirable? What management institutions work for the various combinations of fishery resources, industry structure and political jurisdictions? What factors determine the desirable balance

Fig. 14.3 It is important to understand the philosophy and the operating modes of those involved in a fishery to determine what use rights system is appropriate. It is also crucial that use rights be compatible with the fishery's goals and traditions, the nature of the resource, and the environment in which the fishery operates. Indeed, in many cases, the best use rights system may well be the one already in place!

between resource owners and resource users, from central government control to co-management to self-regulation?

There are certainly no clear answers to these questions. Often, as discussed above, the debate revolves around a dichotomy between community-based and market-based alternatives. Examples of the former include various aboriginal fisheries, as well as the much-studied Northwest Atlantic lobster fisheries (Johannes 1978; National Research Council 1986; Acheson 1989; Berkes 1989; Pinkerton & Weinstein 1995). The market alternative often focuses on applications of individual transferable quotas (ITQs), examples of which were outlined earlier in the chapter.

Under what circumstances are these two broad options most suitable? Berkes (1986: pp. 228–229) provides useful insights into the choice, in the context of Turkish fishery case studies:

- 'local-level management provides a relevant and feasible set of institutional arrangements for managing some coastal fisheries', particularly 'small-scale fisheries in which the community of users is relatively homogeneous and the group size relatively small'.
- 'for offshore fish resources and larger-scale, more mobile fishing fleets, community-level management is less likely to work. In such cases, the often discussed measures for the assignment of exclusive and transferable fishing rights may indeed be more appropriate.'

This can be broadened somewhat into a set of initial hypotheses regarding the desirability of community-based management, the ITQ market-based approach and an intermediate individual quota option.

(1)   Community-based management may best serve sustainability goals in fisheries in which:
   - the structure is small-scale/artisanal;
   - history and tradition play a major role;
   - fishers have clear ties to their coastal communities.
(2)   Individual transferable quota schemes may be most suitable in cases in which:
   - the fishery has a predominantly industrial, capital-intensive orientation;
   - the fishery does not play a major role in supporting coastal communities;
   - the goal of rent generation dominates over community and socioeconomic goals such as employment stability, equity and benefits to the local economy (Charles 1994).
(3)   A non-transferable quota scheme may be a feasible intermediate option, balancing the benefits of individual rights with social and community stability, if there are certain conditions in place, for example:
   - while occasional transfers within a season are permitted, permanent transfers or sales are fully prohibited;
   - the IQs operate within the framework of a community quota, so that rules are set locally, and the fishers in the community have group control over the IQ system.

It should be noted that the 'community-based versus market-based' debate should be differentiated from the 'effort control versus catch control' issue. Unfortunately, these do tend to become confused in that the market approach has been dominated by ITQ quota setting. However, as discussed earlier, hybrids such as community quotas and market-oriented effort controls are feasible and worth examining in some contexts.

Finally, it is important to note that a diversity of use rights approaches can operate side by side within a single fishery jurisdiction. For example, on Canada's Atlantic coast, the ground-fish fisheries (for cod, haddock, flatfish, etc.) operate under various regulatory frameworks.

- The corporations that predominate in the offshore industrial sector operate with *enterprise allocations*; essentially individual non-transferable quotas (with in-season transfers being permitted).
- The midshore medium-sized vessels mostly operate with individual transferable quotas (ITQs), although there are some restrictions on quota ownership.
- The fixed-gear fishers and the small inshore boats operate under limited entry with a range of institutional arrangements, ranging from 'competitive quotas' (parts of the TAC) with certain effort rights, to community quotas managed collectively, to individual quotas that are generally non-transferable.

Meanwhile, the lobster fishery in the same area operates with a combination of limited entry, individual effort rights, somewhat informal territorial use rights, and various biological controls (with no quota controls).

14.17 Summary

The first half of this chapter reviewed the theoretical side of rights in fishery systems, and discussed the concepts of use rights and management rights, and the larger picture of property rights, including the various types of such rights, the range of characteristics incorporated in the rights, and the property regimes in which they arise. It was seen that rights can be viewed as being grouped within 'bundles', with various types and characteristics of rights in a given bundle, depending on the situation. There could be various combinations of access, withdrawal (harvest), management, exclusion and alienation rights, and various groupings of property characteristics. In general, the more of these characteristics the bundle of rights possesses, the stronger the rights. The types of rights in the bundle, and their characteristics, define and influence matters of 'ownership'.

It was argued that, while ownership issues are important in fisheries, much of the policy debate within the context of fishery management is specifically about *use rights* and *management rights*. With this in mind, the second half of the chapter focused on specific use rights options, including TURFs, limited entry, and individual quantitative input (effort) rights and output rights (quotas). Each of these has its problems, among which are a common lack of familiarity with TURFs, capital stuffing with limited entry, difficulties in controlling a diversity of inputs with effort rights, and various anti-conservation incentives such as high-grading and dumping with quota rights. For this reason, it is important to work from a *portfolio* of rights approaches, in order to choose a model that is most suitable, and most acceptable, in a given context. This theme is addressed further in Chapter 15.

In this regard, it is notable that the promotion of one specific option, i.e. market-based individual quota (ITQ) systems, has been so successful in some nations that governments are often unaware of other options, such as successful community-based and input rights alternatives. In some locations, however, this situation seems to be changing, ranging from renewed interest in input rights in some European contexts, to formal actions in the South Pacific to re-establish community-level management rights and traditional input-based management practices. One point is clear: discussions of rights in fishery systems is likely to be a major part of fishery debates in coming years.

Chapter 15

Resilient Fishery Systems and Robust Management

Fishery systems around the world are in a state of transition. Strong external forces, including environmental, technological, economic, social and institutional factors, are driving some of the changes. However, much of the impetus for change is from within, driven by a widespread dissatisfaction with fishery management in the past, widely viewed as a failure. It has become almost obligatory among fishery commentators to call for new approaches to, and even a 're-inventing' of, fishery management. Over the past decade in particular, and certainly reinforced by the collapse of Canada's Atlantic groundfish fishery, there have been hundreds of articles and books on the subject.

Many fundamental issues are involved in the transition to fisheries of the future.

- What constitutes a sustainable fishery, and how can fishery sustainability be assessed?
- What are the implications of the competing visions of fishery use and management?
- What policy directions promote sustainable fisheries within an uncertain environment?
- What research directions are needed for a better understanding of complex fishery systems?
- What impacts do external economic, social and technological forces have on fisheries?

Many of these issues have been examined in Chapters 10–14. In particular, Chapter 11 focused on the great uncertainty in fishery systems, and the need for a precautionary approach to cautious decision making. Chapter 12 focused on the complexity inherent in fisheries, and the need for an ecosystem approach to management; one emphasising a 'systems' perspective. Both the precautionary approach and the ecosystem approach represent efforts to enhance the *resilience* of fishery systems.

This chapter builds on the theme of resilience-enhancing management and policy, focusing here on the idea of *robust management*. In contrast to previous chapters, the discussion herein is notably subjective. The content reflects this author's beliefs concerning priorities for management and policy in moving towards more resilient and sustainable fishery systems. Hopefully, a reasonable case is made here for the directions advocated, but at the same time, readers are encouraged to reflect on their own priorities and preferences.

15.1 Resilience

Resilience is a concept introduced by the ecologist C.S. Holling (1973), to describe the capability of ecosystems to absorb unexpected shocks and perturbations without collapsing, self-destructing or otherwise entering an intrinsically undesirable state (Berkes & Folke 1998). Specifically, Holling (1973: p. 17) wrote:

> 'Resilience determines the persistence of relationships within a system and is a measure of the ability of these systems to absorb changes of state variables, driving variables and parameters, and still persist. In this definition resilience is the property of the system and persistence or probability of extinction is the result.'

Holling drew a strong distinction between the concept of resilience and that of stability, which he defines as 'the ability of a system to return to an equilibrium state after a temporary disturbance. The more rapidly it returns, and with the least fluctuation, the more stable it is.' He highlights the point that 'a system can be very resilient and still fluctuate greatly, i.e. have low stability', and notes examples that suggest that 'the very fact of low stability seems to introduce high resilience'.

Holling's contribution to resilience *theory* has been great, but he went further to draw a key conclusion about *management* in ecological systems, i.e. that management approaches focusing on the pursuit of stability could be detrimental to resilience and lead to critical problems. Specifically, Holling (1973: p. 21) noted that the pursuit of a stable maximum sustained yield for a renewable resource could change the underlying forces operating on the system to such an extent that 'a chance or rare event that previously could be absorbed can trigger a sudden dramatic change and loss of structural integrity of the system.'

It has been realised that the concept of resilience applies beyond ecosystems (see, e.g. the papers in Berkes & Folke 1998). Indeed, for natural resource systems such as fisheries, resilience implies that not only the relevant ecosystem, but also the human and management systems are able to absorb perturbations, such that the system as a whole remains able to sustain (on average) a reasonable flow of benefits over time. In a fishery, then, we can envision resilient management institutions, resilient fishing communities, a resilient economic structure in the fishery, and a resilient ecosystem in which the fish live. In this chapter, we explore possible policy measures and management approaches that contribute to resiliency in fishery systems.

15.2 Robust management

Fishery management must play a key role in seeking to maintain and enhance the resilience of fishery systems. Yet unlike the generally accepted goal of controlling fishery exploitation, the maintenance of resiliency is not often expressed as a fishery manager's objective. How is this goal to be achieved, in a world where the limitations on what is possible through fishery management are beginning to emerge clearly? *Robust* management approaches are needed, i.e. ones that are reasonably successful in meeting societal objectives, even if (a) our current understanding of the fishery (notably the status of the resources), its environment and the processes of change over time turns out to be incorrect, and/or (b) the actual capability to control fishing activity is highly imperfect. In other words, a desirable management system is one in which obtaining reasonable performance from the fishery (some acceptable level of success) is not dependent on having an *accurate* view of the structure and dynamics of the system, and in particular is not reliant on the most uncertain and unobservable of variables. (Of course, if the true state of the natural and human world lies outside what we imagine to be a 'plausible' range, we cannot expect 'perfect' outcomes!)

15.2.1 Rethinking fishery management

The move to robust management requires a rethinking of the philosophy of management, including the adoption of new structural and decision-making tools, notably the precautionary approach and the ecosystem approach. The key point is that the complexities and uncertainties inherent in fishery systems make it risky to rely on management methods that are sensitive to highly uncertain variables or which depend on high levels of controllability. Unfortunately, many fishery management systems have just such a reliance, and thus face two major problems.

- *Illusion of certainty.* While the great uncertainties inherent in fisheries are well documented (Chapter 11), some management systems exhibit a tendency to ignore major elements of uncertainty, so that far from recognising and working within the bounds of this uncertainty, management may create an *illusion of certainty* that leads to the opposite result.
- *Fallacy of controllability.* Humans have managed to steer a spaceship to the moon, walk around there and return home. If that is possible out in space, surely any system on Earth can be as precisely controlled. Yet in reality, this is by no means the case. The fishery is a good example of a system that can only be partially, and imperfectly, controlled. Unfortunately, this point is by no means universally recognised – a *fallacy of controllability* is often in place, reflecting a sense that more can be known, and more controlled, in fishery systems than can be realistically expected.

In reality, there is a need to focus more on the challenge of developing management measures that optimise the overall sustainability of inherently uncontrollable fishery systems. This requires a more flexible, adaptive approach in which fishing plans, and individual 'fishery business plans', are designed to adapt to unexpected changes in the natural world. While clearly a desirable attribute for the management of any highly uncertain system, this is not easily achieved. Possible approaches to help move in the appropriate direction are discussed later in this chapter.

15.2.2 The case of quota management

Quota management involves both (a) determining the total allowable catch (TAC) and (b) subdividing the TAC into *quotas* (fractions of the TAC) and allocating these to specific nations, gear types, vessel categories, and possibly individual fishers or companies. The TAC-setting process is seen as a desirable means to set a quantitative limit on harvests that can be removed from the ocean, while quota setting provides a mechanism to allocate the available TAC. Quota management is among the most common, even dominant, of fishery management methods, particularly in developed countries, and its use is increasing in the world's fisheries. It is therefore a useful case to examine in terms of the robustness of the management tool.

With respect to robustness, three issues are considered here. First, as discussed in Chapters 5 and 14, underlying quota management is the process of TAC setting, which requires estimating the biomass for each fish stock, and determining a total allowable catch (TAC) as a fraction of that biomass. However, biomass estimates, and thus the TAC as well, are highly uncertain, and even biased, due largely to limitations in the two major sources of assessment information. On the one hand, research vessel surveys provide only a temporal snapshot of the stock, and may cover only a fraction of the fishable area, thus failing to capture the spatial diversity of the stocks. This can lead to a possibly biased estimate of the stock, depending on stock structure and migratory patterns. On the other hand, commercial fishery catch rates are often seen erroneously as a direct indicator of strong fish stocks, an assumption that can produce highly misleading conclusions, In fact, good catch rates (a 'successful' fishery) create the illusion of an abundant stock, whereas in reality, catch rates may have little connection with how many fish there are. Even as stocks decline, fishers and their technological tools may well be able to find and catch much of the remaining fish! Furthermore, this can be aggravated by a failure of stock assessments to account consistently for the technological changes that increase the catching power of the fleets. All these problems mean that the TAC is as uncertain (and biased) as is the initial estimate of the fish in the sea, which limits the robustness of the system.

Second, once the TAC is set, the uncertainty is often forgotten and an *illusion of certainty* appears, with the TAC being considered as an unchangeable number. This illusion of certainty is reinforced through the typical subdividing of the TAC into allocations by fleet sector, gear type or vessel size category, and then in some cases down to the individual fisher level: the individual quota. This process creates the impression of the TAC as a well-established 'pie' that can be cut into several precisely determined shares (pieces of the pie that can be calculated to many decimal places!). The management system itself is built around the certainty of a firm TAC, with fishery participants tending to view their shares as sacrosanct, indeed literally 'banking' on these shares. Even if new scientific evidence obtained during a fishing season indicates a potential conservation problem if the current TAC remains in place, efforts to reduce it may well be delayed to the next year, to avoid (a) failures to meet 'business plans' and market commitments already made by fishers, (b) equity problems if fishers are inequitably affected by changes in allowable catches, and (c) potential antagonism facing managers if such changes are made (Charles 1995b).

This perspective, which downplays the inherent uncertainty in the fishery, has led to conservation problems in the past (Angel *et al.* 1994). This is particularly so when the allowable catch is subdivided to individual companies or fishers, since the capability of the fishers to harvest individual quotas whenever they like during the season makes it especially difficult to be equitable in reducing the TAC mid-season (Charles 1995b). Such a system may be built around an illusion of certainty, with the TAC, and thus the ITQ, viewed as a 'sure thing' each year.

Third, the apparent simplicity of quota management (just choose a number (the TAC), divide it up and monitor catch levels as a mere matter of accounting) reflects the *fallacy of controllability*. In reality, (a) it is quite likely that the TAC was set at the wrong level (due to incorrect or biased biomass estimates), and (b) we may monitor the fish brought to shore, but we do not know the total amount of fish killed (e.g. by dumping at sea) to obtain those landings. Among the factors at the root of this situation is the lack of control the manager has over the behaviour of fishers, and by extension, over the quality of the fishery data collected. This may produce direct effects in terms of negative impacts on conservation, as well as faulty assessments of stock status, leading to overestimates of feasible catches and resulting over-exploitation (Charles 1995b, 1998a; Walters & Pearse 1996).

The above three issues may call into question the robustness of a system that relies heavily on quota management. On the other hand, it is unclear whether any other management measure would be more robust in the same situations. What is notable is the reality that, despite the collapse of various quota-managed fisheries, there is relatively little questioning of the growing reliance in the world's fisheries on quota management, nor research into ways to ameliorate its drawbacks. Some possibilities in this regard are discussed below.

15.3 Moving to robust management

If fisheries are to be managed sustainably within an uncertain environment, it is crucial to follow more robust and adaptive methods of management, ones designed to function successfully even given unexpected changes in nature's course, or an ignorance of nature's inherent structure. Movement in this direction requires attention to a number of key themes in fishery management and policy. Five fundamental (and essentially equally important) themes, shown in the box below, are discussed sequentially in the remainder of the chapter.

15.4 A robust, adaptive management portfolio

A wide array of management instruments is available in fishery systems, from use rights arrangements (Chapter 14) to traditional catch, effort and biological controls (Chapter 5). Each instrument has advantages and disadvantages. An overemphasis on any single harvesting or management method is unlikely to provide the desired robustness. There will always be some

---

Robust management for resilient fisheries

- A robust, adaptive management portfolio
- Self-regulatory institutions
- Fishery system planning
- Livelihood diversification
- Using all sources of knowledge

situation in which any such method will fail to ensure sustainability. The drawbacks of relying on quota management were discussed above. Similarly, a reliance on effort management may cause problems if it is not possible to control enough of the various fishery inputs, and a focus on closed seasons or closed areas may lead to excessive exploitation on unprotected parts of the stock, or on distinct sub-stocks. The point is that any single management measure cannot be considered 'safe'. A *portfolio* (set) of appropriate *mutually reinforcing* management tools is needed, selected on a case-by-case basis, and taking into account:

- society's objectives;
- biological aspects of the resource;
- human aspects such as tradition and experience;
- the level of uncertainty and complexity in the fishery;
- the predicted consequences of the various instruments.

Consider, for example, a management system focused on catch controls (TACs). Given the discussion earlier, such a system might be made more robust by supplementing catch controls with other measures to try to ensure that conservation goals are not thwarted. Among the additional measures could be marine protected areas/marine reserves (as discussed in Chapter 12) and effort (input) controls (FRCC 1996). Dual use of catch and effort controls, whether qualitative limits on the 'how, when and where' of fishing (referred to as 'parametric management' by Wilson *et al.* 1994), or quantitative limits on variables such as vessel numbers or amounts of fishing gear or fishing time, provides a double-check that conservation is ensured. This is because, with high levels of over-capacity in the fleets, setting a catch limit in the absence of effort limitation can lead to (a) extensive dumping, discarding and high-grading, and (b) more chance that a quota erroneously set too high will be caught. In the absence of substantial capacity-reduction measures, the key is to ensure that the realisation of this capacity, effort at sea, is suitably controlled while a TAC is being pursued.

Furthermore, while the management of fishing effort has its own challenges (Chapters 5 and 14), its use in a management portfolio may improve robustness, since (a) effort controls are placed on relatively observable variables intrinsic to the fishing fleet and individual vessels, (b) effort controls are less sensitive to uncertainties in the relative proportions within a multi-species fishery, since (in contrast to single-species TACs) they are not species-specific, and (c) effort controls may allow harvests to adapt naturally to the state of the resource, if certain catchability conditions are satisfied. (See Christensen *et al.* (1999) for further discussion of these points.)

In any case, the point here is that, just as a spaceship travelling into space is equipped with a multiplicity of control measures, so that severe problems are avoided if one system fails, so too can risk be reduced if a portfolio of management measures is utilised in the fishery system. The key goal here is for the portfolio to be 'mutually reinforcing' in that the various tools each help to rectify the shortcomings of the others.

It should be added, however, that even given an 'optimal' portfolio of management tools, there remains the need to guard against the *illusion of certainty* and the *fallacy of controllability*, described above. Indeed, there is a risk that after considerable cost and effort to combine together a set of management measures, we come to believe that we are 'safe', that by 'diversifying our portfolio' we no longer need to worry about uncertainty and can manage the fishery system in a controllable fashion. From all that has been discussed in this book, such a belief may well lead to disaster. Thus, the development of a management portfolio is important, but other actions and approaches are needed as well. These include a precautionary approach (Chapter 11), to guide both the philosophy of decision making and the specific decision criteria adopted, and a range of additional policy and management approaches, as discussed below.

15.4.1 Adaptive management

In particular, no matter how successful a management system is in lessening the overall sensitivity to uncertainty, such uncertainties will not disappear. Thus, it remains important to institutionalise processes for (a) continuous learning about the fishery system, through suitable monitoring, and (b) maintaining the capability and willingness to make appropriate adjustments, over both short and long time scales, by adapting in a timely manner to unexpected circumstances, so that conservation (as well as socioeconomic) goals are not compromised. This is what is meant by *adaptive* management. Two manifestations of adaptive management are discussed here.

15.4.2 Monitoring technology

The 'continuous learning' aspect of adaptive management, noted above, involves maintaining information flows about the human side of the fishery system. An important example of this concerns the process of technological change, which has had a major impact on many fisheries, in which fishing techniques, vessel design and notably the use of electronic gear have evolved greatly over the course of the past several decades. Many of the impacts have been negative, and these can often be traced to a lack of monitoring as well as a lack of subsequent understanding of the changes. The need for information flow and adaptive response to changing technologies is evident from an examination of some potential impacts.

- From an ecological perspective, there can be unexpected direct impacts on the fish and the ecosystem. For example, with more powerful vessels, an increasing fraction of the ocean environment comes within reach. Natural refuges for the fish are lost, and vulnerability to excessive harvesting pressure increases.
- Problems also arise when management does not notice technological change, or fails to take it into account in its assessment of the fish stocks and in designing its management plans. For example, in the late 1980s, a new 'turbo trawl' technology was rapidly adopted by trawlers fishing in Canada's Gulf of St. Lawrence. This produced a major increase in catching power, so that even as stocks declined, the fleet was able to maintain its catch per unit effort (CPUE), i.e. the catch level per day of fishing, or per tow of the net. For some time, the technological development was not incorporated into stock assessments, so the latter were overly optimistic (due to a false image of fishers 'doing well'), leading to excessive TACs and resulting in overfishing.
- Technological change in the form of industrialisation can affect what was formerly a stable, labour-intensive, small-scale fishery, diminishing its role as the 'engine' of the local economy and threatening the sustainability of fishing communities. Even if the 'modern-

ised' fleet provides greater total profit to harvesters, the distributional effects may make it detrimental to coastal communities.

While technological advancements have undoubtedly made individual fishers more efficient at finding and killing fish, at the same time, conservation and management have been made more challenging. To improve robustness in management, and to build resilience in the fishery overall, there is a need for adaptive monitoring of the impacts of technology and the processes of technological change, as well as the related examination of technological options with regard to trade-offs between their potential benefits and costs.

15.4.3 *Flexibility*

Adaptive management includes the recognition that in fishing, operating plans must be particularly flexible to allow for the highly uncertain nature of the most crucial input, the fish. This approach does not imply that the fishery manager must respond dramatically to even the smallest apparent change in the stock, but rather that new information must be integrated with existing information on a regular basis, with management actions being reassessed accordingly.

Some fishery management systems, even if they are able to adapt to change over long time frames, seem poor at doing so in the short term, e.g. within the fishing season. This problem, described above in the case of quota management, may be based on a sense that the resource is resilient enough within any given year, and that changes in abundance may be accounted for through annual adjustments rather than in-season changes, or it may arise from a desire to allow fishing firms to keep to their business plans. In any case, while ideally it would be desirable from the fishing industry perspective to follow fixed annual harvest plans and deterministic production methods, we can see that there may be a trade-off between the apparent industry *stability* so obtained and the ecological (and long-term economic) *resilience.*

In some fisheries, management is more inherently adaptive. An example lies in the salmon fisheries of North America's Pacific coast. The nature of the salmon stocks involves migration out of natal rivers to the sea for a period of years, and then a return to the same rivers for spawning. For most salmon stocks (chum, sockeye and pink species), the fish essentially 'disappear' from active monitoring while they are at sea, and then return in highly uncertain numbers. This high level of uncertainty in the stocks has led to interactive management in which not only do harvest patterns change from year to year as stocks vary, but even within a given year, abundances are estimated adaptively as the season progresses. Allowable catches are adjusted incrementally; fisheries are opened and closed, sometimes from day to day, as managers learn more about the stocks (Walters 1986; Parsons 1993). This requires a flexibility that is widely accepted among salmon fishers, despite its inconvenience (Fig. 15.1). Thus, while not without problems (Fraser River Sockeye Public Review Board 1995), the overall approach in the salmon fishery is an example of 'living with uncertainty' (Chapter 11).

Fig. 15.1 Fishers are accustomed to *living with uncertainty* in their work, but fishery management does not always take a similar perspective. These salmon trollers are part of a fishery that has a relatively high level of adaptive management – fishery openings and closings occur on a weekly or even daily basis, as information about fish stock sizes is regularly updated.

15.5 Self-regulatory institutions

The preceding two themes have focused on the portfolio of management tools, the extent to which these tools are inherently robust and adaptive, and the manner by which they are implemented in practice. However, no amount of attention to management tools is likely to make them effective within a dysfunctional institutional environment.

For example, an ineffective management institution will be unable to counter economic incentives that may run counter to societal objectives in a given fishery system. In particular, there is an incentive under effort controls to thwart those controls by increasing uncontrolled inputs, while under quota management, there are incentives to harvest more fish than is allowed in the established catch limits, and to maximise the value of quotas through high-grading, dumping and discarding at sea. Often, the 'solution' to overcome this behaviour is seen to be greater enforcement, but the cost can be high and the performance imperfect. For example, under quota management, excessive harvests may be reduced through an enforcement system with third-party monitoring of landings, but there often remains the means to circumvent the monitoring, such as by 'trans-shipping' of catches to other vessels. There are also problems with such a system if there are many boats, many landing sites, many buyers and many species in the fishery. Dealing with at-sea practices is even more difficult, with enforcement requiring elaborate means such as satellite monitoring or the placing of full-time observers on every vessel. Thus, while enforcement effort is important, it is typically infeasible to monitor and regulate fully all fishing activity at sea.

The above scenario is but one illustration of the consequences that can arise when fisheries are 'managed' with a focus on techniques and methods, in the absence of due attention to underlying institutional arrangements, and to an understanding of what makes such institutions succeed or fail. In this regard, recall that the concept of an 'institution' has a dual meaning. In common usage, an institution might be seen as an organisational arrangement of some sort, by which people interact, pursue society's goals and manage themselves: examples could be a Department of Fisheries, a fisher association or cooperative, or a public education system. Alternatively, underlying this is a more precise yet abstract sense of an institution as a set of rules or 'norms' governing the behaviour of individuals. As North (1990) put it:

> 'Institutions are the rules of the game in a society or, more formally, are the humanly devised constraints that shape human interaction. In consequence they structure incentives in human exchange, whether political, social, or economic.'

These two senses of the term are clearly related. A fishery management agency is an organisation (or an institution in the first sense of the term) that itself reflects what is (according to the second sense of the term) an underlying management institution, i.e. the set of rules adopted by society to manage the natural resource. Other examples of institutions might include the market-place, the legal system and municipal councils. The analysis of institutions has become prominent in fisheries and other resource systems in recent years, particularly within the realm of *common property theory* (e.g. Chapter 14) and fields such as new institutional economics.

The focus here is on the need to create and nurture *self-regulatory institutions*, i.e. those that can effectively and sustainably self-regulate the use of fishery resources. Such institutions seem best able to provide the sort of inherent management needed to overcome the type of problems described above. Three key questions, surely fundamental to success in fishery management, can be posed about self-regulatory institutions.

- Under what circumstances do such institutions develop?
- Under what circumstances is there sufficient 'institutional sustainability' (Chapter 10) and internal resilience to maintain the institution in the face of perturbations and shocks?
- More broadly, under what circumstances is the institution itself effective at managing the fishery and maintaining resilience in the system as a whole?

To address these questions, this section draws heavily on the insights of Elinor Ostrom, who is well known for her efforts to analyse these matters across a wide range of common property regimes (those based on common-pool resources: 'CPR'), including fishery systems.

15.5.1 *Emergence of institutions*

To begin, Ostrom (1992: p. 301) describes four conditions conducive to the development of a suitable institution by those using the resource (the 'appropriators').

> 'Individuals will tend to switch from independent strategies for exploiting a CPR to more costly, coordinated strategies when they share a common understanding that:

(1)   continuance of their independent strategies will seriously harm an important resource for their survival;
(2)   coordinated strategies exist that effectively reduce the risk of serious harm to the
CPR;
(3)   most of the other appropriators from the CPR can be counted on to change strategies if they promise to do so;
(4)   the cost of decision making about future coordinated strategies is less than the
benefits to be derived from the adoption of coordinated strategies.'

Ostrom rephrases these points into an expression of the simultaneous need for (a) a common
understanding of the problem (i.e. the need for management and conservation), (b) a common understanding that there exist coordinated options to solve the problem, (c) a common
perception of 'mutual trust and reciprocity' and (d) a common perception that the benefits of
coordination exceed the costs.

15.5.2 *Sustainability of institutions*

While the above conditions do not guarantee that a suitable management institution will
form, let us suppose that this does occur. The next question in sequence is: will the institution
survive? In other words, is there institutional sustainability and resilience? There is clearly
no point in discussing the self-regulatory nature of an institution unless that institution is not
only able to develop in the first place, but also to maintain itself thereafter. On this matter,
Ostrom (1992) suggests that a local institution of fishery users (specifically, in her terms,
an organisation of appropriators) is more likely to survive over the long term under a set of
conditions paraphrased here as follows.

(1)   Access and use are regulated based on a small number of agreed rules.
(2)   Enforcement of these rules is shared by all users, but is 'supplemented by some "official" observers and enforcers'.
(3)   The relevant organisation operates with 'internally adaptive mechanisms'.
(4)   Fishery users also have legal claims as owners of the resource.
(5)   The relevant organisation is 'nested' within larger organisations, and is seen by the latter as being legitimate.
(6)   Processes of change external to the institution are not excessively rapid.

It is of interest to relate these conditions to earlier discussions. For example, condition (1) may
be seen as informing the discussion on the need for a robust portfolio of management tools,
while condition (3) refers to the need for an adaptive approach, not within management tools,
as discussed above, but now with respect to the management institution itself. Conditions (2)
and (5) relate back to the desired level of co-management (Chapter 13), with an advantage
being seen both in balancing self-enforcement with support from governmental agents, and
in legitimising the fishery management institution within a larger institutional framework.
In a similar vein, condition (4) relates to the discussion of property rights in Chapter 14.
Finally, condition (6) is a recognition that if dynamic change in the external world (popula-

tion, technology, politics, etc.) is too great, impacts on a local institution may occur to such an extent that its adaptive nature is unable to cope.

15.5.3 *Effectiveness of institutions*

Finally, we turn to the third of the original questions stated above. Even if a management institution develops and sustains itself, is it effective in producing and maintaining a sustainable, resilient fishery system? This might seem an odd question, in that surely the institution would disappear if it were not effective over time. Ostrom (1992) notes, however, that for various political and organisational reasons, institutions may persist regardless of their performance, and thus it is useful to consider when an institution is likely to perform effectively. Considerable work has been, and is being, carried out on this topic (see, for example, the papers in Bromley 1992). Ostrom (1995: pp. 35–40) and Ostrom (1990) suggest that effective, robust institutions tend to be characterised by eight 'design principles'. These are paraphrased and summarised below in a fishery context.

(1)   The boundaries of the resource itself and of the set of fishery users are clearly defined.
(2)   Operational (use) rules in the fishery are in keeping with local conditions.
(3)   Those affected by operational (use) rules can participate in modifying those rules.
(4)   Effective monitoring of the resource and fishing activity is carried out by, or is accountable to, the fishery users.
(5)   Sanctions for violations of operational rules are 'graduated', based on the seriousness and context of the offence, and are issued collectively by the fishery users themselves, or by officials accountable to the users.
(6)   Mechanisms for rapid, low-cost local-level conflict resolution are available.
(7)   The institution's right to exist and operate is accepted by government authorities.
(8)   Appropriate management rules are established at each spatial and/or organisational scale of the system, with these arrangements being 'nested' in an effective manner.

These 'design principles', deduced from examination of a wide variety of case studies, may be of considerable use both in developing institutions and in understanding which are likely to be effective. It is beyond the scope of this book to elaborate further on each of these factors (see Ostrom's original work for a detailed discussion).

In assessing the effectiveness of management institutions, a related perspective focuses on the importance of *resilience*. In particular, authors in the volume by Berkes & Folke (1998) highlight resilience as a key characteristic of a successful management institution (as it is also for ecological/resource systems). It was found that in many of the case studies examined in that volume, 'adaptiveness and resilience have been built into institutions so they are capable of responding to and managing processes, functions, dynamics and changes in a fashion that contributes to ecosystem resilience' (Folke & Berkes 1998: p. 5). In other words, resilient institutions seem to be needed to maintain resilient ecosystems in the face of intensive resource exploitation. Indeed, Folke & Berkes (1998: p. 5) continue by noting that in a poorly managed situation, 'as resilience or the buffering capacity of the system gradually declines, flexibility is lost, and the linked social–ecological system becomes more vulnerable to surprise and crisis …'.

Within this perspective, what are the ingredients of a resilient institution? This question has been addressed to some extent (see, for example, Gunderson *et al.* (1995)), but remains an important topic of research. Folke & Berkes (1995: p. 132) argue that to build adaptiveness and resilience into institutions:

> 'The task is to make institutional arrangements more diverse, not less so; to make natural system–social system interactions more responsive to feedbacks; and to make management systems more flexible and accommodating of environmental perturbations.'

They further argue the importance of two fundamental elements (Folke & Berkes 1998):

- a key factor seems to be the concept of 'social/institutional memory', and in particular the role of traditional ecological knowledge or traditional ecological knowledge (TEK) (discussed in detail later in this chapter);
- related to this is the building of resilience through the use of traditional ecological approaches to management, such as (a) embracing small-scale disturbances to avoid major catastrophes, (b) the use of reserves and habitat protection measures, and (c) avoiding reliance on a single species or fishery by encouraging multiple occupations and sources of livelihood (discussed in greater detail below).

15.5.4 Discussion

The 'best' institutional arrangement will certainly vary with the context. It will be dependent not only on natural and human realities, but also on society's objectives and the varying priorities attached to each. Nevertheless, considerable progress is being made in examining case studies and synthesising the ingredients of sustainable, resilient, self-regulatory management institutions for fishery systems, perhaps leading eventually to agreed guidelines along the lines discussed here. The issue is of great importance given that, in the past, poor institutional arrangements in many fisheries led to disastrous conservation failures.

As noted earlier, successful management requires the 'right' institutions; ones that are structured properly, with widespread support, and which are seen as fair and just. Strong self-regulatory institutions enhance the resilience of the fishery system, in part by providing a feedback loop between harvesting and management, since the same people carry out both the fishing and the management functions. This facilitates the creation and reinforcement of incentives to shift the behaviour of fishery users in desired directions. For example, effective community institutions may serve to create social incentives for the sharing of fishery resources and for more responsible behaviour in fishing.

This discussion of self-regulatory institutions has been rather conceptual in nature, yet there are clear practical manifestations. In particular, the many moves toward community-based management and co-management (Chapter 13) involve the creation or reinforcement of such institutions. Fishers as a group commit to a management plan, and may also develop the means for self-policing. While this is not a simple panacea, since there remain many incentives to thwart regulations through illegal or anti-conservationist actions, the focus is on discouraging such actions through social mechanisms and moral suasion; pressure from the community to 'do the right thing'.

Many examples are also emerging in which self-regulatory institutions have led to conservation actions beyond what a central authority could have achieved. As but one example, consider the reopening in 1997 of the cod fishery in the northern Gulf of St. Lawrence (NAFO area 4RS3Pn) on the Atlantic coast of Canada. This was recognised by fishers as a potentially fragile conservation situation, and in developing their harvesting plan, the fishers agreed among themselves to use only fixed gear, rather than bottom trawlers, in the fishery. This focus on technologies with lower potential impact reflected a robust, precautionary approach (Chapter 11); one that politically, the government would not have been able to impose itself.

15.6 Fishery system planning

While poor planning efforts can be, and have been, detrimental to fishery sustainability, careful planning, carried out appropriately, can enhance the fishery system's resilience. (See Lamson 1986b for an interesting effort to apply ecological systems theory and the resilience concept to coastal communities.) Fundamentally, planning for resilience requires a reasonable understanding of the objectives being pursued in the fishery, so that it can be designed to operate efficiently with desirable levels of interaction between the human and natural systems. Two aspects of planning are described here: ideas related to the pursuit of an 'efficient' fishery, and approaches to reducing fishing capacity.

15.6.1 Fishery system efficiency

As discussed in Chapters 13 and 14, the concept of *efficiency* is frequently misused but is inherently simple: to obtain the greatest benefits with the least cost. From this perspective, efficiency can be addressed at the level of a fishing vessel, a fleet, a fishery and society at large. Unfortunately, it is often discussed only at the first of these levels, as 'harvesting efficiency', i.e. seeking the maximum rate of harvest, or profit, obtained by a fisher (or vessel owner) at a given time. This view of efficiency, which is focused on the short term and on the individual, has its place, but it is not sufficient, since there is no reason to believe that what is efficient at such a level implies efficiency for the fishery system, or for communities and society. In particular, calls for 'efficiency' are rarely based on an assessment of the broad objectives being pursued, and of how benefits and costs are to be determined.

In contrast, consider the idea of *fishery system efficiency*. This broader view of efficiency is taken from a long-term perspective. Given a conservation-based constraint on the fish available annually, it makes sense to view an 'efficient' fishery as one which maximises the net benefits obtained *per fish caught*, with increases in efficiency requiring increased benefits without killing more fish. In addition, fishery system efficiency seeks maximum net benefits as measured from the community or coastal economy perspective, rather than that of the individual fisher. The benefits are defined broadly, in order to incorporate all that is valued in society: perhaps a combination of profits and rents, employment, community well-being, ecological resiliency and so on.

From this perspective of fishery system efficiency, what *vision* of the future provides the greatest net benefits for a given available harvest? What policy measures are preferable? What is the preferred *fishery configuration*, i.e. what should the fishery look like in terms of a balance among fishery users, gear types and the like? Specifically, this first requires decisions about the desired 'mix' among multiple user groups (such as commercial, recreational, subsistence and aboriginal fishers), scales of operation (notably small-scale versus large-scale, or artisanal versus industrial), and gear types (e.g. trawlers, long-liners, gill netters, etc.). Second, within any single user group or gear type in the fishery, there is a need to decide on the balance among the various inputs that combine to produce fishing effort, including labour, capital, technology, management and enforcement activity. These decisions all depend on the blend of societal objectives pursued, and the capability of the various fishery players to meet those objectives.

Consider, for example, the key dichotomy highlighted earlier in this book between small-scale and large-scale fishery configurations. In recent decades, while small-scale fisheries have often received verbal support from decision makers, the focus in many jurisdictions has been on (a) fishery industrialisation, emphasising technological advances and market mechanisms, and (b) a corresponding view of fishery resources as 'commodities' within an industry. As a result of this growing 'industrial' view of the fishery, it is often *assumed* that small-scale fisheries are inefficient, indeed, perhaps historical anachronisms in today's world.

However, it is important to ask how small-scale fisheries rate in terms of *fishery system efficiency*, and specifically their efficiency in pursuing resilience and sustainability. What would the outcome be if efficiency were measured appropriately? As yet there seems to be no answer to this question, but it is certainly worthy of study. Several points might be considered. First, although larger vessels are certainly able to catch larger quantities of fish, it is uncertain what scale of vessel is financially superior, even using a traditional financial analysis. Second, when efficiency is viewed from a wider perspective, i.e. that of integrating the harvesting aspects of the fishery with onshore activities and the coastal economy as a whole (of which the small-scale fishery is often the 'engine'), small-scale fisheries may provide the greatest net benefits to the system as a whole. Finally, there is the matter of efficiency in fishery management. While a smaller number of larger vessels may seem more manageable, this ignores the point discussed earlier that in small-scale fisheries, the use in management of local decision making, local knowledge and the community's power of moral suasion can lead to increased efficiency and effectiveness. Indeed, social science research suggests that when fishers (and communities) are more geographically connected to the local resource, and dependent on it for their livelihood, they tend to be more knowledgeable about the resource, and more conservationist.

That such points often fail to be addressed highlights the simplistic view of efficiency that dominates at present, and the strong need for integrated analysis to support the intelligent design of fishery policy and planning, in order to optimise the well-being and sustainability of the overall fishery system.

15.6.2 *Fishery capacity*

A key element in fishery system planning lies in determining a suitable magnitude of interaction between the human system and the natural system: one that achieves overall fishery system efficiency. This challenge is inherent in a topic of concern that has become almost as widespread as that of overfishing, namely the matter of fishing fleet over-capacity (or over-capitalisation).

Over-capacity arises when there is a mismatch between the level of fishing effort *needed* to take the available harvest, and the *actual* level of potential fishing effort that could be exerted in the fishery. The latter is referred to as 'fishing capacity' and reflects the *catching power* of the fishers, i.e. the capability of the fishing fleets to kill fish. Fishing capacity is a complex mix of available 'factor inputs' (fishers, vessels and gear), with a greater availability of any one of these implying greater fleet capacity. Therefore, a situation of over-capacity is one in which there is greater catching power in place than is needed to catch the available fish. This arises through a dynamic process involving the entry of fishers, the expansion of fishing effort, investment in vessel construction, and so on.

Why is over-capacity of such concern? One major answer lies with the economic perspective. Over-capacity can imply a waste of human-made physical capital, with more wood, steel, plastics and electronics used in the fishing fleet than is necessary to catch the available fish. Furthermore, if other productive activities are available for fishers elsewhere in the economy, maintaining a greater number of fishers than is needed to catch the fish can also be seen as wasteful economically. However, if the fishers have few other options for employment, and if the physical capital is already in place in the fishery, then there is less economic need for concern about capacity.

A second theme concerns possible conservation impacts. Indeed, it is often said that over-capacity is among the greatest constraints to sustainability in fisheries. While this may be a valid point, it is not over-capacity *per se* that is the problem. Indeed, a fishery with over-capacity may have abundant fishers, boats and gear, and thus a high *potential* to exert fishing effort and thus to overfish, but a powerful fleet does no damage to the stocks if the boats are tied to the dock! Thus, it is not so much over-capacity itself that is a problem, but its combination with ineffective fishery management that is unable to control the fishing effort of the fleet and the killing of fish. The effect on conservation is indirect, through problems of manageability: (a) a large number of boats is presumably more difficult to manage than a smaller number, and (b) since over-capacity often reflects high investment levels, and correspondingly high debt among fishers, this can increase the pressure to increase harvests in order to pay debts. In the end, then, the key is to plan the desired fishery configuration (the number and types of fishing units) and effectively limit overall fishing effort at sea, through suitable management as well as capacity reduction.

Here, the focus is on the latter point: approaches to capacity reduction. If fishery policy were based on a single goal, a suitable capacity reduction policy might be fairly obvious. For example, a single-minded pursuit of maximum harvesting efficiency might focus on eliminating fishers until one reached the minimum number of fishers capable of harvesting the available catch. However, if fishery policy involves multiple objectives, that is, if society seeks a balance among a range of social, economic and conservation goals, then capacity reduction must similarly be designed to consider impacts on a range of factors, such as conservation, ecological balance, rent generation and income distribution, fishing community welfare and institutional stability. Thus, the matter of how capacity reduction can contribute to achieving the multiple objectives for the fishery set by society is a key matter to be resolved at the outset.

Fundamentally, capacity reduction needs to be part of a planning process that moves the fishery system towards a desired configuration. This implies the possibility that capacity reduction will need to be focused selectively on certain fishery sectors or certain inputs. For example, the optimal capacity reduction scheme may be one that reduces employment to create a more capital-intensive fishery, or one that reduces capital and promotes a shift to a more labour-intensive fishery. Either might be accomplished by changing allocations of catches, or through a 'buy-back' of targeted boats.

Unfortunately, it is rare to see the incorporation of capacity reduction in a planning process based on an objective examination of fishery configurations. Instead, one finds frequent repetitions of a simplistic view of over-capacity – 'too many fishermen, chasing too few fish'. This places the focus of concern on the fishers rather than on over-capitalisation, and can lead to misguided policy measures that reduce resiliency in the fishery. Examples may include (a) capacity reduction programmes favouring specialist/full-time fishers over generalist/part-time fishers, and (b) 'use it or lose it' policies of government that force fishers to fish regularly or risk losing their fishing rights. Both of these policies reward those who place the most pressure on the resource, while perversely penalising fishers who respond to low stock abundance by reducing their impact on the stocks (perhaps by shifting temporarily to other work).

Within the context of fishery planning, three major classes of capacity reduction programmes can be envisioned. These range from those in which the market determines the final configuration of the fishery, to those that deliberately seek to meet certain specified objectives through planned and/or community-based approaches.

Market approaches

In a market-based approach, the market-place determines the eventual configuration of the fishery, which is likely to be based on a combination of individual efficiency and access to financial capital. For example, with individual transferable quota (ITQ) schemes (Chapter 14), fishery participants are assigned or acquire rights to harvest specified quantities of fish, and the rights can then be bought and sold. Markets become established to facilitate the trading of rights, and in theory those who are less efficient from an individual perspective exit the fishery after selling off their rights to others.

Planned approaches

These are targeted, selective approaches that aim to achieve a desired overall fishery configuration, which may take into account such factors as the differential capacity of the various gear types and fleet sectors in the fishery, and the conservation impacts of the harvesting technologies. For example, targeted vessel buy-back mechanisms can be used to reduce fishing capacity directly, whether within basic limited entry (licence limitation) programmes or within individual non-transferable quota (INTQ) programmes.

Community-based approaches

Local-level and/or co-management approaches to capacity reduction produce a fishery configuration that is planned explicitly to meet local objectives. Under suitable circumstances, such approaches can be efficient both at the fishery level (reducing fishing costs by coordinating fishing effort among community participants) and at the regulatory level (e.g. by reducing conflict, increasing self-enforcement and decentralising management).

Whatever approach to capacity reduction is chosen, it should reflect an explicit choice that is based on pursuing the specified fishery objectives, enhancing fishery system efficiency (defined broadly), and fitting within a context of fishery system planning (even if a market-based approach is followed).

15.7 Livelihood diversification

No amount of research or management of fish stocks is likely to produce a resilient fishery system if humans are completely reliant on these resources for their livelihoods, and are unable to survive without over-exploiting them. Yet, of the three fundamental problems present in many fisheries, i.e. over-exploited stocks, over-capacity in the fleets and a lack of economic alternatives beyond the fishery, attention has focused almost exclusively on the first two.

This causes problems when, for example, *rationalisation* policies focus on dealing with over-exploitation and over-capacity by reducing the number of fishers. Smith (1981: p. 22) pointed out that such an action is likely to aggravate problems posed by a lack of livelihood (employment) alternatives since 'Management programmes fail to deal adequately with fishermen who are displaced'. Smith noted that in the absence of non-fishery economic alternatives, rationalisation may fail either because it is (a) non-sustainable and politically infeasible, owing to adverse impacts on those dependent on the fishery, or (b) practically infeasible, since fishers removed from the fishery and without other options will do what they feel is necessary – including illegal fishing – to maintain their livelihood. Furthermore, a reliance on rationalisation policies may lead to declines in community sustainability through inequitable resource allocation and social dislocation.

Thus, it may be possible that in some contexts, as Panayotou (1980: p. 146) suggested in reference to the Thailand fishery, 'the solutions to the problems of small-scale fishermen are to be found outside the fishing sector'. A systems approach is particularly crucial in such situations. It is not enough to focus only within the fishery. Inherent linkages between fishery and non-fishery aspects reinforce the need for an understanding of interactions beyond the fishery system (with other coastal and marine activity such as aquaculture, shipping and tourism) as well as the need for integrated and multidisciplinary approaches (Chapter 13).

Solving the problem of a lack of *livelihood diversity* (economic diversification) is by no means simple. If it were, fishery-based economies would already have become diversified by now, in response to past fishery downturns. Yet in most cases, efforts in this direction seem critical to the success of programmes for sustainable fishery systems, especially in the context of heavily exploited fisheries. Such efforts will typically be composed of within-fishery and non-fishery actions, including those described below.

15.7.1   Encourage multi-species fisheries

Within the fishery system, it is useful to encourage multi-species fishing, in which fishers utilise a range of fish resources, in contrast to policies that lead to the specialisation of fishers in single-species fisheries. By diversifying across sources of fish, the individual fisher reduces risks, and at the same time, the collective pressure to over-exploit is also reduced.

15.7.2   Encourage multiple sources of livelihood for fishers

Still focusing on fishers but looking beyond the fishery *per se*, the existence of 'occupational pluralism', i.e. fishers holding other jobs during non-fishing times, is common as a traditional practice in many seasonal fisheries. Through such practices, fishers avoid total reliance on fishing for their income, reduce the pressure they would otherwise face to obtain a livelihood entirely from the fishery, and thus also reduce pressure on the fish stocks. Encouraging such practices, and by implication, discouraging excessive specialisation by fishers, boosts the resiliency of the fishery system.

15.7.3   Diversify the economy

Finally, there is a need to diversify (broaden the base of) the fishery-dependent economy by creating new, sustainable economic activity outside the fishery sector. From the perspective of the individual, this enhances the range of available livelihood choices, both for current fishers and for young people looking for a job. The process is also likely to increase income levels outside the fishery. All this will tend to make it more attractive for fishers who are so inclined to leave the fishery, and reduce incentives for others to enter the fishery (through an increased *opportunity cost* of remaining in the fishery, as the economist puts it). This leads to an overall reduction in fishing capacity, and reduced pressure on the resources. Thus, economic diversification, combined with conservation-oriented management restrictions within the fishery, can increase resilience as well as community and socioeconomic sustainability (Charles & Herrera 1994). An example of tourism-based diversification is shown in Fig. 15.2.

   Economic diversification may, in many cases, be the single most important need in the pursuit of sustainable, resilient fisheries, but it is also the most challenging of tasks. As noted above, if it were a simple matter, such economic diversification would have been accomplished long ago. Yet there is hope that progress might be made through more *integrated* approaches, as are found in the themes of coastal zone management (e.g. Wells & Ricketts 1994), integrated coastal development (e.g. Arrizaga *et al.* 1989), and the *sustainable livelihood* initiative of the United Nations Development Programme (UNDP). This pursuit might involve (a) a focus on indigenously created employment alternatives within the local region or community, taking advantage of comparative advantages in ocean-related activity, such as the development of alternative fisheries, fish farming, coastal tourism and the like, (b) attention to constraints on local development that may be due to factors at the macroeconomic or macropolitical level, and (c) attention to the need for institutional arrangements that promote effective governance at the local level.

Fig. 15.2 Diversification of the coastal economy – through tourism, for example, as in Cahuita, Costa Rica – may seem beyond the purview of fishery discussions, but in many cases it is among the most crucial policy measures in moving towards a *sustainable fishery*.

15.8 Use all sources of knowledge: traditional ecological knowledge

Over the past few centuries, but particularly in the twentieth century, scientific study of the oceans has expanded dramatically. Oceanographic studies examine ocean currents, tidal action and plankton dynamics. Biological studies deal with food webs, species interactions and resource stock assessments. Geologists carry out research to understand the ocean floor. Meteorologists look to the ocean for insights into global climate phenomena. Social scientists study the users of marine resources, and the communities in which they live. The list of research subjects is impressive. Yet despite this apparent abundance of work, our scientific knowledge base concerning the oceans and those who use them is clearly limited.

Indeed, there may well be connections between the collapse of several major fisheries in recent decades and the state of the knowledge base upon which management decisions have been made. From a biological perspective, we can ask whether we understand enough to know how the biological features of a fish stock can lead to a collapse. Within an economic framework, there arises a question of whether the information flow among fishers produces economic incentives that lead to undesirable results. In human terms, we can ask what underlying attitudes among resource users and managers might lead to poor use, or neglect, of the information available in fisheries. Whatever the perspective, it seems clear that one of the significant contributors to fishery collapses has been the combination of (a) a *lack* of knowledge in some cases, and (b) a failure to use all *available* sources of information and knowledge in other cases.

The information that already exists but has been under-utilised in fishery management typically lies beyond the standard scientific apparatus, in the realm of what is called *traditional ecological knowledge* (TEK). The latter knowledge base incorporates the accumulated information and wisdom concerning the natural world that has been built up over time by fishers and coastal communities, through regular interaction with their environment and the natural resources therein. Berkes (1999: p. 23) defines TEK as follows:

> 'Putting together the most salient attributes of traditional ecological knowledge, one may arrive at a working definition of traditional ecological knowledge as *a cumulative body of knowledge, practice and belief, evolving by adaptive processes and handed down through generations by cultural transmission, about the relationship of living beings (including humans) with one another and with their environment.*'

TEK, which in this context is related to the similar terms 'local knowledge' and 'fishermen's knowledge', must be incorporated into, and nurtured by, fishery science and management. This section discusses some of the forms that TEK takes, and some of the processes involved in integrating TEK into the fishery management system.

First, it is important to note that while knowledge about nature *per se* is a crucial part of TEK, it is not the only part. For example, resource users and coastal communities can hold much wisdom about what resource management arrangements function best within their cultural and belief systems, about workable approaches to improving compliance among ocean users, and about which fishing techniques are most effective, or most conservationist, within the local context. Berkes (1999: p. 33) identifies four interrelated levels of traditional knowledge:

(1)   'local knowledge of animals, plants, soils and landscape' including knowledge of the species available, the biological and geographical nature of those species, and relevant changes over time;

(2)   'a resource management system, one which *uses* local environmental knowledge *and also includes* an appropriate set of practices, tools and techniques. Those ecological practices require an understanding of ecological processes, such as the functional relationships among key species ...';

(3)   'appropriate social institutions, sets of rules-in-use and codes of social relationships' to facilitate 'coordination, cooperation, rule-making to provide social restraints and rule enforcement';

(4)   'the *world view* which shapes environmental perception and gives meaning to observations of the environment' including 'religion, ethics and, more generally, belief systems'.

The first of these is perhaps the most basic, but typically requires the second level to be effective; i.e. success may not be achieved simply through the holding of ecological knowledge, but rather through the use of this within a resource management system. This in turn requires suitable institutions and social relationships. Finally, all this lies within a philosophical 'world view' that underlies the collection and use of TEK. Indeed, such knowledge may be incorporated within the local culture, and even within the language (Kurien 1998).

Fishery conservation measures in the Pacific islands

Johannes (1978) has made the extraordinary point that 'almost every basic fisheries conservation measure devised in the West was in use in the tropical Pacific centuries ago'. The range of conservation measures traditionally used in various Pacific islands include: closed fishing areas, closed seasons, restrictions on catching small fish, effort control (limiting the number of traps) and escapement controls (ensuring enough fish escape the fishery). See Berkes (1999) for a recent summary of the many measures documented by Johannes. Many such measures were components of sophisticated re-source management systems. These systems often failed to survive the commercialisa-tion and industrialisation of the fishery, but are being reintroduced in some Pacific na-tions, as their intrinsic value is being recognised once again.

Thus, it is useful not only to understand the nature of local and/or indigenous knowledge about fish stocks, aquatic ecosystems and the broader environment, but also to examine and learn from TEK arising in the form of resource management systems. Berkes (1999: p. 106) highlighted this point:

'... traditional coastal and lagoon fisheries around the world provide a rich set of local adaptations from which modern management systems can learn. These systems are found not only in isolated parts of the world but also in industrialized areas. Examples include "fisheries brotherhood" systems that act to regulate resource use and manage resource conflicts, the guild-like *prud'homie* system of the French Mediterranean, and the *confreries* of Catalonia, Spain ...'

A further note about TEK is that (despite the 'T') the knowledge base is not just *traditional*, in the sense of being developed over long time periods. The knowledge base of fishers ('fisher-men's knowledge') also includes information acquired recently, such as that obtained in the course of fishing in recent years. Specifically, fishers can provide information about distribu-tions of fish over time and space within a given year, including local abundances, changes in migrations, etc. While such information does not necessarily translate into a comprehensive estimate of overall stock abundance, it can be valuable input for scientists and managers. Each piece of information is potentially important to an understanding of current fish stock levels. Coastal residents may also have important information on environmental conditions in the ecosystem, ranging from changes in ocean currents to the existence of new land-based sources of marine pollution. Thus, while many examples of TEK are drawn from develop-ing countries and indigenous (native) societies, in fact a knowledge base is likely to exist wherever fishers and other coastal dwellers interact with the aquatic environment.

While it is important to acknowledge the existence of TEK and related fishermen's knowl-edge, perhaps the key challenge lies in integrating this knowledge with modern fishery sci-ence and management, which have developed largely without recognition that fishers and others hold valuable ecological and resource management knowledge. There are perhaps two

Fisher knowledge in Newfoundland, Canada

Two related components of local knowledge can be identified among the inshore (coastal) fishers of Newfoundland, on Canada's Atlantic coast. First, there is a base of 'traditional knowledge' (TEK) accumulated from centuries of fishing experience, and manifested, for example, in a detailed understanding of fish stock structure (notably the variety of sub-stocks) and migratory patterns. Second, there is the current knowledge obtained from the regular presence of fishers 'on the water'. This includes, for example, information on near-shore stock abundance and interactions of predators and prey (such as observations of cod 'chasing' caplin, or seals 'chasing' cod). This knowledge base is well known, but generally not well documented. However, since the collapse of the cod fishery in the early 1990s, academic and government researchers have shown an increasing interest in studying TEK and related fisher knowledge.

An interesting example of such a study is that of Neis *et al.* (1996). These researchers carried out interviews with fishers in several coastal communities, gathering responses documenting fisher knowledge of four aspects of the fishery and fish ecology: stock structure, changes over time in fishing effort and catchability, perceptions of current stock abundance, and the potential impact on cod from a renewed caplin fishery (the caplin being a key food source for cod). For example, the study documented an impressively rich knowledge of stock structure, with fishers distinguishing between 'herring fish' (cod that follow herring migrations, and which tend to spawn inshore), 'caplin fish' (cod, mostly from offshore, that follow caplin) and various other categories. In addition, fishers identified spawning grounds and spoke of 'mother fish', i.e. large deep-dwelling cod, the size of which is thought to make them crucial to spawning success, and the harvesting of which older fishers identified as leading to the decline of the cod fishery. Such fisher knowledge, documented in specific research studies such as the one cited above, as well as in governmental stock status reports (based on increasingly regular interactions between scientists and fishers) will probably be important in improving the assessment of fish stocks off Newfoundland.

major facets to this challenge. The first and most important is a sociocultural and institutional issue: developing both the sense of trust and the means of communication between scientists and managers on the one hand, and those holding the knowledge on the other. This process, which clearly requires considerable care, can be facilitated through the establishment of systems of community-based management and co-management (Chapter 13). Such systems encourage participation in decision making (and in research) by those most familiar with, and most attached to, the fish resources and their environment. The latter group typically are those living in coastal communities adjacent to the fish stocks, which implies the additional need for policy measures to maintain the knowledge base by providing such fishers with secure permanent access to the local fishery.

A second, subsequent, and more technical challenge involves the practical steps required to utilise TEK within a management context. This is clearly more than simply adding a new piece of data into a mathematical model. Indeed, as noted above, TEK alone may not provide an abundance estimate similar to that calculated in a stock assessment. Instead, it is necessary to (a) document the knowledge base, (b) explore current and potential uses of this knowledge base, and (c) examine how the knowledge base interacts with (or could interact with) the scientific apparatus surrounding fishery and ocean management. Such efforts are underway in a variety of jurisdictions around the world, and hold considerable potential for improving the functioning of fishery management systems. Indeed, Berkes (1999: p. 54) notes that:

> 'The complementarity of local knowledge and scientific knowledge is an increasingly important theme in resource management. For example, there is an emerging consensus in Oceania that, given the scarcity of scientific knowledge and research resources, alternative coastal fishery models are needed. These models involve the use of local knowledge to substitute for, or complement, scientific knowledge ...'

Given this potential to improve the performance of fishery management, and at the same time to improve the interaction between fishers and communities on the one hand, and scientists and managers on the other, increasing attention can be anticipated (and indeed is urged) to the many strong bases of local knowledge (whether it is considered 'traditional', 'indigenous' or simply 'fisherman's' knowledge) that are now understood to exist around the world. Obviously, such efforts do not guarantee successful management in the future, but at the same time, to neglect any available sources of knowledge is to court disaster.

15.9 Summary and conclusions

This chapter has sought to emphasise the fundamental and increasingly recognised need for resilience in fishery systems, and for robustness in the management system. In doing so, the presentation has drawn on the major themes of this book (notably those of Chapters 10–14) while introducing a set of management and policy approaches to enhance resilience and robustness.

- A robust, mutually reinforcing management portfolio.
- Adaptive management.
- Self-regulatory institutions.
- Fishery system planning.
- Livelihood diversification.
- Using all sources of knowledge.

These approaches focus specifically on the management system, but in most cases also require attention to the human system. It should be noted that this set of measures is by no means exhaustive, but rather illustrative of desired directions. Furthermore, these measures to move fishery systems towards greater resilience must be seen within a context, described in preceding chapters, of broader directions and priorities leading on a path to sustainability. These directions are reviewed and summarised briefly below.

15.9.1 *Focusing on sustainability*

Given the acceptance of fishery *sustainability* as a critical benchmark in evaluating policy and management measures, a multidimensional view of sustainable development is required, incorporating ecological sustainability, as a clear prerequisite for a sustainable fishery, as well as socioeconomic, community and institutional components of sustainability (Chapter 10). This direction is clearly related to the requirement of resilience: fishery strategies must enhance resilience in the ecosystem, the economy, the communities and all relevant institutions. In particular, sustainable fishing communities are of importance both to their residents and to the broader society, and the sustainability of management institutions underlies that of the fishery system as a whole. An integrated process of *sustainability assessment* can provide a mechanism for measuring the various aspects of progress towards sustainability.

15.9.2 *Living with uncertainty*

Recognition that we can never hope to 'overcome' uncertainty in fishery systems has led to an understanding that instead, the goal should be one of *living with uncertainty* (Chapter 11). This requires an understanding of the forms by which uncertainty arises in fisheries, the application of methods for risk assessment and risk management, and the adoption of the precautionary approach to provide guidance to managers on appropriate choices with respect to harvesting levels and methods. This guidance helps managers and other decision makers in the fishery to 'err on the side of conservation'. Living with uncertainty requires overcoming the *illusion of certainty*. This may involve the sort of measures outlined earlier in this chapter, such as (a) reducing risks through adoption of a broader management portfolio, involving mutually reinforcing measures, (b) developing management measures that are robust in the face of uncertainty, and (c) adopting adaptive management approaches.

15.9.3 *Accepting complexity, embracing diversity*

Interest is growing with respect to adopting an ecosystem approach, and in general a more integrated approach to the study and management of fishery systems (Chapter 12). This recognises the complexity and diversity of such systems, as well as the desirability of accepting complexity and 'embracing' diversity. For example, an ecosystem approach implies making explicit links between stock assessment efforts and studies of the ocean environment, and incorporating an analysis and understanding of human objectives and behaviour into fishery management decision making. On the research side, it is promising that multidisciplinary initiatives are expanding, and integrated research methods, while not yet in the mainstream, are attracting more attention. On the management side, moves to 'embrace' complexity and diversity imply overcoming the *fallacy of controllability*, recognising that complete control of a complex system is not a realistic goal.

15.9.4 *Participating in management and research*

Co-management, i.e. the development, implementation and enforcement of management measures by a suitable combination of government, fishers, communities and the public, is rapidly expanding and evolving in fishery systems (Chapter 13). There are various models for this, but the key ingredient is to increase the role of resource users, which serves to lessen the conflict between fishers and managers that has tended to lead to failure in top-down management regimes. As a consequence, there is a clear need to involve fishers, their organisations and their communities in managing local resources, based on sharing decision-making power as well as sharing the responsibility to ensure the fishery's sustainability. Furthermore, it seems reasonable that where fish in the sea are publicly owned, as in most national fisheries, the public should also play a role in developing policy concerning the overall use of those resources. Finally, while participation is important to management, it is equally relevant to fishery research, where the fishery sector and non-governmental organisations are increasingly playing an important role.

15.9.5 *Promoting local control*

It is important to improve the utilisation of indigenous/traditional/local knowledge (TEK), including locally developed resource management systems, in combination with scientific data and methods. Development (or revitalisation) of community-based management approaches (Chapter 13) can provide the means to make use of local resource knowledge and indigenous social and culturally based controls on resource use. While this naturally provides inherent support for community sustainability, the potential of this approach also lies in enhancing other sustainability components (as well as other goals, such as resource conservation and economic efficiency) if local-level control can provide more efficient and effective resource management.

15.9.6 *Seeking appropriate rights and institutions*

It is often said that a crucial step in the management of any activity lies in 'getting the incentives right' and 'getting the institutions right'. In this chapter, it was noted that one of the steps towards greater resilience and robustness was the creation of the 'right' self-regulatory institutions, so that fishers and others would have the 'right' incentives to operate in accordance with the regulations, and in particular to avoid anti-conservationist actions. In Chapter 14, a review was provided of the wide range of rights that can be present, and utilised, in fishery systems. Such rights, when present in the context of effective management institutions, help to clarify the roles and responsibilities of the various players in the fishery, and thereby to steer incentives in the desired directions.

Two key points must be noted here. First, while an *appropriate* rights system, one that fits well into human and management institutions, can enhance a fishery and its sustainability, the imposition of an inappropriate rights systems can lead to undesired consequences, such as a loss of resilience in communities or institutions. Thus, it is crucial to emphasise that it is not just a rights system, but an *appropriate* rights system that must be sought. This point was discussed in some detail within Chapter 14.

Second, institutions provide more than a system of rights. The 'right' institutions will also shape the underlying perspectives of those involved. In particular, this may help to realise a crucial ingredient in a sustainable fishery: a *conservation ethic*. This means a fundamental embracing of conservation as part of the 'belief system' of fishery participants (Charles

1994). It appears to be the case that underlying much of the crisis in the world's fisheries is the problem of *attitude*: how fishery players view the system in which they operate. This matter of conservation attitudes applies not only to those one might first think of in this regard, i.e. the fishers, and the nature of their actions at sea, but equally to other resource users, scientists, managers and policy makers in the fishery system (Charles 1995b). Indeed, it is by looking at underlying attitudes that we may understand *why* fish stocks became depleted in fisheries even where almost everyone involved, from both industry and government, expressed concern for conservation. The 'right' institutions can help create an atmosphere in which old attitudes change, and a conservation ethic emerges.

15.9.7 A closing note

This marks the end of a book that has attempted to cover a rather broad spectrum. An effort has been made throughout to justify the book's focus on *sustainable fishery systems*, i.e. the need to look at fisheries as systems, and to focus on the sustainability and resilience of those systems. An understanding of the structure and dynamics of fisheries (as emphasised in the first half of the book) has proved to be important in examining sustainability and resilience – the major recurring themes of the second half of the book. Within the fishery system, cutting across the various system components are a number of key ingredients which have an impact on sustainability and resilience: uncertainty, complexity, diversity, conflict and fishery rights. These omnipresent issues need to be viewed as much as possible within an integrated, interdisciplinary context.

  This is a time of change in fishery systems. There is a widely recognised and urgent need to restructure fisheries, and fishery management, to promote sustainability of the fish resources, the ecosystem and the coastal communities. New approaches are being sought out, developed and implemented. This book has focused on some of the ideas being debated, the competing visions of how the fishery should be managed, and exactly who is to have the privilege of participating in it. Undoubtedly, a single book cannot answer, or even consider, all the major questions, but we have certainly seen that a variety of positive responses are available in fishery systems. I hope that you, the reader, have found this presentation to be useful, and the overall product to be at least somewhat stimulating and perhaps even provocative.

Appendix

Integrated Biosocioeconomic Analysis of Fishery Systems

The integrated analysis of fishery systems described in Chapter 8 uses a *graphical* approach and is based on a 'bioeconomic' framework incorporating fish stock and fishing fleet variables. In this appendix, we extend the analysis to a 'biosocioeconomic' approach; one which specifically seeks to understand the dynamics of fishers (in the form of fishing effort and labour dynamics) in addition to fish and fleets. As with any such approach, the analysis involves the use of *models*, which may be of two major forms:

- *behavioural* models are designed as a means to explore the behaviour of the system and its components, such as the fishers and the fish stocks;
- *optimisation* models aim to understand how the 'best' outcome possible depends on both the dynamics of the system and the inherent objectives being pursued.

Two examples of biosocioeconomic models are examined here, a behavioural and an optimisation model, each describing certain aspects of the natural and human dynamics within fishery systems. These examples reflect just some of many possible scenarios that can be examined with such models. (Note that the main text in this appendix is non-technical, while optional boxes contain the mathematical and technical details of the models. Thus, the reader preferring to avoid mathematics can read the main text without difficulty, while the more mathematically inclined can also examine the boxed material.)

A behavioural model

Consider a scenario in which a particular fishery forms the dominant core of a fishery-dependent local economy (community), which in turn operates within a larger multi-sector regional economy. In such a situation, the dynamics of the fishery's labour force reflect changes in the overall community population. This is a common occurrence, for example, in northern nations such as Norway and Canada, as well as in many developing countries. Let us further assume a degree of *labour mobility*, meaning that the labour force (workers) within the broader *regional* economy are able to move into and out of the *local* economy (i.e. primarily the fishery), in response to both internal conditions (the state of the fishery) and external conditions (the attractiveness of employment elsewhere in the region). Given this scenario, we examine the dynamics of the natural and human systems together, through a biosocioeconomic analysis. We focus on two dynamic variables of the fishery system, the *fish stock* and the *fishery labour force*. (The fishing fleet is also a key variable, but for simplicity, this additional complicating factor is not included here. See Charles (1989) for a full treatment of the model.)

We will make a number of assumptions here to simplify the discussion. First, with respect to the dynamics of the fish stock, we ignore the complexities of multiple species with diverse age and size structures, and focus on a fish stock which, at any time t, can be described as a single aggregated population or biomass, denoted $x(t)$. This fish stock is assumed to vary in a continuous manner, driven jointly by natural reproductive dynamics and human impact on the resource: the harvesting activity. The latter depends on the level of fishing effort, $E(t)$, which is assumed to be regulated by the fishery management process at each point in time.

Meanwhile, the fishery labour force, $L(t)$, is also assumed to adjust continuously to changing conditions over time. How might these adjustment processes occur? If the local labour force is relatively low (given current conditions in the fishery compared with those elsewhere in the broader economy), we would expect the labour force to expand. Conversely, if the labour force is relatively high (beyond that suggested by local versus external conditions), it

---

Mathematical details of the behavioural model

Fish stock dynamics

The rate of change, dx/dt, in the fish stock is given as $dx/dt = F(x) - h$, i.e. subtracting the rate of harvest per unit time (h) from the resource stock's natural growth rate, $F(x)$. To be specific, we assume here that the latter function reflects logistic population dynamics, $F(x) = sx(1 - x/K)$, where s is an intrinsic growth rate and K a carrying capacity. We further assume that the rate of harvest is $h(t) = qE(t)x(t)$, where $E(t)$ is the level of fishing effort, measured in the same units as the labour force L, and q is a constant called the catchability coefficient. Then the population dynamics are: $dx/dt = sx(1 - x/K) - qEx$.

Fishery labour force dynamics

Labour dynamics are modelled using a differential equation describing the rate of change in the labour force, dL/dt. In analogy with natural population dynamics, we assume that the labour force dynamics are driven by:

- a maximum *per capita* growth rate 'r', reflecting the sum of natural growth and immigration under *ideal* conditions;
- a 'carrying capacity' $\bar{L}(t)$ representing the maximum sustainable level for $L(t)$, given specified internal and external conditions. This carrying capacity can be viewed as a 'natural' or 'target' level that the labour force tends towards gradually over time.

These assumptions lead us to a 'logistic' growth pattern, in which a relatively small labour force $L(t)$, which is less than $\bar{L}(t)$, the 'natural' level, will tend to grow, since the fishery seems relatively attractive to potential fishers (cf. Smith 1968). However, contraction will occur if there are already so many fishers that $L(t) > \bar{L}(t)$. Then the labour force dynamics are such that the actual *per capita* growth rate of the labour force at any time *t* is given by the product of '*r*' and a term involving the ratio $L(t)/\bar{L}(t)$. Thus, the rate of change, dL/dt, is given by the equation $dL/dt = rL(1 - L/\bar{L})$.

Note, however, that unlike a basic logistic equation, in which the state variable tends towards a constant level, in this case as conditions change, so too does the 'natural' labour force, $\bar{L}(t)$. The carrying capacity is in fact a 'moving target' towards which $L(t)$ adjusts continuously, driven by conditions in the fishery and in the external economy. (This is actually analogous to a population dynamics model in which carrying capacity varies with environmental factors.) This carrying capacity, $\bar{L}(t)$, is determined, as discussed in the text, by the state of the external economy (assumed to be constant here) and the attractiveness of the fishery itself. The latter depends on two specific determinants.

- A measure of how lucrative the fishery appears, given by the average per capita income received by fishers, namely the term $(\pi + T)/L$. Here, the fishery rents π are given by total fishery revenue (price per unit harvest multiplied by actual harvest) minus operating costs (the variable costs of supplying fishing effort) minus the opportunity costs of labour in the fishery. If we assume linear costs, with c_E and c_L being unit costs of effort and labour, respectively, and a constant price p, then the rents can be written as $\pi = \pi(x,L,E) = p(qEx) - c_E E - c_L L$ using the relationship $h = qEx$. Furthermore, *T* represents the level of transfers to fishery participants from the external economy (assumed to be constant).
- A measure of the rate of employment in the fishery, represented by $E(t)/L(t)$. Note that if $E(t) = L(t)$, the fishery is operating with full employment of its labour force. However, if $E(t) < L(t)$, the labour force is under-utilised, and if $E(t) > L(t)$, the available workforce is overextended.

The higher the income and employment levels, the more attractive is the fishery to fishers. This is modelled here with a 'Cobb–Douglas' function, in which the attractiveness of the fishery is given by multiplying together the two terms above, each raised to a constant power (α and β, respectively): $\bar{L}(t) = (E/L)^\alpha ([\pi + T]/L)^\beta M$. Here, E, L, π and T are as discussed earlier, and *M* is a constant that incorporates the constant state of the external economy. The labour dynamics described above are fully specified given this expression for the carrying capacity.

Thus, for any given set of initial conditions, the fishery system will evolve over time, driven by the above dynamics of the state variables *x* and *L*, constrained by the manager's choice of fishing effort $E(t)$ at each time *t*. In other words, given the above dynamics and given values for the constant parameters $s, r, K, M, T, p, c_E, c_L, q, \alpha$ and β, the evolution of the fishery will be determined by the choices of effort $E(t)$ made at each point in time.

would be expected to decline. This suggests that labour dynamics are determined by comparing the current level relative to a *carrying capacity*, a 'natural level', based on the two factors described below.

The state of the external economy. The better the employment situation beyond the local economy, the greater the tendency for labour movement away from the fishery, and thus the lower the carrying capacity for labour in the fishery. At the same time, the extent to which labour will exit the fishery may well depend not only on employment and wages (the usual economic elements), but also on such sociocultural factors as local tradition, family ties, employment-sharing, involvement in the 'hidden economy' and attitudes towards income support programmes. For the present discussion, both the state of the external economy and the sociocultural factors are treated as constant, in order to focus on changes within the fishing economy.

Internal conditions in the local economy (fishery). There are assumed to be a set of factors that determine the perceived attractiveness of the fishery to current and potential participants relative to other employment options (and consequently, the aggregate fishing effort). The set of possible factors is diverse, but we focus on two here (see Charles 1989, for other options). First, following Smith (1968), we assume that the expansion of fishing effort is driven in part by how lucrative the fishery appears. Here, we measure this by the average per capita income received by fishers. Second, we assume that fishers find the fishery attractive if there is a high level of employment, i.e. if the level of allowable fishing effort is high relative to the labour available, so that each fisher faces few restrictions on fishing activity.

Assuming suitable population dynamics for the fish stock, and labour dynamics driven by the difference between the current labour force and the 'carrying capacity', we can model the evolution of the fishery system for any given management scenario of allowable fishing effort levels over time. For example, consider the case of a fishery in which the labour force is initially relatively large, while the fish stock is significantly depleted. (See box on p. 240 for details.) Suppose, in addition, that managers seek to rebuild the depleted stock by harvesting at a modest level initially (say 50% for the first 5 years), and then increasing that level thereafter (say, to 70%). Figure A.1 shows the evolution of the resulting fishery system. Note in Fig. A.1 how, as designed within the model, the labour force varies over time, tending continuously towards its constantly shifting *natural* level.

An optimisation model

The idea of the behavioural model above is to mimic, and more importantly, to predict, the evolution of the fishery system over time, given the dynamics of the fish and the people involved, and a *given* harvesting policy. An optimisation model, on the other hand, is meant to determine the *best* harvesting policy. Such a model still needs to incorporate the dynamics of the behavioural model, but must solve for the *optimal harvest* rather than focusing on predicting the consequences of a given harvest policy.

Suppose, then, that the dynamics of the fish stock and the labour force are as in the above model, with the exception that now, while the fishing effort E is a function of time, $E(t)$,

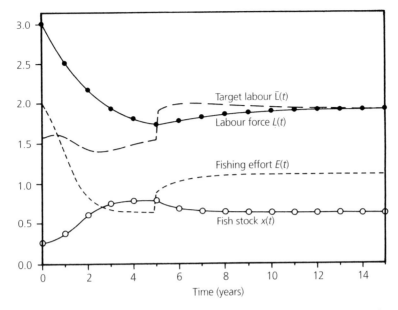

Fig. A.1 Fishery system dynamics for the case of an initially small fish stock and a large labour force. Maximum intrinsic growth rates are $s=3$ for the fish population and $r=0.25$ for the labour force. The harvest increases from a relatively low level for the initial rebuilding period to a higher level from year 5 onwards. Note that the labour force $L(t)$ continuously approaches its target level $\bar{L}$, but that this quantity varies over time depending on profit rates, effort levels and the size of the labour force itself.

Assumptions and parameter values of the behavioural example

The (arbitrarily set) parameter values for the example described in the text are as follows:

$s=3$	Intrinsic growth rate of fish stock
$r=0.25$	Intrinsic growth rate of labour force
$K=1$	Carrying capacity of fish stock
$M=2.12$	Maximum labour force under historical conditions
$T=2$	Net non-fishery transfers into fishery system
$p=1$	Price of fish
$c_E=0.01$	Unit cost of fishing effort
$c_L=0$	Unit opportunity cost of labour
$q=1$	Catchability coefficient
$\alpha=\beta=0.5$	Exponents in the Cobb–Douglas model

Note that simulating a continuous-time fishery requires 'discretising' the model, i.e. changing it to 'discrete time', using a small time increment, in this case $\Delta t=0.1$ years.

it is not pre-set by the manager, but rather must be determined by solving the optimisation problem. In this case, the fishing effort is known as the *control variable* or *decision variable*, because it is the one quantity in the fishery system that is assumed to be (in theory at least) controllable through management decision making.

The optimisation problem of fishery management is to choose the value of fishing effort $E(t)$, and as a consequence the harvest level, at every point in time, but how does the decision maker decide on a suitable level of fishing effort? This is the role of a key component of every optimisation model, the *objective function*. This function is to be maximised subject to a set of constraints, including the inherent dynamics above.

What does the objective function look like? In a dynamic optimisation problem such as this, the essential idea is to determine the *total net social benefits* obtained in each time period, and then to maximise the sum of these over all time periods. For example, one possible goal would be to maximise a sum of annual net benefits obtained each year over all future years. Such a sum is an example of a *present value*, since it gives the total *value* of the flow of future benefits, as seen from the *present*.

Specifically, the present value in the above case, calculated as a simple summation of benefits in future years (whenever they may be received), is based on placing as much importance on the distant future as on the present. An alternative perspective is to place greater weight on immediate rather than future benefits. This is a process known as *discounting*, since future benefits are 'discounted' relative to ones received sooner. This reflects the common behaviour of investors seeking out the greatest rate of return on a financial investment. To allow for both of these possibilities, a *discount rate* is used. If positive, the further into the future the benefits are received, the less they contribute to the present value, while if the discount rate is set at zero, all benefits are counted equally, whenever received. Thus, the fishery optimisation problem can be described as that of choosing a management plan so as to maximise the present value of the discounted sum (over all times t) of the annual benefits. If time is measured in years, this implies adding up N values, one for each year, within a time horizon N.

Now that we have stated the structure of the problem, we need to look more closely at the annual net benefits. These are captured in what economists call a *welfare function*, and defines what is considered important to society in terms of annual socioeconomic benefits. For example, the fishery rents function, discussed earlier, is a special case of a welfare function in which the goal in the fishery is to maximise rents. (Note that the annual net benefits reflect *society's* perspective, in contrast to the measure of fishery 'attractiveness' incorporated in the labour force carrying capacity above, which reflects the *fisher's* perspective. These two measures may or may not be related, since what makes the fishery desirable to fishers does not necessarily correspond to society's perspective.)

So what indicators of fishery performance are important to society? The benefits function must capture the multiple objectives inherent in fishery systems – objectives such as resource conservation and food production (at least in developing countries), as well as specifically socioeconomic goals such as:

- generation of economic wealth;
- generation of reasonable incomes for fishers;
- maintaining employment for fishers;

- maintaining the well-being and viability of fishing communities.

In the present analysis, we focus on this last set of four socioeconomic objectives to illustrate the ideas behind a multi-objective fishery analysis. (Thus, resource conservation is taken not as an objective but as a constraint in this discussion, with the avoidance of stock extinction incorporated into all management options. Similarly, the objective of food production (harvest maximisation) is treated implicitly in the model.) The multiple goals to be pursued are elaborated below.

Economic wealth (rent generation)

This represents the most common objective in traditional bioeconomic fishery models, where it is typically calculated in terms of the resource rent ('economic surplus') accruing directly from the fishery (π), and given as the difference between fishery revenues and fishery costs, taking opportunity costs into account. It is also useful to add a second component of rents, namely those arising from the 'social overhead capital' and infrastructure in fishing communities. These are given in monetary terms, and are produced through economic interactions within the fishing community. This might be modelled as being dependent on the size of the relevant population in the community, or equivalently in this situation the difference between current and historic labour force levels. Note that since the community also has other economic activities, the measure used here should include only that fraction of rents from the community infrastructure that are applicable for inclusion in the fishery analysis.

Fisher incomes

The average fisher income, measured by net benefits per capita, provides a measure of economic well-being at the individual (as opposed to fishery) level. Specifically, to incorporate equity considerations, we focus on per capita income relative to the average income in the overall economy.

Employment

The third objective, employment, is a traditional concern of fishery managers, particularly in small-scale or isolated fisheries. Employment in the fishery system is comprised of fishers in the harvesting sector together with those involved in secondary industry (such as fish processing). We model the employment objective in terms of the employment rate, i.e. the fraction of the labour force involved in providing fishing effort at any point in time. This is a useful measure of well-being from the social perspective of a fishing community. With the fishery being dominant in the local economy, society may desire as great a utilisation of the labour force as possible, i.e. as large a value for this quantity as possible.

Community 'health'

Fishing community viability (or health) is an important factor in determining social welfare, yet its appropriate measurement is by no means clear. One might measure community well-being as total fishery profit levels or even fisher income levels, but here we focus instead on the trend in population of the relevant community. This is based on the notion that a declining population in a community is socially detrimental, while an increasing population tends to reflect a healthy community. However, there are two key points to note. First, an increasing population clearly may in fact lead to greater unemployment and to declining income levels, but these detrimental effects are included explicitly under the second and third objectives above, rather than here. Second, the idea that maintaining a community's population is beneficial may have validity only given unchanging external conditions; for example, improving opportunities in the external economy could lead to a justifiable decline in a fishing community. Thus, *other things being equal*, a growing population is preferred to a shrinking one. This objective is represented by the rate of change of the fishery labour force over time.

Given a set of objectives (four in this case), a typical multi-objective statement of the fishery decision problem involves maximising a social welfare function incorporating each of these goals: rents, fisher incomes, employment and community health. This maximisation is carried out subject to the fishery dynamics and any other constraints. However, a full solution is mathematically complex, so in this example, we seek an approximately optimal solution using simulation methods. This allows us to examine the biosocioeconomic dynamics of a fishery system, driven by the dynamics and objective function above, in this case using a 30-year time horizon. In this example, a simulation is carried out for each 'allowable' harvest strategy (those lying within a set of possibilities), and the strategy is chosen that maximises the objective function. The management plan so determined is thus only optimal relative to the set of possibilities considered.

In the first scenario shown here, it is assumed for simplicity that the objective being pursued is simply maximising the sum of discounted rents, subject to relatively fast adjustment dynamics for both the fish stock and the labour force, and starting with a relatively small initial fish stock size. The resulting 'optimal' fishing effort policy calls for initial recovery of the stock through a low harvest level in the first 5 years, increasing to an equilibrium level thereafter. In fact, however, the actual harvest level in this case lies initially below its optimal level, since fishing effort is constrained by the size of the labour force ($E \leq L$). This leads to an increase in the fish stock (Fig. A.2), while a high profit rate leads to gradual expansion in the labour force and a corresponding increase in the fishing effort level. However, once the fish stock has grown to a certain extent, effort must decrease in order to maintain the desired harvest. When, in year 5, the optimal rate of harvest rises, this produces jumps in both the effort and the natural labour force. Labour and the fish stock then gradually adjust towards a new long-term equilibrium.

The second scenario involves an objective function that places equal weight on rent generation (π) and community health (dL/dt). Thus, the goal is to maximise the present value of this equally weighted sum. There are assumed to be relatively slow dynamics and an initially large fish stock. In this case, the optimal harvest level changes to a greater extent over time, producing fluctuations in both the fish stock and the labour force. Initially, the large fish stock and low labour force produce incentives for entry into the fishery, indicated by the high natural labour force $L(t)$. This leads to expansion of both $L(t)$ and the fishing effort $E(t)$ (Fig. A.3), as well as a corresponding decline in the fish stock, and thus a decrease in the desirability of the

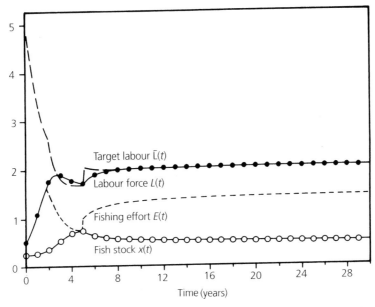

Fig. A.2 Approximately optimal dynamics of the fish stock $x(t)$, effort $E(t)$, labour $L(t)$ and target labour force $\bar{L}(t)$ for a 30-year time horizon, beginning with a relatively low fish stock. Maximum intrinsic growth rates are $s=3$ for the fish population and $r=1$ for the labour force. The optimal harvest is 55% of the biomass carrying capacity for the first 5 years and 75% thereafter. Note that the constraint $E(t)<L(t)$ restricts the fishing effort, and hence the actual harvest level, in the first few years until the labour force has increased sufficiently.

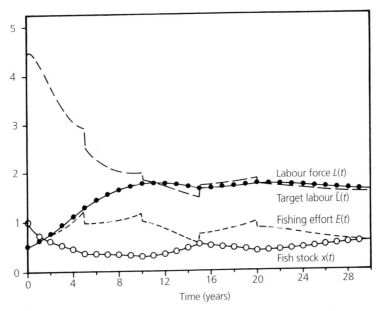

Fig. A.3 Approximately optimal dynamics of the fish stock $x(t)$, effort $E(t)$, labour $L(t)$ and target labour force $\bar{L}(t)$ for a 30-year time horizon, beginning with a relatively strong fish stock. Maximum intrinsic growth rates of $s=1.5$ for the fish stock and $r=0.25$ for the labour force are lower than those used in Fig. A.1, producing a more gradual approach to equilibrium. However, shifts in the desired harvest, which ranges from 45% to 35% of the biomass carrying capacity, lead to periodic jumps in effort E and target labour $\bar{L}$, and corresponding disequilibrium dynamics in the labour force L.

Mathematical details of the multi-objective optimisation

The optimisation challenge is based on balancing the four objectives described in this example, which are given mathematically below.

(1) The fishery rents, denoted by $\pi = \pi(x,L,E) = p(qEx) - c_E E - c_L L$ where the various parameters and variables are as in the boxes of the behavioural example above.
(2) The per capita income in the fishery (and the local economy), given by $(\pi/L - Y)$ where π/L actually represents per capita income net of labour opportunity costs, and Y is the comparable per capita income in the external economy. In general, Y may depend on such external factors as changing unemployment rates, although it is treated as a constant here.
(3) The employment rate E/L, as described in the behavioural example. (Note that other options may be of interest as an employment objective. For example, the level of fishing effort $E(t)$ provides a clear and direct measure of actual fishery employment. An alternative goal relating to equity in employment opportunities involves minimising discrepancies in employment rates between the fishery-dependent economy and the remainder of the economy, perhaps measured by the squared difference between the fishery employment rate and that pertaining in the external economy, $[(E/L) - R]^2$.)
(4) Community health is modelled here by the rate of change of the fishery labour force over time, dL/dt. This is meant to reflect the idea that in a fishery-dependent area, a declining labour force implies a declining population and a less prosperous community. (Of course, this may or may not be realistic in any given case.)

Suppose that, for simplicity, we assume that the objective function is simply a 'weighted sum' of these four objectives, with each multiplied by a constant weighting term (a_1, a_2, a_3 and a_4, respectively) to indicate its importance relative to the other goals. (Since the four objectives are measured in very different units (money, money/unit labour, time and labour), some care is needed in combining them within one objective function; this is discussed in Charles 1989.) Given this form of a multi-objective function, our overall goal is to choose a fishing effort policy $\{E_t\}$ to maximise a discounted sum over time:

$$\underset{\{E_t\}}{\text{Maximise}} \int_{t=0}^{\infty} e^{-\delta t} [a_1(\pi_t) + a_2(\pi/L_t - Y) + a_3(E_t/L_t) + a_4(dL/dt)]$$

where $e^{-\delta t}$ is a discount factor that decreases (from 1 toward 0) as t increases (reflecting the discounting process), and the summation is given by an integral (a continuous form of summation) over the desired time horizon of the optimisation, written here as infinity. Using the method of *optimal control theory*, this optimisation model can be solved to produce a 'modified golden rule' equation (Clark 1985, 1990), the solution of which indicates the optimal equilibrium harvest and stock size. Such equations have become common in the theory of resource management, but their analysis can involve extensive mathematics, and such an analysis is not presented here (see Charles 1989, for details).

Assumptions and parameter values of the optimisation example

In seeking the 'optimal' pattern of fishing effort for the fishery system, the set of possibilities will be subject to certain constraints. In this specific case, the possible options are subject to the constraints listed below.

(1) Only harvest policies resulting in a long-term sustainable fish stock are allowable; sustainability and conservation of the resource is imposed as a firm constraint.

(2) The overall income level must be at least 0 in any period, assuming that some positive income is needed at every time step for a strategy to be acceptable.

(3) The per capita rate of decline in the labour force (and fishery population) cannot exceed 20%, reflecting the idea that too fast a decline is politically unacceptable.

(4) Fishing effort E cannot exceed the available labour force L; $E(t) \leq L(t)$ at each time t.

(5) To simplify the calculations required, the harvest policy is limited to choosing a set of five harvest levels, one for each of years 0–5, years 5–10, years 10–15, years 15–20 and years 20–30. This is specified to avoid having to search over effort levels in each of 30 years.

The scenarios considered focus on varying the intrinsic growth rates (s, r), the initial conditions and the management objectives, holding other parameters fixed at arbitrary values $(K=1, p=1, q=1, c_E=0.01, c_L=0, T=2, M=2.12, \alpha=\beta=0.50, \delta=10\%)$. Note that in the first scenario, it is assumed that only rents are being maximised, so $a_2=a_3=a_4=0$. In addition, there are relatively fast adjustment dynamics for both the fish stock and the labour force $(s=3$ and $r=1)$ and there is a relatively small initial fish stock size, $x(0)=0.25$. In the second scenario, rents and community health are equally weighted $(a_1=a_4)$ and $a_2=a_3=0$. There are relatively slow adjustment dynamics for the fish stock and the labour force $(s=1.5$ and $r=0.25)$ and a high initial stock, $x(0)=1$.

fishery (L). Eventually, the actual labour force (L) rises to meet its target level, and thereafter 'tracks' changes in the target L caused by jumps in the harvest level.

These scenarios illustrate how the labour force L is continually approaching its target level, albeit with this 'tracking' being subject to delay. In addition, since periodic jumps are assumed to occur in the harvest, the optimal fish stock dynamics are not monotonic (always in one direction), but under- and overshooting of the eventual stock equilibrium can occur.

Analysis

The models discussed here illustrate a biosocioeconomic approach to analysing the joint ecological and socioeconomic dynamics inherent in fishery systems. This involves the determination of appropriate adjustment processes to predict the response of fish stocks and of fishers to changing conditions in the fishery, and the use of these dynamics in multi-objective management of fishery harvests. In the models discussed above, objectives of rent maximisation, labour force size and stability, employment in the fishery-dependent economy, and per capita income level are included explicitly in the optimisation process.

While a basic drawback of simulation methods is the inability to draw general conclusions, the above examples do demonstrate the complexity of dynamic behaviour in fishery systems. Computer simulation of these examples shows the dynamics of both the labour force and its target or 'natural' level, together with the gradual adjustment of the former to the latter. The analysis showed, for example, how approximately optimal harvest levels vary over time, and how labour dynamics can depend significantly on the constraints placed on fishing effort, in particular if effort is restricted by the size of the labour force itself.

References

Abdullah, N.M.R. & Kuperan, K. (1997) Fisheries management in Asia: The way forward. *Marine Resource Economics*, **12**, 345–353.

Acheson, J.M. (1975) The lobster fiefs: Economic and ecological effects of territoriality in the Maine lobster fishery. *Human Ecology*, **3**, 183–207.

Acheson, J.M. (1989) Where have all the exploiters gone? Co-management of the Maine lobster fishery. In: *Common Property Resources: Ecology and Community-Based Sustainable Development* (ed. F. Berkes), pp. 199–217. Bellhaven Press, London.

Allen, P.M. & McGlade, J.M. (1986) Dynamics of discovery and exploitation: The case of the Scotian Shelf groundfish fisheries. *Canadian Journal of Fisheries and Aquatic Sciences*, **43**, 1187–1200.

Andersen, P. & Sutinen, J.G. (1984) Stochastic bioeconomics: A review of basic methods and results. *Marine Resource Economics*, **1**, 117–136.

Andersen, R. (1978) The need for human sciences research in Atlantic coast fisheries. *Journal of the Fisheries Research Board of Canada*, **35**, 1031–1049.

Andersen, R. (ed.) (1979) *North Atlantic Maritime Cultures: Anthropological Essays on Changing Adaptations.* Mouton, The Hague.

Anderson, L.G. (1986) *The Economics of Fisheries Management.* Johns Hopkins University Press, Baltimore, MD.

Anderson, L.G. (1987) A management agency perspective of the economics of fisheries regulation. *Marine Resource Economics*, **4**, 123–131.

Angel, J.R., Burke, D.L., O'Boyle, R.N., Peacock, F.G., Sinclair, M. & Zwanenburg, K.C.T. (1994) *Report of the Workshop on Scotia–Fundy Groundfish Management from 1977 to 1993.* Canadian Technical Report of Fisheries and Aquatic Sciences No. 1979.

Anonymous (1997) *Statement of Conclusions. Intermediate Ministerial Meeting on the Integration of Fisheries and Environmental Issues. 13–14 March 1997. Bergen Norway.* Ministry of Environment, Norway. Oslo.

Anonymous (1998) *Workshop on the Ecosystem Approach to the Management and Protection of the North Sea. Oslo, Norway 15–17 June 1998.* TemaNord 1998: 579. Nordic Council of Ministers, Copenhagen.

Apostle, R., Kasdan, L. & Hanson, A. (1985) Work satisfaction and community attachment among fishermen in southwest Nova Scotia. *Canadian Journal of Fisheries and Aquatic Sciences*, **42**, 256–267.

Apostle, R., Barrett, G., Holm, P., Jentoft, S., Mazany, L., McCay, B., *et al.* (1998) *Community, State, and Market on the North Atlantic Rim: Challenges to Modernity in the Fisheries.* University of Toronto Press, Toronto.

Arico, S. (1998) The ecosystem approach: Evolution of the concept and experiences within the context of the Convention on Biological Diversity. In: *Workshop on the Ecosystem Approach to the Management and Protection of the North Sea, Oslo Norway, 15–17 June 1998*, pp. 40–45. Nordic Council of Ministers, Copenhagen.

Arnason, R. (1994) On catch discarding in fisheries. *Marine Resource Economics*, **9**, 189–208.

Arnason, R. (1996) On the ITQ fisheries management system in Iceland. *Reviews in Fish Biology and Fisheries*, **6**, 63–90.

Arrizaga, A., Buzeta, R. & Fierro, W. (1989) The necessity of integrated coastal development. In: *Artisanal Fisheries: Towards Integrated Coastal Development* (ed. A. Arrizaga). International Development Research Centre, Ottawa.

Atkinson, G., Dubourg, R., Hamilton, K., Munasinghe, M., Pearce, D., Young, C. (1997) *Measuring Sustainable Development: Macroeconomics and the Environment*. Edward Elgar, Cheltenham.

Bailey, C. (1988) The political economy of fisheries development. *Agriculture and Human Values* **5**, 35–48.

Bailey, C. (1997) Lessons from Indonesia's 1980 trawler ban. *Marine Policy*, **21**, 225–235.

Baldursson, E.B. (1984) Work stress in the fishing transformation plants. In: *Labor Developments in the Fishing Industry* (ed. J.L. Chaumel), pp. 26–29. Canadian Special Publication of Fisheries and Aquatic Sciences 72. Department of Fisheries and Oceans, Government of Canada, Ottawa.

Ballantine, W.J. (1994) The practicality and benefits of a marine reserve network. In: *Limiting Access to Marine Fisheries: Keeping the Focus on Conservation* (ed. K.L. Gimbel), pp. 205–223. Center for Marine Conservation and World Wildlife Fund. Washington, DC.

Bergh, J.C.J.M. van den (1996) *Ecological Economics and Sustainable Development: Theory, Methods and Applications*. Edward Elgar, Cheltenham.

Berkes, F. (1986) Local-level management and the commons problem: A comparative study of Turkish coastal fisheries. *Marine Policy*, **10**, 215–229.

Berkes, F. (ed.) (1989) *Common Property Resources: Ecology and Community-Based Sustainable Development*. Bellhaven Press, London.

Berkes, F. (1999) *Sacred Ecology: Traditional Ecological Knowledge and Resource Management*. Taylor and Francis, Philadelphia, PA.

Berkes, F. & Farvar, M.T. (1989) Introduction and overview. In: *Common Property Resources: Ecology and Community-Based Sustainable Development* (ed. F. Berkes). Belhaven Press, London.

Berkes, F. & Folke, C. (1992) A systems perspective on the interrelations between natural, human-made and cultural capital. *Ecological Economics*, **5**, 1–8.

Berkes, F. & Folke, C. (eds) (1998) *Linking Social and Ecological Systems: Management Practices and Social Mechanisms for Building Resilience*. Cambridge University Press, Cambridge.

Berkes, F., Feeny, D., McCay, B.J. & Acheson, J.M. (1989) The benefits of the commons. *Nature*, **340**, 91–93.

Beverton, R.J.H. & Holt, S.J. (1957) *On the Dynamics of Exploited Fish Populations*. Ministry of Agriculture, Fisheries and Food, London.

Bogason (1998) Managing fisheries: the Iceland experience. In: *Managing Our Fisheries, Managing Ourselves* (eds L. Loucks, A.T. Charles & M. Butler), pp. 7–15. Gorsebrook Research Institute, Saint Mary's University, Halifax.

Borgese, E.M. (1995) *Ocean Governance and the United Nations*. Centre for Foreign Policy Studies, Dalhousie University, Halifax.

Borgese, E.M. (1998) *The Oceanic Circle: Governing the Seas as a Global Resource*. United Nations University Press, Tokyo.

Botsford, L.W., Castilla, J.C. & Peterson, C.H. (1997) The management of fisheries and marine ecosystems. *Science*, **277**, 509–515.

Bromley, D.W. (1989) *Economic Interests and Institutions: The Conceptual Foundations of Public Policy*. Basil Blackwell, New York.

Bromley, D.W. (ed.) (1992) *Making the Commons Work: Theory, Practice and Policy*. Institute for Contemporary Studies, San Francisco, CA.

Brownstein, J. & Tremblay, J. (1994) Traditional property rights and cooperative management in the Canadian lobster fishery. *Lobster Newsletter*, **7**, 5.

Butterworth, D., Cochrane, K. & De Oliveira, J. (1997) Management procedures: A better way to manage fisheries? The South African experience. In: *Global Trends: Fisheries Management* (eds E.L. Pikitch, D.D. Huppert & M.P. Sissenwine), pp. 83–90. Symposium 20, American Fisheries Society, Bethesda, MD.

Caddy, J.F. (1984) Indirect approaches to regulation of fishing effort. In: *Papers Presented at the Expert Consultation on the Regulation of Fishing Effort (Fishing Mortality)*. FAO Fisheries Technical Report No. 289 (Suppl. 2). Food and Agriculture Organisation, Rome.

Caddy, J.F. & Griffiths, R.C. (1995) *Living Marine Resources and their Sustainable Development: Some Environmental and Institutional Perspectives*. FAO Fisheries Technical Paper No. 353. Food and Agriculture Organisation, Rome.

Caddy, J.F. & Sharp, G.D. (1986) *An Ecological Framework for Marine Fishery Investigations*. FAO Fisheries Technical Paper No. 283. Food and Agriculture Organisation, Rome.

Calhoun, S. (1991) *A Word to Say: The Story of the Maritime Fishermen's Union*. Nimbus, Halifax, Canada.

CCAMLR (Commission for the Conservation of Antarctic Marine Living Resources) (1999) Internet site. (www.ccamlr.org).

Chapman, M.D. (1987) Women's fishing in Oceania. *Human Ecology*, **15**, 267–288.

Charles, A.T. (1983a) Optimal fisheries investment: Comparative dynamics for a deterministic seasonal fishery. *Canadian Journal of Fisheries and Aquatic Sciences*, **40**, 2069–2079.

Charles, A.T. (1983b) Optimal fisheries investment under uncertainty. *Canadian Journal of Fisheries and Aquatic Sciences*, **40**, 2080–2091.

Charles, A.T. (1988) Fishery socioeconomics: A survey. *Land Economics*, **68**, 276–295.

Charles, A.T. (1989) Bio-socio-economic fishery models: labor dynamics and multi-objective management. *Canadian Journal of Fisheries and Aquatic Sciences*, **46**, 1313–1322.

Charles, A.T. (1991a) Small-scale fisheries in North America: Research perspectives. In: *Proceedings of the Conference on Research in Small-Scale Fisheries* (eds J.-R. Durand, J. Lemoalle & J. Weber), pp. 157–184. Editions de l'ORSTOM, Paris.

Charles, A.T. (1991b) *Fisheries Management and Development: A Policy and Strategic Assessment.* International Centre for Ocean Development, Halifax, Canada.

Charles, A.T. (1992a) Uncertainty and information in fishery management models: A Bayesian updating algorithm. *American Journal of Mathematical and Management Sciences*, **12**, 191–225.

Charles, A.T. (1992b) Fishery conflicts: A unified framework. *Marine Policy*, **16**, 379–393.

Charles, A.T. (1992c) Canadian fisheries: Paradigms and policy. In: *Canadian Ocean Law and Policy* (ed. D. VanderZwaag), pp. 3–26. Butterworths, Markham, Canada.

Charles, A.T. (ed.) (1993) *Fishery Enforcement: Economic Analysis and Operational Models*. Oceans Institute of Canada, Halifax, Canada.

Charles, A.T. (1994) Towards sustainability: The fishery experience. *Ecological Economics*, **11**, 201–211.

Charles, A.T. (1995a) Fishery science: The study of fishery systems. *Aquatic Living Resources*, **8**, 233–239.

Charles, A.T. (1995b) The Atlantic Canadian groundfishery: Roots of a collapse. *Dalhousie Law Journal*, **18**, 65–83.

Charles, A.T. (1995c) Sustainability assessment and bio-socio-economic analysis: Tools for integrated coastal development. In: *Philippine Coastal Resources Under Stress* (eds M.A. Juinio-Meñez & G. Newkirk), pp. 115–125. Coastal Resources Research Network, Halifax, Canada, and Marine Science Institute, Quezon City, Philippines.

Charles, A.T. (1997a) Fisheries management in Atlantic Canada. *Ocean and Coastal Management*, **35**, 101–119.

Charles, A.T. (1997b) *Sustainability Indicators: An Annotated Bibliography with Emphasis on Fishery Systems, Coastal Zones and Watersheds*. Strategy for International Fisheries Research, Ottawa.

Charles, A.T. (1997c) The path to sustainable fisheries. In: *Peace in the Oceans: Ocean Governance and the Agenda for Peace. Proceedings of Pacem in Maribus XXIII, Costa Rica, 3–7 December 1995* (ed. E.M. Borgese), pp. 201–213. Intergovernmental Oceanographic Commission Technical Series 47. UNESCO, Paris.

Charles, A.T. (1998a) Living with uncertainty in fisheries: Analytical methods, management priorities and the Canadian groundfishery experience. *Fisheries Research*, **37**, 37–50.

Charles, A.T. (1998b). Beyond the status quo: Re-thinking fishery management. In: *Re-inventing Fisheries Management* (eds T.J. Pitcher, P.J.B. Hart & D. Pauly). Kluwer, Dordrecht.

Charles, A.T. (1998c) *Fisheries in Transition*. Ocean Yearbook 13 (eds E.M. Borgese, A. Chircop, M. McConnell & J.R. Morgan), pp. 15–37. University of Chicago Press, Chicago, IL.

Charles, A.T. & Herrera, A. (1994) Development and diversification: Sustainability strategies for a Costa Rican fishing cooperative. In: *Proceedings of the 6th Conference of the International Institute for Fisheries Economics and Trade* (eds M. Antona, J. Catanzano & J.G. Sutinen), pp. 1315–1324. IIFET/ORSTOM, Paris.

Charles, A.T. & Lavers, A. (2000) *Sustainability Assessment for Nova Scotia's Fisheries: Toward a Genuine Progress Index*. GPI Atlantic, Halifax, Canada.

Charles, A.T. & Leith, J. (1999) *Social and Economic Approaches to Biodiversity Conservation: An Annotated Bibliography with Emphasis on Aquaculture*. International Development Research Centre, Ottawa.

Charles, A.T., Brainerd, T.R., Bermudez, A., Montalvo, H.M. & Pomeroy, R.S. (1994). *Fisheries Socioeconomics in the Developing World: Regional Assessments and an Annotated Bibliography*. International Development Research Centre, Ottawa.

Charles, A.T., Amaratunga, C., Amaratunga, T. & Herrera, A. (1996). *Marine and Coastal Research in Costa Rica: Current State and Policies for the Future.* Publicaciones UNA, Heredia, Costa Rica [in Spanish].

Charles, A.T., Mazany, R.L & Cross, M.L. (1999) The economics of illegal fishing: A behavioral model. *Marine Resource Economics*, **14**, 95–110.

Chaumel, J.-L. (ed.) (1984) *Labor Developments in the Fishing Industry.* Canadian Special Publication of Fisheries and Aquatic Sciences, No. 72. Department of Fisheries and Oceans, Government of Canada, Ottawa.

Chen, Y. & Paloheimo, J.E. (1995) A robust regression analysis of recruitment in fisheries. *Canadian Journal of Fisheries and Aquatic Sciences*, **52**, 993–1006.

Chesson, J. & Clayton, H. (1998) *A Framework for Assessing Fisheries with respect to Ecologically Sustainable Development.* Bureau of Rural Sciences, Canberra.

Christensen, S., Lassen, H., Nielsen, J.R. & Vedsmand, T. (1999) Effort regulation: A viable alternative in fisheries management? In: *Alternative Management Systems for Fisheries* (ed. D. Symes), pp. 179–187. Fishing News Books, Blackwell Science, Oxford.

Christensen, V. & Pauly, D. (1996) Ecological modeling for all. *Naga: The ICLARM Quarterly*, **19**, 25–26.

Christy, F.T. (1982) *Territorial Use Rights in Marine Fisheries: Definitions and Conditions.* FAO Fisheries Technical Paper No. 227. FAO, Rome.

CIDA (Canadian International Development Agency) (1993) *Women and Fisheries Development.* Supply and Services Cat. No. E94-219/1993. Government of Canada, Ottawa.

Clark, C.W. (1973) The economics of overexploitation. *Science*, **181**, 630–634.

Clark, C.W. (1985) *Bioeconomic Modelling and Fisheries Management.* Wiley-Interscience, New York.

Clark, C.W. (1990) *Mathematical Bioeconomics: The Optimal Management of Renewable Resources*, 2nd edn. Wiley-Interscience, New York.

Clark, C.W., Clarke, F.H. & Munro, G.R. (1979) The optimal exploitation of renewable resource stocks: Problems of irreversible investment. *Econometrica*, **47**, 25–49.

Clark, I.N., Majors, P.J. & Mollett, N. (1988) Development and implementation of New Zealand's ITQ management system. *Marine Resource Economics*, **5**, 325–350.

Clark, J.R. (1996) *Coastal Zone Management Handbook.* CRC Lewis, Boca Raton, FL.

Clement, W. (1984) Canada's coastal fisheries: Formation of unions, cooperatives and associations. *Journal of Canadian Studies*, **19**, 5–33.

Cochrane, K.L. (1999) Complexity in fisheries and limitations in the increasing complexity of fisheries management. *ICES Journal of Marine Science*, **56**, 917–926.

Cochrane, K.L., Butterworth, D.S., de Oliveira, J.A.A. & Roel, B.A. (1998). Management procedures in a fishery based on highly variable stocks and with conflicting objectives: Experiences in the South African pelagic fishery. *Reviews in Fish Biology and Fisheries*, **8**, 177–214.

Collie, J.S. & Sissenwine, M.P. (1983) Estimating population size from relative abundance data measured with error. *Canadian Journal of Fisheries and Aquatic Sciences*, **40**, 1871–1879.

Collie, J.S. & Spencer, P.D. (1994) Modeling predator–prey dynamics in a fluctuating environment. *Canadian Journal of Fisheries and Aquatic Sciences*, **51**, 2665–2672.

Commission of the European Communities (ed.) (1994) *European Directory of Research Centers in the Fisheries Sector.* Kluwer, Dordrecht.

Committee on Biological Diversity in Marine Systems (1995). *Understanding Marine Biodiversity: A Research Agenda for the Nation.* Ocean Studies Board, National Research Council. National Academy Press, Washington, DC.

Committee on Fish Stock Assessment Methods (1998) *Improving Fish Stock Assessments.* National Academy Press, Washington, DC.

Committee to Review the Community Development Quota Program (1999) *The Community Development Quota Program in Alaska.* National Research Council. National Academy Press, Washington, DC.

Congress of the Philippines (1998) *An Act providing for the Development, Management, and Conservation of the Fisheries and Aquatic Resources, Integrating all Laws Pertinent thereto, and for other Purposes.* Republic Act No. 8550, Congress of the Philippines, Republic of the Philippines. Metro Manila, Philippines.

Copes, P. (1986) A critical review of the individual quota as a device in fisheries management. *Land Economics*, **62**, 278–291.

Copes, P. (1994) Individual fishing rights: Some implications of transferability. In: *Proceedings of the Sixth Confer-*

ence of the International Institute of Fisheries Economics and Trade (eds M. Antona, J. Catanzano & J.G. Sutinen), pp. 907–916. IIFET/ORSTOM, Paris.

Copes, P. (1995) Problems with ITQs in fisheries management, with tentative comments on relevance for Faroe Islands fisheries. In: *Nordiske Fiskersamfund i fremtiden (Nordic Fishing Communities in the Future)* (ed. S.T.F. Johansen), pp. 19–40. Tórshavn, Faroe Islands. Nordic Council of Ministers, Copenhagen.

Copes, P. (1997a) Adverse impacts of individual quota systems on conservation and fish harvest productivity. *Proceedings of the Eighth Conference of the International Institute of Fisheries Economics and Trade.* Keynote address, Marrakech, Morocco, July 1996. Institut scientifique des pêches maritimes, Casablanca, Morocco.

Copes, P. (1997b) Social impacts of fisheries management regimes based on individual quotas. *Proceedings of the Workshop on Social Implications of Quota Systems in Fisheries.* Vestman Islands, Iceland, May 1996. Nordic Council of Ministers, Copenhagen.

Costanza, R., d'Arge, R., de Groot, R., Farber, S., Grasso, M., Hannon, B., *et al.* (1997a) The value of the world's ecosystem services and natural capital. *Nature,* **387,** 253–260. (Also reprinted in *Ecological Economics,* **25,** 3–15, 1998.)

Costanza, R., Cumberland, J., Daly, H., Goodland, R. & Norgaard, R. (1997b) *An Introduction to Ecological Economics.* St. Lucie Press, CRC Press, Boca Raton, FL.

Coull, J.R. (1993) *World Fisheries Resources.* Routledge, London.

Crean, K. & Symes, D. (eds) (1996) *Fisheries Management in Crisis.* Fishing News Books, Blackwell Science, Oxford.

Criddle, K.R. (1996) Predicting the consequences of alternative harvest regulations in a sequential fishery. *North American Journal of Fisheries Management,* **16,** 30–40.

Cunningham, S., Dunn, M.R. & Whitmarsh, D. (1985) *Fisheries Economics: An Introduction.* Mansell, London.

Cushing, D. (1975) *Fisheries Resources of the Sea and their Management.* Oxford University Press, Oxford.

Daan, N. (1987) Multispecies versus single-species assessment of North Sea fish stocks. *Canadian Journal of Fisheries and Aquatic Sciences,* **44** (Suppl. 2), 360–370.

Daan, N. (1997) Multispecies assessment issues for the North Sea. In: *Global Trends: Fisheries Management* (eds E.L. Pikitch, D.D. Huppert & M.P. Sissenwine), pp. 126–133. Symposium 20, American Fisheries Society, Bethesda, MD.

Daan, N. & Sissenwine, M.P. (1991) *Multispecies Models Relevant to Management of Living Resources.* ICES Marine Sciences Symposium, Vol. 193. International Council for the Exploration of the Sea (ICES), Copenhagen.

Danielsson, A. (1997) Fisheries management in Iceland. *Ocean & Coastal Management* **35,** 121–135.

Davis, A. (1984) Property rights and access management in the small-boat fishery: A case study from southwest Nova Scotia. In: *Atlantic Fisheries and Coastal Communities: Fisheries Decision-Making Case Studies* (eds C. Lamson & A. Hanson), pp. 133–164. Dalhousie Ocean Studies Program, Halifax, Canada.

Davis, D.L. (1988) 'Shore skippers' and 'grass widows': Active and passive women's roles in a Newfoundland fishery. In: *To Work and To Weep: Women in Fishing Economies* (eds J. Nadel-Klein & D.L. Davis), pp. 211–229. Social and Economic Papers No. 18, Institute of Social and Economic Research, Memorial University of Newfoundland, St. John's.

Department of Fisheries and Oceans (1992) *Crosbie Announces First Steps in Northern Cod (2J3KL) Recovery Program.* Press release No. NR-HQ-92-58E, 2 July 1992. Ottawa.

Department of Fisheries and Oceans (1996) *Preliminary Landed Values Canada 1995* (unpublished report updated 15 October 1996). DFO Marketing and Economics, Ottawa.

Deriso, R.B. (1980) Harvesting strategies and parameter estimation for an age-structured model. *Canadian Journal of Fisheries and Aquatic Sciences,* **37,** 268–282.

Dewees, C.M. (1998) Effects of individual quota systems on New Zealand and British Columbia fisheries. *Ecological Applications,* **8** (Suppl.), S133–S138.

Doeringer, P.B. & Terkla, D.G. (1995) *Troubled Waters: Economic Structure, Regulatory Reform, and Fisheries Trade.* University of Toronto Press, Toronto.

Doeringer, P.B., Moss, P.I. & Terkla, D.G. (1986) *The New England Fishing Economy: Jobs, Income, and Kinship.* University of Massachusetts Press, Amherst, MA.

Dubbink, W. & van Vliet, M. (1996) Market regulation versus co-management? Two perspectives on regulating fisheries compared. *Marine Policy,* **20,** 499–516.

Dudley, N. & Waugh, G. (1980) Exploitation of a single-cohort fishery under risk: a simulation–optimization approach. *Journal of Environmental Economics and Management*, **7**, 234–255.

Dugan, J.E. & Davis, G.E. (1993) Applications of marine refugia to coastal fisheries management. *Canadian Journal of Fisheries and Aquatic Sciences*, **50**, 2029–2042.

Dunn, E. (1996) Fisheries and the development of sustainability indicators. *North Sea Monitor*, June 1996, 13–19.

Durand, J.R., Lemoalle, J. & Weber, J. (eds) (1991*) La Recherche Face à la Pêche Artisanale (Research and Small-Scale Fisheries).* Editions de l'ORSTOM, Paris.

Dyer, C.L. & McGoodwin, J.R. (1994) *Folk Management in the World's Fisheries: Lessons for Modern Fisheries Management.* University Press of Colorado, Niwot, CO.

Eglington, A., Israel, R. & Vartanov, R. (1998) Towards sustainable development for the Murmansk region. *Ocean & Coastal Management*, **41**, 257–271.

European Commission (1998) *The European Union Fishing Fleet in 1995.* Directorate General XIV, Fisheries. European Commission, Brussels.

European Commission (1999) Directorate General XIV, Fisheries. European Commission, Brussels (http://europa.eu.int/comm/dg14).

Eurostat (1998) *European Fisheries in Figures.* Directorate for Agriculture, Environment and Energy Statistics, Eurostat, Luxembourg.

FAO (Food and Agriculture Organization) (1983) *Report of the Expert Consultation on the Regulation of Fishing Effort (Fishing Mortality).* Fisheries Technical Report No. 289. FAO, Rome.

FAO (Food and Agriculture Organization) (1984) *Report of the FAO World Conference on Fisheries Management and Development.* FAO, Rome.

FAO (Food and Agriculture Organization) (1985) *Expert Consultation on the Acquisition of Socio-economic Information in Fisheries.* FAO Fisheries Report No. 344. FAO, Rome.

FAO (Food and Agriculture Organization) (1995a) *Precautionary Approach to Fisheries. Part 1. Guidelines on the Precautionary Approach to Capture Fisheries and Species Introductions.* FAO Fisheries Technical Paper No. 350, Part 1. FAO, Rome.

FAO (Food and Agriculture Organization) (1995b) *Code of Conduct for Responsible Fisheries.* FAO, Rome.

FAO (Food and Agriculture Organization) (1997a) *Review of the State of World Fishery Resources: Marine Fisheries.* FAO Fisheries Circular No. 920 FIRM/C920. Marine Resources Service, Fishery Resources Division, Fisheries Department, FAO, Rome (www.fao.org/fi/publ/circular/c920/intro.asp#A2).

FAO (Food and Agriculture Organization) (1997b) *Fisheries Management.* FAO Technical Guidelines for Responsible Fisheries No. 4. Fishery Resources Division and Fishery Policy and Planning Division, FAO, Rome.

FAO (Food and Agriculture Organization) (1999a) *FAO Glossary.* Fisheries Department, FAO, Rome (www.fao.org/fi/glossary/glossary.asp).

FAO (Food and Agriculture Organization) (1999b) *Fishery Softwares: BEAM 4 Analytical Bioeconomic Simulation of Space-Structured Multispecies and Multifleets Fisheries.* FAO of the United Nations, Rome (www.fao.org/fi/statist/fisoft/beam4.asp).

FAO (Food and Agriculture Organization) (1999c) *Major Issues Affecting the Performance of Regional Fishery Bodies.* Meeting of FAO and non-FAO regional fishery bodies or arrangements, Rome, 11–12 February 1999. FI:RFB/99/2. FAO of the United Nations, Rome.

FAO (Food and Agriculture Organization) (1999d) *Indicators for Sustainable Development of Marine Capture Fisheries.* FAO Technical Guidelines for Responsible Fisheries No. 8. Fishery Resources Division, FAO, Rome.

Ferrer, E.M., dela Cruz, L.P. & Domingo, M.A. (eds) (1996) *Seeds of Hope: A Collection of Case Studies on Community-Based Coastal Resources Management in the Philippines.* College of Social Work and Community Development, University of the Philippines, Quezon City, Philippines.

Ferris, S. & Plourde C. (1982) Labour mobility, seasonal unemployment insurance, and the Newfoundland inshore fishery. *Canadian Journal of Economics*, **15**, 426–441.

Flaaten, O., Salvanes, A.G.V., Schweder, T. & Ulltang, Ø. (eds) (1998) Objectives and uncertainties in fisheries management with emphasis on three North Atlantic ecosystems: A selection of papers presented at an international symposium in Bergen, Norway, 3–5 June 1997. Special Issue, *Fisheries Research,* **37**, 1–310.

Fogarty, M.J., Hilborn, R. & Gunderson, D. (1997) Rejoinder: Chaos and parametric management. *Marine Policy*, **21**, 187–194.

Folke, C. & Berkes, F. (1995) Mechanisms that link property rights to ecological systems. In: *Property Rights and the Environment: Social and Ecological Issues* (eds S. Hanna & M. Munasinghe), pp. 121–137. Beijer International

Institute of Ecological Economics and the World Bank, Washington, DC.

Folke, C. & Berkes, F. (1998) *Understanding Dynamics of Ecosystem–Institution Linkages for Building Resilience.* Beijer Discussion Paper Series No.112. Beijer International Institute of Ecological Economics, Royal Swedish Academy of Sciences, Stockholm.

Folke, C., Hammer, M., Costanza, R. & Jansson, A. (1994) Investing in natural capital — why, what and how? In: *Investing in Natural Capital: The Ecological Economics Approach to Sustainability* (eds A. Jansson, M. Hammer, C. Folke & R. Costanza), pp. 1–20. Island Press, Washington, DC.

Folke, C., Kautsky, N., Berg, H., Jansson, A. & Troell, M. (1998) The ecological footprint concept for sustainable seafood production: A review. *Ecological Applications*, **8** (Suppl.), S63–S71.

Francis, R.I.C.C. (1992) Use of risk analysis to assess fishery management strategies: a case study using Orange Roughy (*Hoplostethus atlanticus*) on the Chatham Rise, New Zealand. *Canadian Journal of Fisheries and Aquatic Sciences*, **49**, 922–930.

Francis, R.I.C.C. & Shotton, R. (1997) 'Risk' in fisheries management: A review. *Canadian Journal of Fisheries and Aquatic Sciences*, **54**, 1699–1715.

Frangoudes, K. (1996) Fishermen households and fishing communities in Greece: A case study of Nea Michaniona. In: *Fisheries Management in Crisis* (eds K. Crean & D. Symes), pp. 119–128. Fishing News Books, Blackwell Science, Oxford.

Fraser River Sockeye Public Review Board (1995) *Fraser River Sockeye 1994: Problems and Discrepancies.* Public Works and Government Services Canada, Cat. No. Fs23-263/1995E, Ottawa.

FRCC (Fisheries Resource Conservation Council) (1994) *Conservation: Stay the Course – Report to the Minister of Fisheries and Oceans on 1995 Conservation Requirements for Atlantic Groundfish.* FRCC 94.R4E, Minister of Supply and Services, Ottawa (www.ncr.dfo.ca/frcc).

FRCC (Fisheries Resource Conservation Council) (1995) *A Conservation Framework for Atlantic Lobster. Report to the Minister of Fisheries and Oceans.* FRCC95.R.1, Cat. No. Fs 23-278/1995E, Supply and Services Canada, Ottawa (www.ncr.dfo.ca/frcc).

FRCC (Fisheries Resource Conservation Council) (1996) *Quota Controls and Effort Controls: Conservation Considerations.* Discussion Paper FRCC.96.TD.3. FRCC, Ottawa (www.ncr.dfo.ca/frcc).

FRCC (Fisheries Resource Conservation Council) (1997a) *A Groundfish Conservation Framework for Atlantic Canada.* FRCC.97.R.3. Ottawa (www.ncr.dfo.ca/frcc).

FRCC (Fisheries Resource Conservation Council) (1997b) *Georges Bank: 1997 Conservation Requirements for Georges Bank Groundfish Stocks. Report to the Minister of Fisheries and Oceans.* FRCC.97.R.2. FRCC, Ottawa (www.ncr.dfo.ca/frcc).

Fricke, P. (1985) Use of sociological data in the allocation of common property resources. *Marine Policy*, **9**, 39–52.

Friis, P. (1996) The European fishing industry: Deregulation and the market. In: *Fisheries Management in Crisis* (eds K. Crean & D. Symes), pp. 175–186. Fishing News Books, Blackwell Science, Oxford.

Fuller, S. & Cameron, P. (1998). *Marine Benthic Seascapes: Fishermen's Perspectives.* Ecology Action Centre, Halifax, Canada.

Furlong, W.J. (1991) The deterrent effect of regulatory enforcement in the fishery. *Land Economics*, **67**, 116–129.

Gallucci, V.F., Saila, S.B., Gustafson, D.J. & Rothschild, B.J. (eds) (1996) *Stock Assessment: Quantitative Methods and Applications for Small-Scale Fisheries.* CRC Lewis, Boca Raton, FL.

Gandhi, M. (1947) *India of My Dreams* (ed. R K. Prabhu). Bombay, India. (See p. 1 In: *The Oceanic Circle: Governing the Seas as a Global Resource* by E.M. Borgese. United Nations University Press, Tokyo, 1998.)

Garcia, S.M. (1994) The precautionary principle: Its implications in capture fisheries management. *Ocean and Coastal Management*, **22**, 99–125.

Garcia, S.M. & Newton, C. (1994) Current situation, trends and prospects in world capture fisheries. In: *Global Trends: Fisheries Management* (eds E.K. Pikitch, D.D. Huppert & M.P. Sissenwine). American Fisheries Society Symposium 20, Bethesda, MD.

Gimbel, K.L. (ed.) (1994) *Limiting Access to Marine Fisheries: Keeping the Focus on Conservation.* Center for Marine Conservation and World Wildlife Fund, Washington, DC.

Gleick, J. (1988) *Chaos: The Amazing Science of the Unpredictable.* Minerva, Random House, London.

Gonzalel, E. (1996) Territorial use rights in Chilean fisheries. *Marine Resource Economics* **11**, 211–218.

Gonzalez, A.L., Herrera, A.F., Villalobos, U.L., Breton, C.Y., Lopez, E., Breton, E.E., *et al.* (1993) *Comunidades*

Pesquero-Artesanales en Costa Rica. Editorial de la Universidad Nacional, Heredia, Costa Rica.

Gordon, H.S. (1954) The economic theory of a common property resource: The fishery. *Journal of Political Economy*, **62**, 124–142.

Gough, J. (1993) A historical sketch of fisheries management in Canada. In: *Perspectives on Canadian Marine Fisheries Management* (eds L.S. Parsons & W.H. Lear), pp. 5–53. Canadian Bulletin of Fisheries and Aquatic Sciences No. 226, National Research Council of Canada, Ottawa.

Government of Australia (1999) *Australia's Fisheries Resources*. Bureau of Rural Sciences, Government of Australia (www.brs.gov.au/fish/fishery.html#afz).

Government of Iceland (1999) *Fish Processing*. Ministry of Fisheries, Government of Iceland (brunnur.stjr.is/interpro/sjavarutv/english.nsf/pages/front).

Graham, J. (1998) *An evolving dynamic: Community participation in community-based coastal resource management in the Philippines*. MA thesis, Dalhousie University, Halifax, Canada.

Gulland, J.A. (ed.) (1977) *Fish Population Dynamics*. Wiley-Interscience, New York.

Gulland, J.A. (1982) The management of tropical multispecies fisheries. In: *Theory and Management of Tropical Fisheries, ICLARM Conference Proceedings 9* (eds D. Pauly & G.I. Murphy), pp. 287–298. International Center for Living Aquatic Resources Management, Manila, Philippines.

Gulland, J.A. (1983) *Fish Stock Assessment: A Manual of Basic Methods*. Wiley-Interscience, New York.

Gulland, J.A. (1987) The management of North Sea fisheries. *Marine Policy*, **11**, 259–272.

Gunderson, L., Holling, C.S. & Light, S. (eds) (1995) *Barriers and Bridges to the Renewal of Ecosystems and Institutions*. Columbia University Press, New York.

Gunnlaugsdottir, S.S. (1984) Restraint in workers' situation as determined by organization of the labor process in the fishing industry in Iceland. In: *Labor Developments in the Fishing Industry* (ed. J.L.Chaumel), pp. 23–25. Canadian Special Publication of Fisheries and Aquatic Sciences 72. Department of Fisheries and Oceans, Government of Canada, Ottawa.

Halliday, R.G., Peacock, F.G. & Burke, D.L. (1992) Development of management measures for the groundfish fishery in Atlantic Canada: A case study of the Nova Scotia inshore fleet. *Marine Policy*, **16**, 411–426.

Hammer, M. (1998) The adaption of an ecosystem approach to the conservation and sustainable use of biological diversity – the Malawi principles. In: *Workshop on the Ecosystem Approach to the Management and Protection of the North Sea, Oslo, Norway, 15–17 June 1998*, pp. 51–53. Nordic Council of Ministers, Copenhagen.

Hammer, M., Jansson, A. & Jansson, B. (1993) Diversity change and sustainability: Implications for fisheries. *Ambio*, **22**, 97–105.

Hammond, A., Adriaanse, A., Rodenburg, E., Bryant, D. & Woodward, R. (1995) *Environmental Indicators: A Systematic Approach to Measuring and Reporting on Environmental Policy Performance in the Context of Sustainable Development*. World Resources Institute, Washington, DC.

Hancock, D.A., Smith, D.C., Grant, A. & Beumer, J.P. (eds) (1997) *Developing and Sustaining World Fisheries Resources: The State of Science and Management. 2nd World Fisheries Congress Proceedings*. CSIRO, Collingwood, Australia.

Hanna, S. & Munasinghe, M. (eds) (1995a) *Property Rights and the Environment: Social and Ecological Issues*. Beijer International Institute of Ecological Economics and the World Bank, Washington, DC.

Hanna, S. & Munasinghe, M. (eds) (1995b) *Property Rights in a Social and Ecological Context: Case Studies and Design Applications*. Beijer International Institute of Ecological Economics and the World Bank, Washington, DC.

Hannesson, R. (1993) *Bioeconomic Analysis of Fisheries*. Halsted Press (Wiley), New York.

Hannesson, R. (1996) *Fisheries Mismanagement: The Case of the North Atlantic Cod*. Fishing News Books, Blackwell Science, Oxford.

Hardin, G. (1968) The tragedy of the commons. *Science*, **162**, 1243–1247.

Hatcher, A.C. (1997) Producers' organizations and devolved fisheries management in the United Kingdom: Collective and individual quota systems. *Marine Policy*, **21**, 519–533.

Healey, M.C. (1984) Multiattribute analysis and the concept of optimum yield. *Canadian Journal of Fisheries and Aquatic Sciences*, **41**, 1393–1406.

Herr, R. (1990) *The Forum Fisheries Agency: Achievements, Challenges and Prospects*. Institute of Pacific Studies, University of the South Pacific, Suva, Fiji.

Herrera, A. & Valerín, N. (1992) *Legislación pesquera de principal interés para los pescadores artesanales del Golfo de Nicoya*. Publicaciones UNA, Universidad Nacional, Heredia, Costa Rica.

Hersoug, B., Holm P. & Ranes S.A. (1999) Three challenges to the future of fisheries management in Norway: ITQs, regional co-management and eco-labelling. In: *Alternative Management Systems for Fisheries* (ed. D. Symes), pp. 136–144. Fishing News Books, Blackwell Science, Oxford.

Hightower, J.E. & Grossman, G.D. (1985) Comparison of constant effort harvest policies for fish stocks with variable recruitment. *Canadian Journal of Fisheries and Aquatic Sciences*, **42**, 982–988.

Hilborn, R. (1985) Fleet dynamics and individual variation: Why some fishermen catch more fish than others. *Canadian Journal of Fisheries and Aquatic Sciences*, **42**, 2–13.

Hilborn, R. (1987) Living with uncertainty in resource management. *North American Journal of Fisheries Management*, **7**, 1–5.

Hilborn, R. & Walters, C.J. (1992) *Quantitative Fisheries Stock Assessment: Choice, Dynamics and Uncertainty*. Chapman & Hall, New York.

Holland, D.S. & Brazee, R.J. (1996) Marine reserves for fisheries management. *Marine Resource Economics*, **11**, 157–171.

Holling, C.S. (1973) Resilience and stability of ecological systems. *Annual Review of Ecology and Systematics*, **4**, 1–23.

Holling, C.S. (ed.) (1978) *Adaptive Environmental Assessment and Management*. Wiley, New York.

Honoré, A.M. (1961) Ownership. In: *Oxford Essays in Jurisprudence* (ed. A.G. Guest), Chapter 5. Oxford University Press, Oxford.

Hourston, A.S. (1978) *The Decline and Recovery of Canada's Pacific Herring Stocks*. Fisheries and Marine Technical Report No. 784. Minister of Supply and Services, Cat. No. Fs 97-6/784, Ottawa.

Hutchings, J.A. & Myers, R.A. (1994) What can be learned from the collapse of a renewable resource? Atlantic cod, *Gadus morhua*, of Newfoundland and Labrador. *Canadian Journal of Fisheries and Aquatic Sciences*, **51**, 2126–2146.

Hutchings, J.A. & Myers, R.A. (1995) The biological collapse of Atlantic cod off Newfoundland and Labrador: An exploration of historical changes in exploitation, harvesting technology, and management. In: *The North Atlantic Fisheries: Successes, Failures and Challenges* (eds R. Arnason & L. Felt). Institute of Island Studies, Charlottetown, Canada.

Hutchings, J.A., Walters, C. & Haedrich, R.L. (1997) Is scientific inquiry incompatible with government information control? *Canadian Journal of Fisheries and Aquatic Sciences*, **54**, 1198–1210.

IIRR (1998) *Participatory Methods in Community-Based Coastal Resource Management*. International Institute for Rural Reconstruction, Silang, Philippines.

Independent World Commission on the Oceans (1998) *The Ocean, Our Future*. Cambridge University Press, Cambridge.

Inter-American Institute for Cooperation on Agriculture (1991) *Toward a Working Agenda for Sustainable Agricultural Development*. Paper DP-25. Inter-American Institute for Cooperation on Agriculture, San Jose, Costa Rica.

International Ocean Institute (1998) *The Halifax Declaration on the Ocean*. International Ocean Institute, Halifax, Canada.

International Whaling Commission (1994) The revised management procedure (RMP) for Baleen whales. *Reports of the International Whaling Commission*, **44**, 145–152.

Iverson, E.S. (1996) *Living Marine Resources: Their Utilization and Management*. Chapman & Hall, New York.

Jackson, A.R.W. & Jackson, J.M. (1996) *Environmental Science: The Natural Environment and Human Impact*. Longman, Harlow, UK.

Jansson, A., Hammer, M., Folke, C. & Costanza, R. (eds) (1994) *Investing in Natural Capital: The Ecological Economics Approach to Sustainability*. Island Press, Washington, DC.

Jentoft, S. (1985) Models of fishery development: The cooperative approach. *Marine Policy*, **9**, 322–331.

Jentoft, S. (1989) Fisheries co-management. *Marine Policy*, **13**, 137–154.

Jentoft, S. (1998) Social science in fisheries management: A risk assessment. In: *Re-inventing Fisheries Management* (eds T.J. Pitcher, P.J.B. Hart & D. Pauly), pp. 177–184. Kluwer, Dordrecht.

Jentoft, S. & McCay, B.J. (1995) User participation in fisheries management: Lessons drawn from international experience. *Marine Policy*, **19**, 227–246.

Jentoft, S. & Mikalsen, K.H. (1994) Regulating fjord fisheries: Folk management or interest group politics? In: *Folk*

Management in the World Fisheries (eds C.L. Dyer & J.R. McGoodwin), pp. 287–316. University of Colorado Press, Boulder, CO.

Jentoft, S., McCay, B.J. & Wilson, D.C. (1998) Social theory and fisheries co-management. *Marine Policy*, **22**, 423–436.

Johannes, R.E. (1978) Traditional marine conservation methods in Oceania and their demise. *Annual Review of Ecology and Systematics*, **9**, 349–364.

Jónsson, Ö.D. (1996) The geopolitics of fish: The case of the North Atlantic. In: *Fisheries Management in Crisis* (eds K. Crean & D. Symes), pp. 187–194. Fishing News Books, Blackwell Science, Oxford.

Josupeit, H. (1998) *Value-Added Products in Europe*. Food and Agriculture Organization of the United Nations, GLOBEFISH, Rome.

Kalland, A. (1996) Marine management in coastal Japan. In: *Fisheries Management in Crisis*. (eds K. Crean & D. Symes), pp. 71–83. Fishing News Books, Blackwell Science, Oxford.

Karagiannakos, A. (1995) *Fisheries Management in the European Union*. Avebury, Aldershot, UK.

Karagiannakos, A. (1996) Total allowable catch (TAC) and quota management system in the European Union. *Marine Policy* **20**, 235–248.

Kay, J.J. & Schneider, E. (1994) Embracing complexity: The challenge of the ecosystem concept. *Alternatives*, **20**, 32–39.

Kearney, J. (1984) The transformation of the Bay of Fundy herring fisheries, 1976–1978: An experiment in fishermen–government co-management. In: *Atlantic Fisheries and Coastal Communities: Fisheries Decision-Making Case Studies* (eds C. Lamson & A.J. Hanson), pp. 165–203. Dalhousie Ocean Studies Program, Halifax, Canada.

Kearney, J. (1992) *Diversity of labour processes, household forms, and political practice. A social approach to the inshore fishing communities of Clare, Digby Neck and the islands.* PhD thesis, Université Laval.

King, L.R. (1986) Science, politics and the Sea Grant College program. *Ocean Development and International Law*, **17**, 37–57.

King, M. (1995) *Fisheries Biology: Assessment and Management*. Fishing News Books, Blackwell Science, Oxford.

King, P.A., Elsworth, S.G. & Baker, R.F. (1994) Partnerships – The route to better communication. In: *Coastal Zone Canada '94, Cooperation in the Coastal Zone: Conference Proceedings* (eds P.G. Wells & P.J. Ricketts), pp. 596–611. Coastal Zone Canada Association, Dartmouth, Canada.

Kirby, M.J.L. (1983) *Navigating Troubled Waters: A New Policy for the Atlantic Fisheries*. Report of Task Force on Atlantic Fisheries. Supply and Services Canada, Ottawa.

Kirkwood, G.P. (1992) Background to the development of revised management procedures. Annex I. *Reports of the International Whaling Commission,* **42**, pp. 236–243.

Krauthamer, J.T., Grant, W.E. & Griffen, W.L. (1987) A sociobio-economic model: The Texas inshore shrimp fishery. *Ecological Modelling*, **35**, 275–307.

Kuik, O. & Verbruggen, H. (1991) *In Search of Indicators of Sustainable Development*. Kluwer, Dordrecht.

Kuperan, K. & Abdullah, N.M.R. (1994) Small-scale coastal fisheries and co-management. *Marine Policy*, **18**, 306–313.

Kurien, J. (1998) Traditional ecological knowledge and ecosystem sustainability: New meaning to Asian coastal proverbs. *Ecological Applications*, **8** (Suppl.), S2–S5.

Laevastu, T., Alverson, D.L. & Marasco, R.J. (1996) *Exploitable Marine Ecosystems: Their Behaviour and Management*. Fishing News Books, Blackwell Science, Oxford.

Lamson, C. (1986a) On the line: Women and fish plant jobs in Atlantic Canada. *Industrial Relations*, **41**, 145–156.

Lamson, C. (1986b) Planning for resilient coastal communities: Lessons from ecological systems theory. *Coastal Zone Management Journal*, **13**, 265–280.

Lamson, C. & Hanson, A.J. (eds) (1984) *Atlantic Fisheries and Coastal Communities: Fisheries Decision-Making Case Studies*. Dalhousie Ocean Studies Program, Halifax, Canada.

Lane, D.E. (1988) Investment decision making by fishermen. *Canadian Journal of Fisheries and Aquatic Sciences*, **45**, 782–796.

Lane, D.E. (1989) Operational research and fisheries management. *European Journal of Operational Research* **42**, 229–242.

Lane, D.E. & Stephenson, R.L. (1995) Fisheries management science: The framework to link biological, economic, and social objectives in fisheries management. *Aquatic Living Resources*, **8**, 215–221.

Lane, D.E. & Stephenson, R.L. (1998) A framework for risk analysis in fisheries decision-making. *ICES Journal of Marine Science*, **55**, 1–13.

Langstraat, D. (1999) The Dutch co-management system for sea fisheries. In: *Alternative Management Systems for Fisheries* (ed. D. Symes), pp. 73–78. Fishing News Books, Blackwell Science, Oxford.

Larkin, P.A. (1977) An epitaph for the concept of maximum sustainable yield. *Transactions of American Fisheries Society*, **106**, 1–11.

Larkin, P.A. (1996) Concepts and issues in marine ecosystem management. *Reviews in Fish Biology and Fisheries*, **6**, 139–164.

Lawson, R. (1984) *Economics of Fisheries Development*. Frances Pinter, London.

Levin, S.A. (ed.) (1993) Forum: Perspectives on sustainability. *Ecological Applications*, **3**, 545–589.

Liss, W.J. & Warren, C.E. (1980) Ecology of aquatic systems. In: *Fisheries Management* (eds R.T. Lackey & L.A. Nielsen), pp. 41–80. Wiley, New York.

Longhurst, A.R. (ed.) (1981) *Analysis of Marine Ecosystems*. Academic Press, London.

Loucks, L., Charles, A.T. & Butler, M. (eds) (1998) *Managing Our Fisheries, Managing Ourselves*. Gorsebrook Research Institute, Halifax, Canada.

Ludwig, D. & Walters, C.J. (1982) Optimal harvesting with imprecise parameter estimates. *Ecological Modeling*, **14**, 273–292.

Ludwig, D. & Walters, C.J. (1989) A robust method for parameter estimation from catch and effort data. *Canadian Journal of Fisheries and Aquatic Sciences*, **46**, 137–144.

Ludwig, D., Hilborn, R. & Walters, C.J. (1993) Uncertainty, resource exploitation, and conservation: Lessons from history. *Science*, **260**, 17–18.

Major, P. (1994) A government perspective on New Zealand's experience with ITQs. In: *Global Trends: Fisheries Management* (eds E.K. Pikitch, D.D. Huppert & M.P. Sissenwine), pp. 264–269. American Fisheries Society Symposium 20, Bethesda, MD.

Mangel, M. (1985) *Decision and Control in Uncertain Resource Systems*. Academic Press, New York.

Mann, K.H. (1982) *Ecology of Coastal Waters: A Systems Approach*. University of California Press, Berkeley, CA.

Mann, K.H. & Lazier, J.R. (1996) *Dynamics of Marine Ecosystems: Biological–Physical Interactions in the Oceans*. 2nd edn. Blackwell Science, Boston, MA.

Marchak, P., Guppy, N. & McMullan, J. (1987) *Uncommon Property: The Fishing and Fish-Processing Industries in British Columbia*. Methuen, Toronto.

McCay, B.J. (1979) Fish is scarce: Fisheries modernization on Fogo Island, Newfoundland. In: *North Atlantic Maritime Cultures* (ed. R. Andersen). Mouton, The Hague.

McCay, B.J. & Acheson, J. (1987) *The Question of the Commons: Anthropological Contributions to Natural Resource Management*. University of Arizona Press, Tuscon, AZ.

McGlade, J.M. (1989) Integrated fisheries management models: Understanding the limits to marine resource exploitation. *American Fisheries Society Symposium*, **6**, 139–165.

Mercer, M.C. (ed.) (1982) *Multispecies Approaches to Fisheries Management Advice*. Canadian Special Publication of Fisheries and Aquatic Sciences 59, Ottawa.

Ministry of Agriculture, Forestry and Fisheries, Government of Japan (1999) *Summary of the Survey on Trends in Fisheries Labor Force, as of November 1, 1997*. Internet document listed under the heading: 'Preliminary Statistical Report on Agriculture, Forestry and Fisheries' (www.maff.go.jp/ettitle.html).

Moloney, D.G. & Pearse, P.H. (1979) Quantitative rights as an instrument for regulating commercial fisheries. *Journal of the Fisheries Research Board of Canada*, **36**, 859–866.

Moore, W.P., Walsh, D., Worden, I. & MacDonald, J.D. (1993) *The Fish Processing Sector in Atlantic Canada: Industry Trends and Dynamics*. Prepared for the Task Force on Incomes and Adjustment in the Atlantic Fishery. Department of Fisheries and Oceans, Ottawa.

Munasinghe, M. & Shearer, W. (1995) *Defining and Measuring Sustainability: The Biogeophysical Foundations*. United Nations University and the World Bank, Washington, DC.

Munro, G.R. (1990) Applications to policy problems: Labor mobility in the fishery. In: *Mathematical Bioeconomics: The Optimal Management of Renewable Resources*, 2nd edn (by C.W. Clark), pp. 76–85. Wiley-Interscience, New York.

Nadel-Klein, J. & Davis, D.L. (eds) (1988) *To Work and To Weep: Women in Fishing Economies*. Social and Economic Papers No. 18. Institute of Social and Economic Research, Memorial University of Newfoundland, St. John's.

National Research Council (1986) *Proceedings of the Conference on Common Property Resource Management.* National Academy Press, Washington, DC.

National Research Council (1999) *Sustaining Marine Fisheries.* National Academy Press, Washington, DC.

Native Council of Nova Scotia (1994) *Mi'kmaq Fisheries Netukulimk: Towards a Better Understanding.* Native Council of Nova Scotia, Truro, Canada.

Neher, P.A., Arnason, R. & Mollett, N. (eds) (1989) *Rights Based Fishing.* Kluwer, Dordrecht, Boston, MA.

Neis, B., Felt, L., Schneider, D.C., Haedrich, R., Hutchings, J. & Fischer, J. (1996) *Northern Cod Stock Assessment: What can be Learned from Interviewing Resource Users?* Atlantic Fisheries Research Document 96/45. Department of Fisheries and Oceans, Ottawa.

Nelson, J.S. (1994) *Fishes of the World,* 3rd edn. Wiley, New York.

Newell, D. & Ommer, R.E. (1999) Introduction: traditions and issues. In: *Fishing Places, Fishing People: Traditions and Issues in Canadian Small-Scale Fisheries* (eds D. Newell & R.E. Ommer), pp. 3–12. University of Toronto Press, Toronto.

Newkirk, G. (1995) Transforming stressed waters. In: *Philippine Coastal Resources Under Stress* (eds M.A. Juinio-Meñez & G. Newkirk), pp. 91–102. Coastal Resources Research Network, Halifax, Canada, and Marine Science Institute, Quezon City, Philippines.

Nielsen, J.R. & Vedsmand, T. (1997) Fishermen's organisations in fisheries management: Perspectives for fisheries co-management based on Danish fisheries. *Marine Policy,* **21,** 277–288.

North, D.C. (1990) *Institutions, Institutional Change and Economic Performance.* Cambridge University Press, Cambridge.

Norwegian Fishermen's Association (1996) *Facts about the Norwegian Fishing Industry.* Norwegian Fishermen's Association, Trondheim.

NRTEE (1998) *Sustainable Strategies for Oceans: A Co-Management Guide.* National Roundtable on Environment and Economy, Ottawa.

O'Boyle, R. (1993) Fisheries management organizations: A study of uncertainty. In: *Risk Evaluation and Biological Reference Points for Fisheries Management* (eds S.J. Smith, J.J. Hunt & D. Rivard), pp. 423–436. Canadian Special Publication of Fisheries and Aquatic Sciences No. 120. National Research Council of Canada, Ottawa.

Odum, E.P. (1974) *The Fundamentals of Ecology.* Saunders, New York.

Odum, E.P. (1983) *Basic Ecology.* Saunders, New York.

OECD (Organisation for Economic Co-operation and Development) (1993) *The Use of Individual Quotas in Fisheries Management.* OECD, Paris.

OECD (Organisation for Economic Co-operation and Development) (1997) *Towards Sustainable Fisheries. Economic Aspects of the Management of Living Marine Resources.* OECD, Paris.

O'Neill, R.V., DeAngelis D.L., Waide, J.B. & Allen, T.F.H. (1986). *A Hierarchical Concept of Ecosystems.* Princeton University Press, Princeton, NJ.

Opaluch, J.J. & Bockstael, N.E. (1984) Behavioral modelling and fisheries management. *Marine Resource Economics,* **1,** 105-115.

Orbach, M.K. (1980) The human dimension. In: *Fisheries Management* (eds R.T. Lackey & L.A. Nielsen), pp. 149–163. Wiley, New York.

Ostrom, E. (1990) *Governing the Commons.* Cambridge University Press, Cambridge.

Ostrom, E. (1992) The rudiments of a theory of the origins, survival, and performance of common-property institutions. In: *Making the Commons Work: Theory, Practice and Policy* (eds D.W. Bromley, D. Feeny, M.A. McKean, P. Peters, J.L. Gilles, R.J. Oakerson, *et al.*), pp. 293–318. Institute for Contemporary Studies, San Francisco, CA.

Ostrom, E. (1995) Designing complexity to govern complexity. In: *Property Rights and the Environment: Social and Ecological Issues* (eds S. Hanna & M. Munasinghe), pp. 33–45. Beijer International Institute of Ecological Economics and the World Bank, Washington, DC.

Ostrom, E. & Schlager, E. (1996) The formation of property rights. In: *Rights to Nature: Ecological, Economic, Cultural and Political Principles of Institutions for the Environment* (eds S. Hanna, C. Folke & K.G. Mäler), pp. 127–156. Island Press, Washington, DC.

Padilla, J.E. & Charles, A.T. (1994) Bioeconomic modeling and the management of capture and culture fisheries.

Naga: The ICLARM Quarterly, **17**, 18–20.

Panayotou, T. (1980) Economic conditions and prospects of small-scale fishermen in Thailand. *Marine Policy*, **4**, 142–146.

Panayotou, T. (1982) *Management Concepts for Small-Scale Fisheries: Economic and Social Aspects*. FAO Fisheries Technical Paper No. 228. FAO, Rome.

Panayotou, T. (ed.) (1985) *Small-Scale Fisheries in Asia: Socioeconomic Analysis and Policy*. International Development Research Centre, Ottawa.

Panayotou, T. & Panayotou, D. (1986) *Occupational and Geographical Mobility In and Out of Thai Fisheries*. FAO Fisheries Technical Paper No. 271. FAO, Rome.

Parsons, L.S. (1993) *Management of Marine Fisheries in Canada*. Canadian Bulletin of Fisheries and Aquatic Sciences No. 225. National Research Council of Canada, Ottawa.

Passino, D.R.M. (1980) Biology of fishes. In: *Fisheries Management* (eds R.T. Lackey & L.A. Nielsen), pp. 81–109. Wiley, New York.

Pauly, D. & Tsukayama, L. (eds) (1987) *The Peruvian Anchoveta and its Upwelling Ecosystem: Three Decades of Change*. ICLARM Studies and Reviews, No. 15. International Center for Living Aquatic Resources Management, Manila, Philippines.

Pauly, D., Silvestre, G. & Smith, I.R. (1989) On development, fisheries and dynamite: A brief review of tropical fisheries management. *Natural Resource Modeling*, **3**, 307–329.

Pauly, D., Christensen, V., Dalsgaard, J., Froese, R. & Torres, F. Jr. (1998) Fishing down marine food webs. *Science*, **279**, 860–863.

Pearce, D., Barbier, E. & Markandya, A. (1990) *Sustainable Development: Economics and Environment in the Third World*. Edward Elgar, Aldershot, UK.

Pearse, P.H. (1980) Property rights and regulation of commercial fisheries. *Journal of Business Administration*, **11**, 185–209.

Pearse, P.H. (1982) *Turning the Tide: A New Policy for Canada's Pacific Fisheries. Final Report of the Commission on Pacific Fishery Policy*. Canadian Government Publishing Centre, Ottawa.

Pearse, P.H. & Walters, C.J. (1992) Harvesting regulation under quota management systems for ocean fisheries: Decision making in the face of natural variability, weak information, risks and conflicting incentives. *Marine Policy*, **16**, 167–182.

Peet, J. & Peet, K. (1990) With people's wisdom: Community-based perspectives on sustainable development. Paper presented at the *Ecological Economics of Sustainability Conference*, 21–23 May 1990. World Bank, Washington, DC.

Pena-Torres, J. (1997) The political economy of fishing regulation: The case of Chile. *Marine Resource Economics* **12**, 253–280.

Pezzy, J. (1989) *Economic Analysis of Sustainable Growth and Sustainable Development. Environment Department Working Paper 15*. World Bank, Washington, DC.

Pikitch, E.L., Huppert, D.D. & Sissenwine, M.P. (eds) (1997) *Global Trends: Fisheries Management*. Symposium 20. American Fisheries Society, Bethesda, MD.

Pinkerton, E.W. (1987) The fishing-dependent community. In: *Uncommon Property: The Fishing and Fish-Processing Industries in British Columbia* (eds P. Marchak, N. Guppy & J. McMullan), pp. 293–325. Methuen, Toronto.

Pinkerton, E.W. (1989) *Cooperative Management of Local Fisheries*. University of British Columbia Press, Vancouver.

Pinkerton, E.W. (1994) Local fisheries co-management: A review of international experiences and their implications for salmon management in British Columbia. *Canadian Journal of Fisheries and Aquatic Sciences*, **51**, 2363–2378.

Pinkerton, E. & Weinstein, M. (1995) *Fisheries that Work: Sustainability through Community-Based Management*. David Suzuki Foundation, Vancouver.

Pitcher, T.J. & Hart, P.J.B. (1982) *Fisheries Ecology*. Croom Helm, London.

Pitcher, T.J., Hart, P.J.B. & Pauly, D. (eds) (1998) *Re-Inventing Fisheries Management*. Kluwer, Dordrecht.

Pollnac, R.B. & Littlefield, S.J. (1983) Sociocultural aspects of fisheries management. *Ocean Development and International Law*, **12**, 209–246.

Polovina, J.J. (1984) Model of a coral reef ecosystem. I. The ECOPATH model and its application to French Frigate Shoals. *Coral Reefs*, **3**, 1–11.

Pomeroy, R.S. (1995) Community-based and co-management institutions for sustainable coastal fisheries management in Southeast Asia. *Ocean and Coastal Management*, **27**, 143–162.

Pomeroy, R.S. & Berkes, F. (1997) Two to tango: The role of government in fisheries co-management. *Marine Policy*, **21**, 465–480.

Pringle, J.D. (1985) The human factor in fishery resource management. *Canadian Journal of Fisheries and Aquatic Sciences*, **42**, 389–392.

Pringle, J.D. & Burke D.L. (1993) The Canadian lobster fishery and its management, with emphasis on the Scotian Shelf and the Gulf of Maine. In: *Perspectives on Canadian Marine Fisheries Management* (eds L.S. Parsons & W.H. Lear), pp. 91–122. Canadian Bulletin of Fisheries and Aquatic Sciences No. 226. National Research Council of Canada and Department of Fisheries and Oceans, Ottawa.

Ralston, S. & Polovina, J.J. (1982) A multispecies analysis of the commercial deep-sea handline fishery in Hawaii. *Fisheries Bulletin*, **80**, 435–448.

Reed, W.J. (1979) Optimal escapement levels in stochastic and deterministic harvesting models. *Journal of Environmental Economics and Management*, **6**, 350–363.

Rees, W.E. (1988) *Defining sustainable development. Background Paper for the Conference Planning for Sustainable Development.* School of Community and Regional Planning, University of British Columbia, Vancouver.

Rees, W.E. & Wackernagel, M. (1994) Ecological footprints and appropriated carrying capacity: Measuring the natural capital requirements of the human economy. In: *Investing in Natural Capital: The Ecological Economics Approach to Sustainability* (eds A. Jansson, M. Hammer, C. Folke & R. Costanza), pp. 362–390. Island Press, Washington, DC.

Regier, H.A. & Grima, A.P. (1985) Fishery resource allocation: An exploratory essay. *Canadian Journal of Fisheries and Aquatic Sciences*, **42**, 845–859.

Renard, Y. & Koester, S.K. (1995) Resolving conflicts for integrated coastal management: The case of Soufrière, St. Lucia. *Caribbean Park and Protected Area Bulletin*, **5**, 5–7.

Rettig, R.B. & Ginter, J.C. (eds) (1978) *Limited Entry as a Fishery Management Tool.* University of Washington Press, Seattle, WA.

Ricker, W.E. (1958*) Handbook of Computations for Biological Statistics of Fish Populations.* Fisheries Research Board of Canada, Ottawa.

Ricker, W.E. (1975) *Computation and Interpretation of Biological Statistics of Fish Populations.* Bulletin 191. Supply and Services Canada, Ottawa.

Robinson, J., Francis, G., Legge, R. & Lerner, S. (1990) Defining a sustainable society: Values, principles and definitions. *Alternatives*, **17**, 36–46.

Rodrigues, A.G. (ed.) (1990) *Operations Research and Management in Fishing.* Kluwer, Dordrecht.

Rogers, R.A. (1995) *The Oceans are Emptying: Fish Wars and Sustainability.* Black Rose Books, Montreal.

Rosenberg, A.A. & Brault, S. (1993) Choosing a management strategy for stock rebuilding when control is uncertain. In: *Risk Evaluation and Biological Reference Points for Fisheries Management* (eds S.J. Smith, J.J. Hunt & D. Rivard), pp. 243–249. Canadian Special Publication of Fisheries and Aquatic Sciences No. 120. National Research Council of Canada, Ottawa.

Rowley, R.J. (1994) Marine reserves in fisheries management. *Aquatic Conservation: Marine and Freshwater Ecosystems*, **4**, 233–254.

Royce, W.F. (1996) *Introduction to the Practice of Fishery Science.* Academic Press, San Diego, WA.

Ruddle, K. (1989a) The continuity of traditional management practices: The case of Japanese coastal fisheries. In: *Traditional Marine Resource Management in the Pacific Basin: An Anthology* (eds K. Ruddle & R.E. Johannes), pp. 263–285. Contending with Global Change Study No. 2. UNESCO/ROSTSEA, Jakarta.

Ruddle, K. (1989b) Solving the common-property dilemma: Village fisheries rights in Japanese coastal waters. In: *Common Property Resources: Ecology and Community-Based Sustainable Development* (ed. F. Berkes), pp. 168–184. Bellhaven Press, London.

Ruddle, K. (1994) Local knowledge in the folk management of fisheries and coastal marine environments. In: *Folk Management in the World's Fisheries: Lessons for Modern Fisheries Management* (eds C.L. Dyer & J.R. McGoodwin), pp. 161–206. University Press of Colorado, Niwot, CO.

Ruddle, K. & Johannes, R.E. (1989) *Traditional Marine Resource Management in the Pacific Basin: An Anthology.* Contending with Global Change Study No. 2. UNESCO/ROSTSEA, Jakarta.

Ruddle, K., Hviding, E. & Johannes, R.E. (1992) Marine resources management in the context of customary tenure.

Marine Resource Economics, **7**, 249–273.

Russ, G.R. & Alcala, A.C. (1994) Sumilon Island Reserve: 20 years of hopes and frustration. *Naga: The ICLARM Quarterly*, **17**, 8–12.

Saila, S.B. & Gallucci, V.F. (1996) Overview and background. In: *Stock Assessment: Quantitative Methods and Applications for Small-Scale Fisheries* (eds V.F. Gallucci, S.B. Saila, D.J. Gustafson & B.J. Rothschild), pp. 1–8. CRC Lewis, Boca Raton, FL.

Saila, S.B., McKenna, J.E., Formacion, S., Silvestre, G.T. & McManus, J.W. (1996) *Stock Assessment: Quantitative Methods and Applications for Small-Scale Fisheries*. CRC Lewis, Boca Raton, FL.

Sainsbury, J.C. (1996) *Commercial Fishing Methods: An Introduction to Vessels and Gears*. Fishing News Books, Blackwell Science, Oxford.

Salz, P. (1986) *Policy Instruments for Development of Fisheries*. Publication No. 574. Agricultural Economics Research Institute, The Hague.

Sampson, D.B. (1994) Fishing tactics in a two-species fisheries model: The bioeconomics of bycatch and discarding. *Canadian Journal of Fisheries and Aquatic Sciences*, **51**, 2688–2694.

Schaefer, M.B. (1954) Some aspects of the dynamics of populations important to the management of the commercial marine fisheries. *Bulletin of the Inter-American Tropical Tuna Commission*, **1**, 27–56.

Schaefer, M.B. (1957) Some considerations of population dynamics and economics in relation to the management of marine fisheries. *Journal of the Fisheries Research Board of Canada*, **14**, 669–681.

Schramm, H.L. & Hubert, W.A. (1996) Ecosystem management: Implications for fisheries management. *Fisheries*, **21**, 6–11.

Scott, A.D. (1955) The fishery: The objectives of sole ownership. *Journal of Political Economy*, **63**, 116–124.

Scott, A.D. (1988) Development of property in the fishery. *Marine Resource Economics*, **5**, 289–311.

Scott, A.D. & Johnson, J. (1985) Property rights: Developing the characteristics of interests in natural resources. In: *Progress in Natural Resource Economics* (ed. A. Scott), pp. 376–403. Clarendon Press, Oxford.

Sea Fisheries Research Directorate (1999) *Directorate Structure*. Sea Fisheries Research Directorate, Government of South Africa, Cape Town (www.gov.za/sfri/sfri.htm).

Seijo, J.C., Defeo, O. & Salas, S. (1998) *Fisheries Bioeconomics: Theory, Modelling and Management*. FAO Fisheries Technical Paper No.368. Food and Agriculture Organization, Rome.

Sen, S. & Nielsen, J.P. (1996) Fisheries co-management: A comparative analysis. *Marine Policy*, **20**, 405–418.

Sherman, K. (1993) Large marine ecosystems as global units for marine resources management – An ecological perspective. In: *Large Marine Ecosystems: Stress, Mitigation, and Sustainability* (eds K. Sherman, L.M. Alexander & B.D. Gold), pp. 3–14. AAAS Press, Washington, DC.

Sherman, K. (1998) Large marine ecosystems as science and management units. In: *Workshop on the Ecosystem Approach to the Management and Protection of the North Sea, Oslo, Norway, 15–17 June 1998*, pp. 46–50. Nordic Council of Ministers, Copenhagen.

Sinclair, P.R. (1983) Fishermen divided: The impact of limited entry licensing in northwest Newfoundland. *Human Organization*, **42**, 307–313.

Sissenwine, M.P. (1984) The uncertain environment of fishery scientists and managers. *Marine Resource Economics*, **1**, 1–30.

Sivasubramaniam, K. (1993) The biosocioeconomic way: A new approach to management of small-scale fisheries in the Bay of Bengal region. *Bay of Bengal News*, **52**, 4–15.

Slocombe, S.D. (1993) Environmental planning, ccosystem science and ecosystem approaches for integrating environment and development. *Environmental Management*, **17**, 289–298.

Smith, I.R. (1979) *A Research Framework for Traditional Fisheries*. ICLARM Studies and Reviews No. 2. International Center for Living Aquatic Resources Management, Manila, Philippines.

Smith, I.R. (1981) Improving fishing incomes when resources are overfished. *Marine Policy*, **5**, 17–22.

Smith, S.J., Hunt, J.J. & Rivard, D. (eds) (1993) *Risk Evaluation and Biological Reference Points for Fisheries Management*. Canadian Special Publication of Fisheries and Aquatic Sciences No. 120. National Research Council of Canada, Ottawa.

Smith, T.D. (1994) *Scaling Fisheries: The Science of Measuring the Effects of Fishing, 1855–1955*. Cambridge University Press, Cambridge.

Smith, V.L. (1968) Economics of production from natural resources. *American Economic Review*, **58**, 409–431.

Sobel, J. (1993) Conserving biological diversity through marine protected areas. *Oceanus*, **36**, 19–26.

Solórzano, R., De Camino, R., Woodward, R., Tosi, J., Watson, V., Vásquez, A., *et al.* (1991) *Accounts Overdue: Natural Resource Depreciation in Costa Rica.* Tropical Science Center and World Resources Institute, Washington, DC.

Soufrière Marine Management Authority (1996) *Report on the Meeting to Review the First Year of Operation of the Soufrière Marine Management Authority.* La Haut, Soufrière, St. Lucia.

Soufrière Regional Development Foundation (1994) *Soufrière Marine Management Area: Agreement on the Use and Management of Marine and Coastal Resources in the Soufrière Region, St. Lucia.* Soufrière Regional Development Foundation, Soufrière, St. Lucia.

Sparre, P., Ursin, E. & Venema, S.C. (1989) *Introduction to Tropical Fish Stock Assessment. Part 1: Manual.* FAO Fisheries Technical Paper No. 306. FAO, Rome.

Squires, D. & Kirkley, J. (1996) Individual transferable quotas in a multiproduct common property industry. *Canadian Journal of Economics*, **29**, 318–342.

Sutinen, J.G. & Andersen, P. (1985) The economics of fisheries law enforcement. *Land Economics*, **61**, 387–397.

Sutinen, J.G. & Hennessey, T.M. (1986) Enforcement: The neglected element in fishing management. In: *Natural Resources Economics and Policy Applications* (eds E. Miles, R. Pealy & R. Stokes), pp. 185–213. University of Washington Press, Seattle, WA.

Sutinen, J.G., Rieser, A. & Gauvin, J.R. (1990) Measuring and explaining noncompliance in federally managed fisheries. *Ocean Development and International Law*, **21**, 335–372.

Sutton, M. (1998) Harnessing market forces and consumer power in favour of sustainable fisheries. In: *Re-Inventing Fisheries Management* (eds T.J. Pitcher, P.J.B. Hart & D. Pauly), pp. 125–135. Kluwer, Dordrecht.

Swartzman, G.L., Getz, W.M. & Francis, R.C. (1987) Binational management of Pacific hake (*Merluccius productus*): A stochastic modeling approach. *Canadian Journal of Fisheries and Aquatic Sciences*, **44**, 1053–1063.

Symes, D. (1997) The European Community's Common Fisheries Policy. *Ocean and Coastal Management*, **35**, 137–155.

Symes, D. (ed.) (1998a) *Property Rights and Regulatory Systems in Fisheries.* Fishing News Books, Blackwell Science, Oxford.

Symes, D. (1998b) Property rights, regulatory measures and the strategic response of fishermen. In: *Property Rights and Regulatory Systems in Fisheries* (ed. D. Symes), pp. 3–16. Fishing News Books, Blackwell Science, Oxford.

Taggart, C.T., Anderson, J., Bishop, C., Colbourne, E., Hutchings, J., Lilly, G., *et al.* (1994) Overview of cod stocks, biology, and environment in the Northwest Atlantic region of Newfoundland, with emphasis on northern cod. *ICES Marine Science Symposium*, **198**, 140–157.

Tait, R.V. & Dipper, F. (1998) *Elements of Marine Ecology.* Butterworth–Heinemann, Oxford.

Terkla, D.G., Doeringer, P.B. & Moss, P.I. (1985) *Common Property Resource Management with Sticky Labor: The Effects of Job Attachment on Fisheries Management.* Discussion Paper No. 108. Department of Economics, Boston University, Boston, MA.

Tettey, E.O. & Griffin, W.L. (1984) Investment in Gulf of Mexico shrimp vessels, 1965–77. *Marine Fisheries Review*, **46**, 49–52.

Townsend, R.E. (1985) The right to fish as an external benefit of open access. *Canadian Journal of Fisheries and Aquatic Sciences*, **42**, 2050–2053.

Townsend, R.E. (1990) Entry restrictions in the fishery: A survey of the evidence. *Land Economics*, **66**, 359–378.

Townsend, R.E. (1995) Fisheries self-governance: Corporate or cooperative structures? *Marine Policy*, **19**, 39–45.

Townsend, R.E. & Charles, A.T. (1997) User rights in fishing. In: *Northwest Atlantic Groundfish: Perspectives on a Fishery Collapse* (eds J. Boreman, B.S. Nakashima, J.A. Wilson & R.L. Kendall), pp. 177–184. American Fisheries Society, Bethesda, MD.

Troadec, J.-P. (1982) *Introduction to Fisheries Management: Advantages, Difficulties and Mechanisms.* FAO Fisheries Technical Paper No. 224. FAO, Rome.

Twomley, B. (1994) License limitation in Alaskan salmon fisheries. In: *Limiting Access to Marine Fisheries: Keeping the Focus on Conservation* (ed. K.L. Gimbel), pp. 59–66. Center for Marine Conservation and World Wildlife Fund. Washington, DC.

Tyler, A.V. & Gallucci, V.F. (1980) Dynamics of fished stocks. In: *Fisheries Management* (eds R.T. Lackey & L.A. Nielsen), pp. 111–147. Wiley, New York.

Ullah, M. (1985) Fishing rights, production relations, and profitability: A case study of Jamuna fishermen in Bangladesh. In: *Small-Scale Fisheries in Asia: Socioeconomic Analysis and Policy* (ed. T. Panayotou). International Development Research Centre, Ottawa.

United Nations (1999) Division for Ocean Affairs and the Law of the Sea. United Nations, New York (www.un.org/Depts/los).

United Nations Industrial Development Organization (UNIDO) (1987) *Industrial Development Strategies for Fishery Systems in Developing Countries*, Vol. 1. Sectoral Studies Series No. 32. United Nations Industrial Development Organization, Vienna.

Ursin, E. (1982) Multispecies fish stock and yield assessment in ICES. In: *Multispecies Approaches to Fisheries Management Advice* (ed. M.C. Mercer), pp. 39–47. Canadian Special Publication of Fisheries and Aquatic Sciences 59. Department of Fisheries and Oceans, Government of Canada, Ottawa.

Von Benda-Beckmann, F. (1995) Anthropological approaches to property law and economics. *European Journal of Law and Economics*, **2**, 309–336.

Walters, C.J. (1980) Systems principles in fisheries management. In: *Fisheries Management* (eds R.T. Lackey & L.A. Nielsen), pp. 167–183. Wiley, New York.

Walters, C.J. (1981) Optimum escapements in the face of alternative recruitment hypotheses. *Canadian Journal of Fisheries and Aquatic Sciences*, **38**, 678–689.

Walters, C.J. (1986) *Adaptive Management of Renewable Resources*. MacMillan, New York.

Walters, C.J. (1998) Designing fisheries management systems that do not depend upon accurate stock assessment. In: *Re-Inventing Fisheries Management* (eds T.J. Pitcher, P.J.B. Hart & D. Pauly), pp. 279–288. Kluwer, Dordrecht.

Walters, C.J. & Hilborn, R. (1976) Adaptive control of fishing systems. *Journal of the Fisheries Research Board of Canada*, **33**, 145–159.

Walters, C.J. & Hilborn, R. (1978) Ecological optimization and adaptive management. *Annual Review of Ecology and Systematics*, **9**, 157–188.

Walters, C.J. & Maguire, J.J. (1996) Lessons for stock assessment from the Northern cod collapse. *Reviews in Fish Biology and Fisheries*, **6**, 125–137.

Walters, C. & Pearse, P.H. (1996) Stock information requirements for quota management systems in commercial fisheries. *Reviews in Fish Biology and Fisheries*, **6**, 21–42.

Warming, J. (1911) Om grundrente af fiskegrunde (On rent of fishing grounds). *National okonomisk Tidsskrift*, **49**, 499–505.

Wells, P.G. & Ricketts, P.J. (1994) *Coastal Zone Canada '94, 'Cooperation in the Coastal Zone': Conference Proceedings*. Coastal Zone Canada Association, Bedford Institute of Oceanography, Dartmouth, Canada.

Wiber, M.G. & Kearney, J.F. (1996) Stinting the commons: Property, policy or power struggle? Comparing quota in the Canadian dairy and fisheries sectors. In: *The Role of Law in Natural Resource Management* (eds J. Spiertz & M.G. Wiber), pp. 145–165. VUGA Uitgeverij, 's-Gravenhage, the Netherlands.

Wilson, J.A. (1982) The economical management of multispecies fisheries. *Land Economics*, **58**, 414–433.

Wilson, J.A., Acheson, J.M., Metcalfe, M. & Kleban, P. (1994) Chaos, complexity and community management of fisheries. *Marine Policy*, **18**, 291–305.

World Book (1999) *World Book Millennium 2000*. Multimedia Encyclopedia (CD-Rom). World Book and IBM (www.worldbook.com).

World Commission on Environment and Development (1987) *Our Common Future*. Oxford University Press, Oxford.

Yang, C.W. (1989) *A methodology for multispecies stock assessment with application to analysing population dynamics of the groundfish community on the Scotian Shelf*. PhD thesis, Dalhousie University, Halifax, Canada.

de Young, B., Peterman, R.M., Dobell, A.R., Pinkerton, E., Breton, Y., Charles, A.T., *et al.* (1999) *Canadian Marine Fisheries in a Changing and Uncertain World*. Canadian Special Publication of Fisheries and Aquatic Sciences 129. NRC Research Press, Ottawa.

Zhong, Y. & Power, G. (1997) Fisheries in China: progress, problems, and prospects. *Canadian Journal of Fisheries and Aquatic Sciences*, **54**, 224–238.

Index